8th International Symposium on Radiation from Relativistic Electrons in Periodic Structures 2009

(RREPS 09)

Journal of Physics: Conference Series Volume 236

Zvenigorod, Moscow, Russia
7-11 September 2009

Editors:

Mikhail N. Strikhanov Yury L. Pivovarov

ISBN: 978-1-61738-659-6
ISSN: 1742-6588

Printed from e-media with permission by:

Curran Associates, Inc.
57 Morehouse Lane
Red Hook, NY 12571

Some format issues inherent in the e-media version may also appear in this print version.

Printed by Curran Associates, Inc. (2010)

For permission requests, please contact the Institute of Physics at the address below.

Institute of Physics
Dirac House, Temple Back
Bristol BS1 6BE UK

Phone: 44 1 17 929 7481
Fax: 44 1 17 920 0979

techtracking@iop.org

Additional copies of this publication are available from:

Curran Associates, Inc.
57 Morehouse Lane
Red Hook, NY 12571 USA
Phone: 845-758-0400
Fax: 845-758-2634
Email: curran@proceedings.com
Web: www.proceedings.com

TABLE OF CONTENTS

This volume contains the papers presented at 8th International Symposium on Radiation from Relativistic Electrons in Periodic Structures (RREPS'09), which was held in Zvenigorod, Moscow Region, Russia, from 7 to 11 September 2009, organized jointly by National Research Nuclear University MEPhI (Moscow) and Tomsk Polytechnic University (Tomsk), Russia.

University MEPhI (Moscow) and Tomsk Polytechnic University (Tomsk), Russia. RREPS was founded in September 1993 by an initiative of the Nuclear Physics Institute at Tomsk Polytechnic University, Russia, with the intention of strengthening basic and applied research focused on radiation from relativistic particles in natural and artificial periodic structures. Since then, the symposium has developed into a forum attracting scientists from different fields and from many countries all over the world. RREPS'09 followed previous successful series of biennial RREPS symposia at Tomsk (1993, 1995, 1997, 2003), Baikal Lake (1999), Aya Lake (Altai, Russia, 2001) and Czech Technical University in Prague (Czech Republic, 2007). Five NIMB topical issues (V 145 No 1-2, October 1998; V 173 No 1-2, January 2001; V 201(1) January 2003; V 227, Issues 1-2, January 2005; V 266, Issue 17, September 2008) have been published as outgrowth of these symposia.

Traditionally, the RREPS program includes following topics:

- General Properties of Electromagnetic Radiation from Relativistic Particles
- Transition Radiation
- Parametric X- Radiation
- Diffraction Radiation and Smith-Purcell Effect
- Coherent Bremsstrahlung and Channeling Radiation
- Crystal- Assisted Processes
- Applications of Monochromatic X- and Gamma- Beams Produced at Electron Accelerators

The present RREPS'09 Symposium was dedicated to the modern problems in radiation from relativistic electrons in crystals and other periodic structures, as well as to new applications of photon and electron beams. During the last few decades, electromagnetic radiation from relativistic particles, both in external fields and in matter, has always been an interesting field for investigation. Every kind of radiation reflects specific processes of fundamental atomic physics, classical or quantum electrodynamics and might have specific applications in accelerator physics (beam diagnostics), nuclear physics (hard photon sources), material science and medicine (X-Ray sources). Nowadays, electromagnetic radiation studies cover electron energies from a few MeV up to hundreds of GeV in many laboratories throughout the world. The goal is to study the physics of generation of various kinds of radiation and their interplay or combined effects and to find successful applications for them. New photon sources, which use new types of radiation at new accelerators (e.g. tabletop synchrotrons), may be considered complementary to conventional photon sources based on synchrotron radiation, undulator radiation and free electron lasers.

We express our thanks to the members of the International Program Committee for their suggestions during the preparation of the scientific program of the workshop.

We warmly thank the National Research Nuclear University MEPhI (Moscow) and the Tomsk Polytechnic University (Tomsk) for the financial and administrative support. We also acknowledge the valuable financial contributions by Russian Fund for Basic Research and "Dynasty" Foundation.

Editors

Mikhail N. Strikhanov
National Research Nuclear University MEPhI, Moscow, Russia

Yury L. Pivovarov
Tomsk Polytechnic University, Tomsk, Russia

On spreading of high energy wave packets and development of radiation process by relativistic electron in space and time

N F Shul'ga[1], V V Syshchenko[2] and S N Shul'ga[1]

[1] Akhiezer Institute for Theoretical Physics of the NSC "KIPT", Akademicheskaya Street, 1, Kharkov 61108, Ukraine

[2] Belgorod State University, Pobedy Street, 85, Belgorod 308015, Russian Federation

E-mail: `shulga@kipt.kharkov.ua`, `syshch@yandex.ru`

Abstract. The problem of the motion of high energy wave packets combined of free electromagnetic waves is considered. It is demonstrated that the transformation of such packets to the packet of spherically diverging waves happens on long distances along the packet's motion direction, that substantially exceed the radiated wavelength. The transition radiation by the "half-bare" ultrarelativistic electron is considered. It is demonstrated that the transition radiation by such electron on the targets located inside and outside the coherence length of the radiation process would be substantially different.

1. Introduction

Moving electron is the charge and the eigenfield (Coulomb field) moving together with it. Changing the electron's trajectory disturbs that field. The disturbance of the field could be treated as a packet of free plane electromagnetic waves. On large distances from the region where the acceleration had happened the packet transforms to the packet of diverging waves (the radiation field). For non-relativistic particles that happens on the distances of order of the length λ of radiated wave [1]. High energies make the stabilizing influence to wave packets that leads to a substantial increase of the length on which the packet's transformation takes place. This length could have macroscopic size, exceeding not only interatomic distances in matter, but also the size of the target and just the size of the experimental installation (detector). Hence it is important to know the behavior of such high-energy packets of electromagnetic waves in the region where that transformation happens. The present article is devoted to the examination of this problem.

Primarily, we consider the motion of Gaussian packet combined of plane waves with directions of the wave vector $\mathbf{k}$ close to each other. It is demonstrated that the shape of such a packet changes on the lengths that substantially exceed the wavelength $\lambda = 1/|\mathbf{k}|$ corresponding to the given absolute value of the wave vector $|\mathbf{k}|$.

Then we consider the motion of the wave packet that coincides in the time moment $t = 0$ with the eigenfield of the ultrarelativistic electron. It is demonstrated that the last packet also conserves its shape for a long time interval. Fourier component of this packet with the wavelength λ changes only on the distances z along the packet's direction of motion that exceed the length

$2\gamma^2\lambda$, where γ is the electron's Lorentz factor. This length coincides with the coherence length of the radiation process of the relativistic electron $l = 2\gamma^2\lambda$ [2, 3].

The problem of special interest is the radiation under sharp (at the time moment $t = 0$) changing of the ultrarelativistic electron's velocity [3 - 5]. We demonstrate that the packets of electromagnetic waves arising in this case are close in their structure to the packets considered above. However, their manifestations in the direction of the initial and final motion of the electron are substantially different. Namely, on the distances $z < 2\gamma^2\lambda$, Fourier components with the wavelength λ of the packet moving along the direction of the initial electron motion will practically coincide with the Fourier components of the initial packet and, consequently, to the Fourier components of the Coulomb field of the electron moving in the initial direction without scattering. Oppositely, in the final electron's direction of motion, the field of the packet of free waves will screen the particle's eigenfield. The electron under such conditions was called in [4] as "half-bare particle", that is the particle whose specific Fourier components of the surrounding field are practically absent for a long time. We put attention to that the transition radiation by such particles and wave packets on the targets placed on the distances from the point of scattering larger and smaller than $2\gamma^2\lambda$ would be substantially different. The corresponding experiment would permit to observe direct manifestation of the "half-bare" electron and the process of its dressing.

Let us note that for the charged particle the Gauss theorem is applicable, according to which the number of force lines of the electromagnetic field surrounding the electron does not change with time [1]. Under this the radiation process by electron can be presented as bending of these force lines [6 - 10]. Such a concept of radiation process relates to the complete electromagnetic field surrounding the electron. However, it does not contain such characteristics of the radiation process as coherence length and wave zone which are connected with specific Fourier components of this field. The term "half-bare electron" relates also to a specific Fourier component of the field surrounding the electron which is defined by the wavelength λ. So, the analysis of a space-time evolution of these Fourier components (wave packets) gives us a supplement for the picture of evolution of complete field surrounding the electron which is in accelerated motion.

We use the system of units in which the speed of light in vacuum is taken equal to the unit: $c = 1$.

2. Gaussian packet

The scalar potential of the packet of free electromagnetic waves could be expressed in the form of the following Fourier decomposition:

$$\varphi(\mathbf{r}, t) = \int \frac{d^3q}{(2\pi)^3} e^{i(\mathbf{qr} - qt)} C_q, \tag{1}$$

where C_q are the coefficients of the decomposition, $q = |\mathbf{q}|$. Consider at first the behavior of the packet combined at $t = 0$ of plane waves with the wave vectors $\mathbf{k}$ directed closely to some given direction (the z axis). Supposing for simplicity that the distribution of the waves over directions of the vector $\mathbf{k}$ is Gaussian at $t = 0$, let us write the potential (1) in the form

$$\varphi_k(\mathbf{r}, 0) = \frac{1}{\pi\Delta^2} \int d^2\vartheta e^{-\vartheta^2/\Delta^2} e^{i\mathbf{kr}}, \tag{2}$$

where ϑ is the angle between $\mathbf{k}$ and the z axis, and Δ^2 is the mean square value of the angle ϑ, $\Delta^2 \ll 1$. Coefficients $C_\mathbf{q}$ of such packet have the form

$$C_\mathbf{q} = (2\pi)^3 \int \frac{d^2\vartheta}{\pi\Delta^2} e^{-\vartheta^2/\Delta^2} \delta(\mathbf{k} - \mathbf{q}), \tag{3}$$

where $\delta(\mathbf{k}-\mathbf{q})$ is the delta-function. In this case, according to (1),

$$\varphi_k(\mathbf{r},t) = \frac{1}{1+ikz\Delta^2/2}\exp\left\{ik(z-t) - \frac{(k\rho\Delta/2)^2}{1+i(kz\Delta^2/2)}\right\}, \tag{4}$$

where ρ is the transverse (in relation to the z axis) component of $\mathbf{r}$.

Equation (4) demonstrates that under $kz\Delta^2/2 \ll 1$

$$\varphi_k(\mathbf{r},t) \approx \exp\left\{ik(z-t) - (k\rho\Delta/2)^2\right\}, \tag{5}$$

and under the condition $kz\Delta^2/2 \gg 1$

$$\varphi_k(\mathbf{r},t) \approx -\frac{2i}{kz\Delta^2}\exp\left\{ik(z-t) + ik\frac{\rho^2}{2z} - \frac{\rho^2}{z^2\Delta^2}\right\}. \tag{6}$$

In the case $z \gg \rho$ the last formula could be written in the form of diverging wave:

$$\varphi_k(\mathbf{r},t) \approx -\frac{2i}{kr\Delta^2}\exp\left\{ik(r-t) - \frac{\rho^2}{z^2\Delta^2}\right\}, \tag{7}$$

where $r = \sqrt{\rho^2+z^2} \approx z + \rho^2/2z$.

So, on the distances z from the center of the packet that satisfy the condition

$$kz\Delta^2/2 \ll 1, \tag{8}$$

the shape of the packet (4) coincides with the packet's shape at $t=0$. Only on the distances z that satisfy the condition

$$kz\Delta^2/2 > 1, \tag{9}$$

the transformation of the packet of plane waves (4) into the packet of diverging spherical waves happens.

In the theory of radiation of electromagnetic waves, the spatial region where the field of moving charges acquires the form of spherically diverging waves, is called as *wave zone* (see, e.g. [1, 11]). Particularly, for non-relativistic charged particles the wave zone begins just on the distances from the radiating system that exceed the radiated wavelength (see [1]). Condition (9) demonstrates, however, that under $\Delta^2 \ll 1$ the formation of the wave zone takes place not on the distances $z > \lambda$, like in the problem of radiation of the non-relativistic particle, but on the distances

$$z > 2\lambda/\Delta^2, \tag{10}$$

which are much larger than the wavelengths $\lambda = 1/k$, of which the packet is composed (4). For small values of Δ^2 the length $z = 2\lambda/\Delta^2$ could reach macroscopic sizes.

3. Approximation of Coulomb field by the packet of plane waves

Such problem arises in the equivalent photons method (or the method of virtual quanta) when the Coulomb field of relativistic electron is replaced at some specific time moment ($t=0$) by the packet of free electromagnetic waves. Indeed, the Fourier decomposition of the electron's Coulomb field could be written in the form

$$\varphi_c(\mathbf{r},t) = \mathrm{Re}\int \frac{d^3k}{(2\pi)^3} e^{i(\mathbf{kr}-\mathbf{kv}t)} C_k^c, \tag{11}$$

where $\mathbf{v}$ is the electron's velocity directed along the z axis, and

$$C_k^c = \frac{8\pi e\Theta(k_z)}{k_\perp^2 + k_z^2/\gamma^2}. \tag{12}$$

Here γ is the electron's Lorentz factor, k_z and $\mathbf{k}_\perp$ are the components of the vector $\mathbf{k}$, parallel and orthogonal to the z axis, $\Theta(k_z)$ is the Heaviside's step function.

It is supposed in the equivalent photons method that at $t = 0$ the packet (1) composed of free electromagnetic waves coincides with the electron's Coulomb field moving with the velocity $\mathbf{v}$ [11 - 13]. That corresponds to Fourier decomposition (1) with the coefficients $C_q = C_k^c$.

For $\gamma \gg 1$ the main contribution to (1) would be made by the values $\mathbf{q} = \mathbf{k}$ which directions are close to the direction of the electron's velocity $\mathbf{v}$. Taking this into account, the packet (1) could be written in the form

$$\varphi(\mathbf{r}, t) = \mathrm{Re} \int_0^\infty dk\, \varphi_k(\mathbf{r}, t), \tag{13}$$

where

$$\varphi_k(\mathbf{r}, t) = \frac{2}{\pi} \exp\left[ik(z-t)\right] \int_0^\infty \frac{\vartheta d\vartheta}{\vartheta^2 + \gamma^{-2}} J_0(k\rho\vartheta)\, e^{-ikz\vartheta^2/2}. \tag{14}$$

Here ϑ is the angle between $\mathbf{k}$ and $\mathbf{v}$ ($\vartheta \ll 1$), and $J_0(x)$ is the Bessel function.

The function $\varphi_k(\mathbf{r}, t)$ has the same structure as the function (4) corresponding to Gaussian distribution of the vectors $\mathbf{k}$ over the angles ϑ. Namely, if $kz\vartheta^2/2 \ll 1$, the main contribution to the integral (14) is made by the values $\vartheta \sim \gamma^{-1}$ and

$$\varphi_k(\mathbf{r}, t) \approx \frac{2}{\pi} K_0(k\rho/\gamma)\, e^{ik(z-t)}, \tag{15}$$

where $K_0(x)$ is the modified Bessel function of the third kind. In this case after integration over k in (13) we find that

$$\varphi(\mathbf{r}, t) \approx \frac{e}{\sqrt{(z-t)^2 + \rho^2/\gamma^2}}. \tag{16}$$

The main contribution to (13) is made by the values $k \sim \gamma/\rho$, hence equation (16) is valid in the range of coordinates ρ and z that satisfy the condition $z < \gamma\rho$. In this range of coordinates the packet under consideration moves with the velocity of light in the z axis direction.

So, on the distances $z \lesssim 2\gamma^2\lambda$ the considered wave packet practically coincides with the initial one (at $t = 0$). Substantial transformation of the packet would happen only on the distances

$$z > 2\gamma^2\lambda. \tag{17}$$

In this case for the evaluation of the integral in (14) over ϑ one could apply the method of stationary phase. As a result we find that

$$\varphi_k(\mathbf{r}, t) = -\frac{2i}{\pi} \frac{1}{\vartheta_0^2 + \gamma^{-2}} \frac{1}{kr} e^{ik(r-t)}, \tag{18}$$

where $r \approx z + \rho^2/2z$ and $\vartheta_0 = \rho/z$ is the point of stationary phase of the integral (14). We see that the components (18) of our packet have in the case under consideration the form of diverging spherical waves. Under this condition the angle ϑ_0 corresponds to the direction of radiation, and the function before the diverging wave describes the angular distribution of the radiation. So, the condition (17) draws out the wave zone in application to given problem.

The value $2\gamma^2\lambda$ presenting in the condition (17) is known in the theory of radiation by ultrarelativistic particles as the *formation length* or the *coherence length* [2, 3].

4. Transition radiation by a "half-bare electron"

High-energy packets of electromagnetic waves considered above manifest themselves in many problems connected with bremsstrahlung and diffraction radiation (see, e.g., [5, 14, 15]). Let us pay attention to some manifestations of such packets in the problem of transition radiation arising after sharp scattering of the high energy electron on large angle.

The retarded solution for the potential of the electromagnetic field after the scattering of the electron at the time moment $t = 0$ on large angle could be expressed in the following form [3]:

$$\varphi(\mathbf{r}, t) = \Theta(r - t)\varphi_{\mathbf{v}}(\mathbf{r}, t) + \Theta(t - r)\varphi_{\mathbf{v}'}(\mathbf{r}, t), \tag{19}$$

where $\varphi_{\mathbf{v}}(\mathbf{r}, t)$ and $\varphi_{\mathbf{v}'}(\mathbf{r}, t)$ are potentials of the Coulomb field of the electrons moving all the time with the velocity $\mathbf{v}$ along the z axis and with the velocity $\mathbf{v}'$ along the z' axis, respectively. Equation (19) demonstrates that after scattering of the electron at $t = 0$ its eigenfield strips out and after that transforms into the radiation field. In the direction of the final particle's motion the electron's eigenfield arises only in the region $r < t$ which is achieved by the signal about the scattering act at $t = 0$ (see figure 1, where the isolines of the scalar potential (19) are presented).

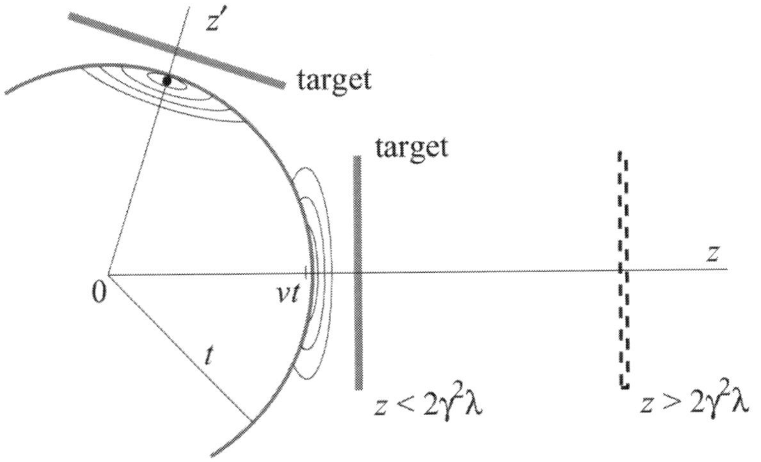

Figure 1. Equipotential surfaces of (19) and possible positions of targets for producing the transition radiation

Consider the Fourier decomposition of (19):

$$\varphi(\mathbf{r}, t) = \frac{e}{2\pi^2} \operatorname{Re} \int \frac{d^3k}{k} e^{i\mathbf{k}\mathbf{r}} \left\{ \frac{1}{k - \mathbf{k}\mathbf{v}} e^{-ikt} + \frac{1}{k - \mathbf{k}\mathbf{v}'} \left[1 - e^{-i(k - \mathbf{k}\mathbf{v}')t}\right] e^{-i\mathbf{k}\mathbf{v}'t} \right\}. \tag{20}$$

The first term in this formula has the form of the packet of free waves moving along initial direction of the electron's velocity $\mathbf{v}$. This packet coincides with the electron's eigenfield at $t = 0$. According to (17), (18), the transformation of the Fourier components of this packet with the wavelength λ to the packet of diverging waves would happen on $z > 2\gamma^2\lambda$. On smaller distances the packet of waves with the given value of $|\mathbf{k}|$ would be close to the initial one.

The length $l = 2\gamma^2\lambda$ on which the formation of the wave zone takes place could have macroscopic size. For example, for the electrons of energy 50 MeV in the range of wavelengths $\lambda \sim 10^{-1}$ cm this length is about 20 meters (the measuring technique in such conditions is developed today — see, e.g. [15, 16]). So in the frames of that length one could arrange a thin target (see the target on figure 1 which is arranged along the z-axis at $z < 2\gamma^2\lambda$) and examine the "transition radiation" of the considered packet (reflection of the waves, their passage through target etc.). The characteristics of such "transition radiation" practically would not differ from the characteristics of the transition radiation of the electron moving in the same direction (however, the electron in the packet under consideration is absent). But if the target would be located on the distance $z > 2\gamma^2\lambda$ (see dashed-line box on figure 1), the features of the considered "transition radiation" would change due to the changing of the packet's shape (formation of the diverging waves).

The second term in (20) describes the field surrounding the electron after its scattering at $t = 0$, when its velocity became equal to $\mathbf{v}'$. This field consists of the electron's eigenfield moving with the velocity $\mathbf{v}'$ (the first term in square brackets in (20)) and the packet of free waves moving in the direction of $\mathbf{v}'$ coinciding at $t = 0$ with the opposite sign with Coulomb field of the electron (the second term in square brackets).

As it was demonstrated above, transformation of the packet of plane waves to the packet of diverging waves takes place on the distances $z' \sim 2\gamma^2\lambda$, where the axis z' is directed along $\mathbf{v}'$. During the time interval t over which the electron passes that distance, the substantial cancellation of the terms in the square brackets in (20) takes place. This mean that the electron stays on that distance in a "half-bare" state: the Fourier components with the wave vector $\mathbf{k}$ of its surrounding field would be suppressed comparing to the case $z' > 2\gamma^2\lambda$. Transition radiation of the electron with such field ("half-bare" electron) on the target located on the distance $z' < 2\gamma^2\lambda$ from the point of scattering (see the target on figure 1 which is arranged along the z'-axis) would be suppressed in comparison to the case $z' > 2\gamma^2\lambda$. The results obtained are valid for sharp scattering of an electron at a large angle. Sharp scattering means that it takes place on the length which is much smaller than the coherence length.

Note that the "half-bare" state of the electron manifests itself in the suppression of bremsstrahlung arising under collisions with atoms in the frames of the coherence length [5, 17, 18]. That leads to such phenomena as Landau-Pomeranchuk-Migdal effect, the effect of suppression of the coherent bremsstrahlung in crystals and the effect of suppression of the radiation in thin layers of substance (see, e.g., [19 - 23]). Examination of the process of transition radiation by "half-bare" electron creates one more opportunity for study of manifestations of such electron under its interaction with matter.

This work is supported in part by the internal grant of Belgorod State University.

References

[1] Landau L D and Lifshitz E M 1987 *The Classical Theory of Fields* (Oxford: Pergamon)
[2] Ter-Mikaelyan M L 1972 *High-Energy Electromagnetic Processes in Condenced Media* (New York: Wiley)
[3] Akhiezer A I and Shul'ga N F 1996 *High Energy Electrodynamics in Matter* (Amsterdam: Gordon and Breach)
[4] Feinberg E L 1966 *Sov. Phys. JETP* **23** 132
[5] Shul'ga N F and Fomin S P 1986 *Phys. Lett.* A **114** 148
[6] Purcell E M 1965 *Electricity and Magnetism (Berkeley Physics Cource, v.2)* (New York: McGrow-Hill)
[7] Tsien R Y 1972 *Am. J. Phys.* **40** 46
[8] Ohanian H C 1980 *Am. J. Phys.* **48** 170
[9] Arutunian S G 1986 *Sov. Phys. Usp.* **150** 445
[10] Bolotovskii B M and Serov A V 1997 *Phys. Usp.* **40** 1055
[11] Jackson J D 1999 *Classical Electrodynamics* (New York: Wiley)
[12] Weizsäcker C 1934 *Z. Phys.* **88** 612
[13] Williams E J 1935 *Dan. Vid. Selsk. Mat. Phys. Medd.* **13** 4
[14] Shul'ga N F and Dobrovol'sky S N 2000 *JETP* **90** 579;
2003 *Nucl. Instrum. and Methods* B **201** 123
[15] Naumenko G, Artru X, Potylitsyn A, Popov Yu and Sukhikh L *Preprint* arXiv:0901.2630v1 [physics.acc-ph]
[16] Shibata Y *et al* 1994 *Phys. Rev.* E **49** 785
[17] Akhiezer A I and Shul'ga N F 1987 *Sov. Phys. Usp.* **30** 197
[18] Feinberg E L 1979 *Sov. Phys. Usp.* **22** 479
[19] Blankenbecler R and Drell S 1996 *Phys. Rev.* D **53** 6265
[20] Akhiezer A I, Shul'ga N F and Fomin S P 2005 *Landau-Pomeranchuk-Migdal Effect* (Cambridge Sci. Pub.)
[21] Klein S 1999 *Rev. of Mod. Phys.* **71** 1501
[22] Uggerhøj U I 2005 *Rev. of Mod. Phys.* **77** 1131
[23] Tompsen H D *et al* 2009 *Phys. Lett.* B **672** 323

Angular momentum and torque radiated by a relativistic charged spin particle

V. A. Bordovitsyn, O. A. Konstantinova , E. A. Nemchenko

Tomsk State University, Lenin Ave. 36, 634050 Tomsk, Russia

E-mail: vabord@sibmail.com, olgakonst87@mail.ru, katrin_nemchenko@mail.ru

Abstract. In this paper two alternative definitions of density of angular momentum of electromagnetic field by Ivanenko-Sokolov and by Teitelboim are discussed. It is shown that both definitions give identical integral characteristics of radiation for an arbitrary moving relativistic charged spin particle. The total power of orbital momentum radiation is proportional to the radiation torque. The angular momentum of radiation is proportional to the kinematical Thomas spin precession. Thus, at present the dynamical interpretation of Thomas precession is obvious. As an example the angular momentum of spin light from a point magnetic momentum moving with constant velocity is studied.

1. Introduction

In 1897 A. I. Sadowsky put forward a hypothesis that circularly polarized light must posses the proper angular momentum [1]. In 1935-1936, B. A. Beth in USA [2] and A. N. S. Holborn in England [3] experimentally proved that circularly polarized light does possess angular momentum. These precise experiments not only confirmed Sadowsky's hypothesis, but also allowed to estimate Planck's constant accurate within 10%.

Observation of considerably greater torque became possible with the emergence of lasers (see, for example, [4]). At the present time there is no doubt about the existence of the angular momentum of circularly polarized electromagnetic waves.

However, the general definition of angular momentum of the electromagnetic field (AMEF) is argued over among physicists even nowadays [4-9].

2. Two alternative methods of the AMEF definition

It may be distinguished two alternative ways of the relativistically covariant methods for the description of AMEF [10-12].

The first method was proposed by Ivanenko and Sokolov [10] on the basis of the principle of least action for the electromagnetic field. This principle leads to the conservation laws of the density of AMEF in the differential form

$$D_\lambda \mathcal{M}^{\mu\nu\lambda} = 0$$

with derivative

$$D_\lambda = \frac{\partial}{\partial R_\lambda} = \tilde{\partial}_\lambda + \frac{\tilde{r}_\lambda}{\tilde{r}_\rho \upsilon^\rho} \frac{d}{d\tau},$$

where a retardation of the radiation field per time $\Delta t = \tilde{t} - t$ is taken into account.
Here

$$\mathcal{M}^{\mu\nu\lambda} = «\mathcal{A}^{\mu\nu\lambda}» + «\mathcal{T}^{\mu\nu\lambda}»$$

is the density tensor of the total AMEF split into «orbital» and «spin» parts:

$$«\mathcal{A}^{\mu\nu\lambda}» = \frac{1}{c}\left(R^{\mu}\mathcal{T}_{c}^{\nu\lambda} - R^{\nu}\mathcal{T}_{c}^{\mu\lambda}\right), \qquad «\mathcal{T}^{\mu\nu\lambda}» = \frac{1}{4\pi c}\left(A^{\mu}H^{\nu\lambda} - A^{\nu}H^{\mu\lambda}\right)$$

respectively. In addition,

$$\mathcal{T}_{c}^{\nu\lambda} = -\frac{1}{4\pi}\left(\frac{1}{4}g^{\nu\lambda}H_{\alpha\beta}H^{\alpha\beta} + D^{\nu}A^{\rho}H_{\rho}{}^{\lambda}\right) \neq \mathcal{T}_{c}^{\lambda\nu}$$

with $D_{\lambda}\mathcal{T}_{c}^{\nu\lambda} = 0$ is the canonical tensor of the energy density. The metrics used everywhere is $g^{\mu\nu} = diag(-1,1,1,1)$.

The purpose of the quotes presence in our notations is due to the fact that the derivative $D_{\lambda}«\mathcal{A}^{\mu\nu\lambda}» = \mathcal{T}_{c}^{\nu\mu} - \mathcal{T}_{c}^{\mu\nu}$ is nonzero. This happens because the canonical tensor $\mathcal{T}_{c}^{\nu\lambda}$ isn't symmetrical and, therefore, the separation of $\mathcal{M}^{\mu\nu\lambda}$ into «orbital» and «spin» parts isn't relativistically invariant. Worse than that, $\mathcal{T}_{c}^{\nu\lambda}$ isn't gauge invariant, so that $«\mathcal{T}^{\mu\nu\lambda}»$ isn't gauge invariant either and this means that it can't claim to be the real observed value. In this connection the definition of the proper AMEF or spin

$$\boldsymbol{\Pi} = \frac{1}{4\pi c}\int[\boldsymbol{E}\boldsymbol{A}]dV$$

is required in additional substantiation.

Alternative pathway is the density tensor of the total AMEF proposed by C. Teitelboim [11]

$$\mathcal{M}^{\mu\nu\lambda} = \frac{1}{c}\left(R^{\mu}\mathcal{T}^{\nu\lambda} - R^{\nu}\mathcal{T}^{\mu\lambda}\right)$$

on basis of the symmetric density tensor of the electromagnetic field energy

$$\mathcal{T}^{\nu\lambda} = -\frac{1}{4\pi}\left[\frac{1}{4}g^{\nu\lambda}H_{\alpha\beta}H^{\alpha\beta} + H^{\nu\rho}H_{\rho}{}^{\lambda}\right] = \mathcal{T}^{\lambda\nu}$$

which possesses gauge-invariance in contrast to the $\mathcal{T}_{c}^{\nu\lambda}$. Tensor $\mathcal{T}^{\nu\lambda}$ is calculated according to C. Teitelboim in terms of radiation theory of relativistic particles in which $H^{\mu\nu} = \overline{H}^{\mu\nu} + \widetilde{H}^{\mu\nu}$, where $\overline{H}^{\mu\nu} \sim 1/\tilde{r}^{2}$ is the well-known strength tensor of the attached fields in the near-charge zone and $\tilde{H}^{\mu\nu} \sim 1/\tilde{r}$ is in the long-distance region which is called the radiation zone. The four-dimensional vector $\tilde{r}^{\mu}(\tau,\tilde{t}) = R^{\mu}(\tilde{t}) - r^{\mu}(\tau)$ traced from the trajectory radius-vector $r^{\mu}(\tau)$ in proper time τ to the spectator point of four-dimensional vector $R^{\mu}(\tilde{t})$. Now the decomposition of the density of total AMEF into orbital and spin parts arise automatically. Tensor $\mathcal{T}^{\nu\lambda}$ decomposes into three parts in agreement with the general theory of relativistic radiation $\mathcal{T}^{\nu\lambda} = \overline{\mathcal{T}}^{\nu\lambda} + \overset{\simeq}{\mathcal{T}}{}^{\nu\lambda} + \widetilde{\mathcal{T}}^{\nu\lambda}$, where $\overline{\mathcal{T}}^{\nu\lambda}$ is the energy density tensor of the convection field in the near-charge zone, $\overset{\simeq}{\mathcal{T}}{}^{\nu\lambda}$ is the energy density tensor of the mixed fields, $\widetilde{\mathcal{T}}^{\nu\lambda}$ is the energy density tensor of the radiation field in the wave zone.

In the following we will consider the AMEF only in the wave zone, for which

$$\widetilde{\mathcal{A}}^{\mu\nu\lambda} = \frac{1}{c}\left(r^{\mu}\widetilde{\mathcal{T}}^{\nu\lambda} - r^{\nu}\widetilde{\mathcal{T}}^{\mu\lambda}\right), \quad \widetilde{\mathcal{M}}^{\mu\nu\lambda} = \frac{1}{c}\left(\tilde{r}^{\mu}\widetilde{\widetilde{\mathcal{T}}}^{\nu\lambda} - \tilde{r}^{\nu}\widetilde{\widetilde{\mathcal{T}}}^{\mu\lambda}\right).$$

In terms of the derivative of a retarded function it can be shown that $D_{\lambda}\mathcal{T}^{\nu\lambda} = 0$, and also $D_{\lambda}\left(\overline{\mathcal{T}}^{\nu\lambda} + \widetilde{\widetilde{\mathcal{T}}}^{\nu\lambda}\right) = 0$, $D_{\lambda}\widetilde{\mathcal{T}}^{\nu\lambda} = 0$. From this one can obtain $D_{\lambda}\widetilde{\mathcal{A}}^{\mu\nu\lambda} = 0$, $D_{\lambda}\widetilde{\mathcal{M}}^{\mu\nu\lambda} = 0$, and as a consequence $D_{\lambda}\mathcal{M}^{\mu\nu\lambda} = 0$.

Thus, the decomposition of the density of the total AMEF onto orbital and spin parts is now gauge and relativistically invariant.

3. Relativistic theory of the AMEF radiation

First we write expressions for the density tensors of orbital momentum in explicit form with regard to the wave zone and the density tensor of the spin momentum.Now we can obtain orbital and spin momenta of the radiation field on basis of Gauss's integral theorem in the form

$$\widetilde{L}^{\mu\nu} = \oint \mathcal{A}^{\mu\nu\lambda} d\sigma_{\lambda}, \quad \widetilde{\Pi}^{\mu\nu} = \oint \mathcal{M}^{\mu\nu\lambda} d\sigma_{\lambda}.$$

Here $d\sigma_{\lambda} = e_{\lambda}\varepsilon^{2} c\, d\tau\, d\Omega_{0}$ is an element of spacelike hypersurface with the normal unit vector

$$e_{\lambda} = -c\frac{\tilde{r}_{\lambda}}{\tilde{r}_{\rho}\upsilon^{\rho}} - \frac{\upsilon_{\lambda}}{c}$$

with $\upsilon^{\mu} = dr^{\mu}/d\tau$, and $d\Omega_{0}$ is a solid angle element in the rest frame of the particle. Integrating $\widetilde{L}^{\mu\nu}$ and $\widetilde{\Pi}^{\mu\nu}$ over the angles in the rest frame by well-known (see [12]) integrals we can obtain expressions for the power of the orbital and spin AMEF radiation in the wave zone

$$\frac{d\widetilde{L}^{\mu\nu}}{d\tau} = \frac{2}{3}\frac{e^{2}}{c^{5}} w_{\rho} w^{\rho}\left(r^{\mu}\upsilon^{\nu} - r^{\nu}\upsilon^{\mu}\right), \quad \frac{d\widetilde{\Pi}^{\mu\nu}}{d\tau} = \frac{2}{3}\frac{e^{2}}{c^{3}}\left(\upsilon^{\mu}w^{\nu} - \upsilon^{\nu}w^{\mu}\right)$$

where $w^{\mu} = d\upsilon^{\mu}/d\tau$.

Our formulas have clear-cut physical interpretation. If we put the well-known charge power of the radiation into operation [13] $d\widetilde{P}^{\mu}/d\tau = \left(2e^{2}/3c^{5}\right) w_{\rho} w^{\rho}\upsilon^{\mu}$, then we can get the expression for the orbital momentum of radiation

$$\frac{d\widetilde{L}^{\mu\nu}}{d\tau} = r^{\mu}\frac{d\widetilde{P}^{\nu}}{d\tau} - r^{\nu}\frac{d\widetilde{P}^{\mu}}{d\tau} = \frac{d}{d\tau}(r^{\mu}\widetilde{P}^{\nu} - r^{\nu}\widetilde{P}^{\mu}),$$

which is exactly same relation between the torque tensor of the particle and its angular momentum tensor derivative as in the relativistic mechanics $\tilde{T}^{\mu\nu} = r^{\mu}\tilde{F}^{\nu} - r^{\nu}\tilde{F}^{\mu} = d\tilde{L}^{\mu\nu}/d\tau$ Thus, the radiation of the orbital AMEF adds up to the radiation of the field momentum of the force.

The power of the radiation of the spin momentum is proportional to the Thomas's precession frequency $\Omega_{Th}^{\mu\nu}$ which is defined by this equation in the relativistic spin theory (see. [12])

$$\left(\frac{d\pi^{\mu}}{d\tau}\right)_{Th} = \frac{1}{c^{2}}\left(\upsilon^{\mu}w^{\nu} - \upsilon^{\nu}w^{\mu}\right)\pi_{\nu} = \Omega_{Th}^{\mu\nu}\,\pi_{\nu},$$

where π^{μ} is a well-known four-dimensional space-like spin vector. This shows that the Thomas precession, which until now was known as purely kinematic relativistic effect, becomes obvious dynamic interpretation.

4. Spin light radiation

The spin light [13] corresponds to the electromagnetic radiation of a relativistic spin particle with a proper magnetic momentum. Such radiation arises only from the electrically neutral particles (neutron [14], neutrino [15]). If these particles have also an electrical charge besides the magnetic momentum, the interference of radiations from the accelerated charge and from the proper magnetic momentum arises. This phenomenon is the most noticeably observed just by extreme electron's energies in the synchrotron accelerators [16]. The radiation associated with the spin leads to the radiative self-polarization of relativistic electrons known as Sokolov-Ternov effect [17, 18]. In what follows the problem of identification of the spin light against the background of powerful synchrotron radiation, of recoil effects and of other relativistic phenomena was studied in details on the basis of the semiclassical theory of relativistic radiation [13, 19].

At the present time spin light as a special type of electromagnetic radiation is discussed in the case not only of synchrotron and undulator radiation of electrons [13, 19-21], but also in the case of high energy neutrino radiation in matter and in gravitational fields [21-24].

For the construction of our theory we will use a well-known method of the classical radiation theory of relativistic particles [11, 13 and 19].

As starting relations for the consideration of AMEF of the spin light we use already known four-dimensional vector-potentials $A^{\mu}{}_{e}$ and $A^{\mu}{}_{m}$, which obtained on the basis of the Lienard-Viehert potentials for electric charge and the Hertz's tensor-potential for the proper magnetic dipole of the particle

$$A^{\mu}{}_{e} = -e\frac{\upsilon^{\mu}}{\tilde{r}_{\rho}\upsilon^{\rho}}, \quad A^{\mu}{}_{m} = \overline{A}^{\mu}{}_{m} + \widetilde{A}^{\mu}{}_{m},$$

$$\overline{A}^{\mu}{}_{m} = c^{3}\frac{\Pi^{\mu\lambda}\tilde{r}_{\lambda}}{\left(\tilde{r}_{\rho}\upsilon^{\rho}\right)^{3}}, \quad \widetilde{A}^{\mu}{}_{m} = -c\left[\frac{\overset{\circ}{\Pi}{}^{\mu\lambda}\tilde{r}_{\lambda}}{\left(\tilde{r}_{\rho}\upsilon^{\rho}\right)^{2}} - \frac{\Pi^{\mu\lambda}\tilde{r}_{\lambda}}{\left(\tilde{r}_{\rho}\upsilon^{\rho}\right)^{2}}\tilde{r}_{\sigma}w^{\sigma}\right].$$

Here index e refers to the charge and m - to the magnetic momentum respectively. Symbol $\circ$ means a derivative with respect to proper time τ.

The stress tensor of electromagnetic field is defined according to the general rule

$$H^{\mu\nu} = D^{\mu}A^{\nu} - D^{\nu}A^{\mu}.$$

With given expressions of vector potentials and in accordance with definitions of the stress tensor and retarded derivative it's possible to find the field strengths

$$H^{\mu\nu}_{e} = \overline{H}^{\mu\nu}{}_{e} + \widetilde{H}^{\mu\nu}{}_{e},$$

$$\overline{H}^{\mu\nu}{}_{e} = ec^{2}\frac{\tilde{r}^{[\mu}\upsilon^{\nu]}}{\left(\tilde{r}_{\rho}\upsilon^{\rho}\right)^{3}}, \quad \widetilde{H}^{\mu\nu}{}_{e} = e\left\{-\frac{\tilde{r}^{[\mu}w^{\nu]}}{\left(\tilde{r}_{\rho}\upsilon^{\rho}\right)^{2}} + \frac{\tilde{r}^{[\mu}\upsilon^{\nu]}}{\left(\tilde{r}_{\rho}\upsilon^{\rho}\right)^{3}}\tilde{r}_{\sigma}w^{\sigma}\right\}$$

- for the charge, and

$$H^{\mu\nu}_{m} = \overline{H}^{\mu\nu}{}_{m} + \widetilde{\widetilde{H}}^{\mu\nu}{}_{m} + \widetilde{H}^{\mu\nu}{}_{m},$$

$$\overline{H}^{\mu\nu}_{\ m} = \frac{c^3}{\left(\tilde{r}_\rho \upsilon^\rho\right)^3}\left\{-2\Pi^{\mu\nu} + 3\frac{\Pi^{[\mu\lambda}\tilde{r}_\lambda \upsilon^{\nu]}}{\tilde{r}_\rho \upsilon^\rho} + 3c^2\frac{\Pi^{[\mu\lambda}\tilde{r}_\lambda \tilde{r}^{\nu]}}{\left(\tilde{r}_\rho \upsilon^\rho\right)^2}\right\},$$

$$\widetilde{H}^{\mu\nu}_{\ m} = \frac{c}{\tilde{r}_\rho \upsilon^\rho}\left\{\frac{\overset{\circ\circ}{\Pi}{}^{[\mu\lambda}\tilde{r}_\lambda \tilde{r}^{\nu]}}{\left(\tilde{r}_\rho \upsilon^\rho\right)^2} - \frac{\Pi^{[\mu\lambda}\tilde{r}_\lambda \tilde{r}^{\nu]}\tilde{r}_\sigma \overset{\circ}{w}{}^\sigma + 3\overset{\circ}{\Pi}{}^{[\mu\lambda}\tilde{r}_\lambda \tilde{r}^{\nu]}\tilde{r}_\sigma w^\sigma}{\left(\tilde{r}_\rho \upsilon^\rho\right)^3} + 3\frac{\Pi^{[\mu\lambda}\tilde{r}_\lambda \tilde{r}^{\nu]}\left(\tilde{r}_\sigma w^\sigma\right)^2}{\left(\tilde{r}_\rho \upsilon^\rho\right)^4}\right\}$$

$$\overset{\approx}{H}{}^{\mu\nu}_{\ m} = \frac{c}{\left(\tilde{r}_\rho \upsilon^\rho\right)^2}\left\{2\overset{\circ}{\Pi}{}^{\mu\nu} - 2\Pi^{\mu\nu}\frac{\tilde{r}_\sigma w^\sigma}{\tilde{r}_\rho \upsilon^\rho} - \frac{\overset{\circ}{\Pi}{}^{[\mu\lambda}\upsilon_\lambda \tilde{r}^{\nu]} + 2\overset{\circ}{\Pi}{}^{[\mu\lambda}r_\lambda \upsilon^{\nu]} + \Pi^{[\mu\lambda}\tilde{r}_\lambda w^{\nu]}}{\tilde{r}_\rho \upsilon^\rho} - \right.$$

$$\left. -3c^2\frac{\overset{\circ}{\Pi}{}^{[\mu\lambda}\tilde{r}_\lambda \tilde{r}^{\nu]} - \Pi^{[\mu\lambda}\tilde{r}_\lambda \upsilon^{\nu]}\tilde{r}_\sigma w^\sigma}{\left(\tilde{r}_\rho \upsilon^\rho\right)^2} + 6c^2\frac{\Pi^{[\mu\lambda}\tilde{r}_\lambda \upsilon^{\nu]}\tilde{r}_\sigma w^\sigma}{\left(\tilde{r}_\rho \upsilon^\rho\right)^3}\right\},$$

- for the magnetic moment. Here a bracket denotes antisymmetrization: $a^{[\mu}b^{\nu]} = a^\mu b^\nu - a^\nu b^\mu$.
Unlike to the charge strength, the tensor of the magnetic momentum consists of three parts: the convection field with $\overline{H}^{\mu\nu}_{\ m} \sim 1/\tilde{r}^3$, the radiation field with $\widetilde{H}^{\mu\nu}_{\ m} \sim 1/\tilde{r}$ and the mixed field $\overset{\approx}{H}{}^{\mu\nu}_{\ m} \sim 1/\tilde{r}^2$.
Since the fields caused by the electrical charge are well-known in the literature (see. [11, 13 and 19]), so in the following we'll pay more attention to the tensor $H^{\mu\nu}_m$.
In spite with the complexity of the fields given above, it can be verified to satisfy the first pair of Maxwell's equations. In the retarded form these equations have the following form

$$D^\mu H^{\rho\nu}_m + D^\nu H^{\mu\rho}_m + D^\rho H^{\nu\mu}_m = 0.$$

The second pair of Maxwell's equations are, again, valid on the world line of the particle

$$D_\nu H^{\mu\nu}_m = 0.$$

It should be emphasized that the first pair of Maxwell's equations is valid independently over all terms with the same derivatives with respect to τ in contrast to the second pair. It should be noted that both of the invariants of magneton electromagnetic field in the wave zone are vanished (as expected in the case of the radiation field of charge)

5. Symmetrical stress tensor of spin light
A dynamic characteristic to define the stress tensor of AMEF is a symmetrical stress tensor of energy of spin light [25]

$$\mathcal{T}^{\nu\lambda} = -\frac{1}{4\pi}\left(\frac{1}{4}g^{\nu\lambda}H_{\alpha\beta}H^{\alpha\beta} + H^{\nu\rho}H_\rho^{\ \lambda}\right).$$

In general this tensor for the charge fields and for the magnetic momentum fields consists of five terms

$$\mathcal{T}^{\mu\nu} = {}_{(-6)}\mathcal{T}^{\mu\nu} + {}_{(-5)}\mathcal{T}^{\mu\nu} + {}_{(-4)}\mathcal{T}^{\mu\nu} + {}_{(-3)}\mathcal{T}^{\mu\nu} + {}_{(-2)}\mathcal{T}^{\mu\nu}$$

In the following we will consider the pure spin light and the charge radiation will be taken into account only in the mixed fields not far from the wave zone of radiation. With respect to this assumption the components of the stress tensor of the spin light we need are the follows

$$_{(-3)}\mathcal{T}^{\mu\nu} = -\frac{1}{4\pi}\left[\widetilde{H}^{\mu\rho}{}_{m}\widetilde{\widetilde{H}}_{\rho m}{}^{\nu} + \overline{H}^{\mu\rho}{}_{e}\widetilde{H}_{\rho m}{}^{\nu} + \widetilde{H}^{\mu\rho}{}_{e}\widetilde{\widetilde{H}}_{\rho m}{}^{\nu} + (\nu,\mu)\right],$$

$$_{(-2)}\mathcal{T}^{\mu\nu} = -\frac{1}{4\pi}\left[\widetilde{H}^{\mu\rho}{}_{m}\widetilde{H}_{\rho m}{}^{\nu} + \widetilde{H}^{\mu\rho}{}_{e}\widetilde{H}_{\rho m}{}^{\nu} + (\nu,\mu)\right],$$

All the invariants of the electromagnetic field which symmetrical stress tensor contains

$$\widetilde{H}_{\alpha\beta m}\widetilde{\widetilde{H}}{}^{\alpha\beta}{}_{m},\ \overline{H}_{\alpha\beta e}\widetilde{H}{}^{\alpha\beta}{}_{m} + \widetilde{H}_{\alpha\beta e}\widetilde{H}{}^{\alpha\beta}{}_{m},\ \widetilde{H}_{\alpha\beta e}\widetilde{\widetilde{H}}{}^{\alpha\beta}{}_{m},\ \widetilde{H}_{\alpha\beta m}\widetilde{H}{}^{\alpha\beta}{}_{m},\ \widetilde{H}_{\alpha\beta e}\widetilde{H}{}^{\alpha\beta}{}_{m}$$

are zeros according to our hypothesis.

With the field strengths in the case of the uniform motion of the charged magneton ($\upsilon^{\mu} = const, w = \overset{\circ}{w} = 0$) we find $_{(-2)}\mathcal{T}^{\mu\nu}{}_{e} = 0$ and

$$_{(-3)}\mathcal{T}^{\mu\nu} = {}_{(-3)}\mathcal{T}^{\mu\nu}{}_{m} + {}_{(-3)}\mathcal{T}^{\mu\nu}{}_{em},$$

where

$$_{(-3)}\mathcal{T}^{\mu\nu}{}_{m} = -\frac{1}{4\pi}\left[\widetilde{H}^{\mu\rho}{}_{m}\widetilde{\widetilde{H}}_{\rho m}{}^{\nu} + (\nu,\mu)\right],\ {}_{(-3)}\mathcal{T}^{\mu\nu}{}_{em} = -\frac{1}{4\pi}\left[\overline{H}^{\mu\rho}{}_{m}\widetilde{\widetilde{H}}_{\rho m}{}^{\nu} + (\nu,\mu)\right],$$

$$_{(-2)}\mathcal{T}^{\mu\nu}{}_{m} = -\frac{1}{4\pi}\widetilde{H}^{\mu\rho}{}_{m}\widetilde{H}_{\rho m}{}^{\nu}.$$

In the explicit form these formulas correspond to

$$_{(-3)}\mathcal{T}^{\mu\nu}{}_{m} = -\frac{c^2}{2\pi}\frac{1}{(\tilde{r}_{\rho}\upsilon^{\rho})^3}\left\{\frac{\tilde{r}^{(\mu}\overset{\circ}{\Pi}{}^{\nu)\alpha}}{\tilde{r}_{\rho}\upsilon^{\rho}} + \frac{\tilde{r}^{(\mu}\upsilon^{\nu)}\overset{\circ}{\Pi}{}^{\alpha\rho}\tilde{r}_{\rho}}{(\tilde{r}_{\rho}\upsilon^{\rho})^2} + 3c^2\frac{\tilde{r}^{\mu}\tilde{r}^{\nu}\overset{\circ}{\Pi}{}^{\alpha\rho}\tilde{r}_{\rho}}{(\tilde{r}_{\rho}\upsilon^{\rho})^3}\right\}\frac{\overset{\circ\circ}{\Pi}_{\alpha\beta}\tilde{r}^{\beta}}{\tilde{r}_{\rho}\upsilon^{\rho}},$$

$$_{(-3)}\mathcal{T}^{\mu\nu}{}_{em} = \frac{ec^3}{4\pi}\frac{\tilde{r}^{(\mu}\overset{\circ}{\Pi}{}^{\nu)\alpha}\tilde{r}_{\alpha}}{(\tilde{r}_{\rho}\upsilon^{\rho})^5},\quad {}_{(-2)}\mathcal{T}^{\mu\nu}{}_{m} = -\frac{c^2}{4\pi}\frac{\tilde{r}_{\alpha}\overset{\circ\circ}{\Pi}{}^{\alpha\rho}\overset{\circ\circ}{\Pi}_{\rho\beta}\tilde{r}^{\beta}}{(\tilde{r}_{\rho}\upsilon^{\rho})^6}\tilde{r}^{\mu}\tilde{r}^{\nu}.$$

Here round brackets mean symmetrisation $a^{(\mu}b^{\nu)} = a^{\mu}b^{\nu} + a^{\nu}b^{\mu}$. It is noteworthy that in all these cases $_{(-3)}\mathcal{T}^{\mu\nu}\tilde{r}_{\nu} = {}_{(-2)}\mathcal{T}^{\mu\nu}\tilde{r}_{\nu} = 0$.

6. Orbital and proper AMEF of spin light

With respect to Teitelboim's assumption [11] the full spin stress tensor AMEF is defined like

$$\mathcal{M}^{\mu\nu\lambda} = \frac{1}{c}\left(R^{\mu}\mathcal{T}^{\nu\lambda} - R^{\nu}\mathcal{T}^{\mu\lambda}\right).$$

The division of the full AMEF into orbital and spin parts has gauge and relativistic invariance in opposed to the traditional formulation (see Sec.2):

$$\mathcal{M}^{\mu\nu\lambda} = \mathcal{A}^{\mu\nu\lambda} + \mathcal{J}\mathcal{T}^{\mu\nu\lambda}.$$

Then we will consider AMEF located at the wave zone, where

$$\tilde{\mathcal{A}}^{\mu\nu\lambda} = \frac{1}{c}\left(r^{\mu}\underset{(-2)}{\mathcal{T}}^{\nu\lambda} - r^{\nu}\underset{(-2)}{\mathcal{T}}^{\mu\lambda}\right), \quad \tilde{\mathcal{I}}^{\mu\nu\lambda} = \frac{1}{c}\left(\tilde{r}^{\mu}\underset{(-3)}{\mathcal{T}}^{\nu\lambda} - \tilde{r}^{\nu}\underset{(-3)}{\mathcal{T}}^{\mu\lambda}\right).$$

Using $\underset{(-2)}{\mathcal{T}}^{\mu\nu}$ spin stress tensor $\tilde{\mathcal{I}}^{\mu\nu\lambda}$ vanishes and the first nonzero term is $\underset{(-3)}{\mathcal{T}}^{\mu\nu}$.

The formulas written above are very cumbersome that is why we'll deal with the uniform motion of the charged magneton. Moreover, according to relations $\upsilon^{\mu} = const, w = \mathring{w} = 0$ may be written $\Pi^{\mu\nu}\upsilon_{\nu} = \mathring{\Pi}^{\mu\nu}\upsilon_{\nu} = \overset{\circ\circ}{\Pi}^{\mu\nu}\upsilon_{\nu} = 0,$ as follows from the spacelikeness of tensor $\Pi^{\mu\nu}$.

Then we can find the stress tensor of the orbital momentum. In a similar fashion we can consider also the spin AMEF. For the uniform motion of the magnetic momentum in the wave zone the stress tensor of the orbital and spin AMEF is defined in accordance with

$$\widetilde{\mathcal{A}}^{\mu\nu\lambda}{}_{m} = -\frac{c}{4\pi}\frac{\tilde{r}_{\alpha}\overset{\circ\circ}{\Pi}{}^{\alpha\rho}\,\overset{\circ\circ}{\Pi}_{\rho\beta}\,\tilde{r}^{\beta}}{\left(\tilde{r}_{\rho}\upsilon^{\rho}\right)^{6}} r^{[\mu}\tilde{r}^{\nu]}\tilde{r}^{\lambda}, \quad \widetilde{\mathcal{I}}^{\mu\nu\lambda}{}_{m} = -\frac{c}{2\pi}\frac{1}{\left(\tilde{r}_{\rho}\upsilon^{\rho}\right)^{5}}\left(\tilde{r}^{[\mu}\mathring{\Pi}^{\nu]\alpha} + \tilde{r}^{[\mu}\upsilon^{\nu]}\frac{\mathring{\Pi}^{\alpha\rho}\,\tilde{r}_{\rho}}{\tilde{r}_{\rho}\upsilon^{\rho}}\right)\overset{\circ\circ}{\Pi}_{\alpha\beta}\,\tilde{r}^{\beta}\tilde{r}^{\lambda}.$$

It is easy to see that in the wave zone $D_{\lambda}\widetilde{\mathcal{A}}^{\mu\nu\lambda}{}_{m} = 0$, $D_{\lambda}\widetilde{\mathcal{I}}^{\mu\nu\lambda}{}_{m} = 0$ and from this all the radiation of the orbital AMEF off the world line moves through the spacelike plane with a surface element $d\sigma_{\lambda}$ (see Sec.2).

Integrating in this relation over $d\sigma_{\lambda}$, taking well-known angular integrals into account (see Sec.3) we can get a radiation power of the orbital and spin AMEF independent on the reference frame

$$\frac{d\widetilde{L}^{\mu\nu}{}_{m}}{d\tau} = \frac{1}{3c^{5}}\overset{\circ\circ}{\Pi}_{\alpha\beta}\,\overset{\circ\circ}{\Pi}{}^{\alpha\beta}\left(r^{\mu}\upsilon^{\nu} - r^{\nu}\upsilon^{\mu}\right). \quad \frac{d\widetilde{\Pi}^{\mu\nu}{}_{m}}{d\tau} = -\frac{2}{3c^{2}}\left(\mathring{\Pi}^{\mu\rho}\,\overset{\circ\circ}{\Pi}{}^{\nu}{}_{\rho} - \mathring{\Pi}^{\nu\rho}\,\overset{\circ\circ}{\Pi}{}^{\mu}{}_{\rho}\right).$$

It is obvious that this magnitude is proportional to the tensor of angular momentum of this very particle $P^{\mu\nu} = r^{\mu}\upsilon^{\nu} - r^{\nu}\upsilon^{\mu}$. In the case of the uniform motion all the components are vanished. Thus $d\widetilde{L}^{\mu\nu}{}_{m}/d\tau = 0$. Note that in the case of the mixed radiation of the orbital AMEF is also vanish due to $\widetilde{H}^{\mu\nu}{}_{e} = 0$.

Similarly we can construct the stress tensor AMEF for mixed radiation

$$\widetilde{\mathcal{I}}^{\mu\nu\lambda}{}_{em} = \frac{ec^{2}}{4\pi}\frac{\tilde{r}^{[\mu}\,\mathring{\Pi}^{\nu\rho]}}{\left(\tilde{r}_{\rho}\upsilon^{\rho}\right)^{5}}\tilde{r}_{\rho}\tilde{r}^{\lambda},$$

and again from $D_{\lambda}\widetilde{\mathcal{I}}^{\mu\nu\lambda}{}_{em} = 0,$ the total AMEF power of interferential radiation is

$$\frac{d\widetilde{\Pi}^{\mu\nu}{}_{em}}{d\tau} = \frac{2e^{2}}{3c^{3}}\overset{\circ\circ}{\Pi}{}^{\mu\nu}.$$

7. Conclusion

Thus, in this work two alternative definitions of density of intrinsic angular momentum of electromagnetic field by Ivanenko-Sokolov and by Teitelboim are discussed. It is shown that in spite of the absence of the outward sign of similarity both definitions give identical integral characteristics of radiation for an arbitrary moving relativistic charge. The total power of orbital momentum radiation is proportional to the relativistic angular momentum of the particle, and corresponds to the radiation torque. The spin angular momentum of radiation is proportional to the kinematical Thomas spin precession, which becomes obvious dynamical interpretation.
To sum up, it can be stated that this work has discovered a new trend in the relativistic theory of radiation and the spin properties of relativistic particles. In this paper we also demonstrate a new phenomenon of nature – the proper angular momentum of electromagnetic of spin light radiation.

This work was supported by RF President grant № SS 871.2008.2, and by the Russian Science and Innovations Federal Agency under contract № 02.740.11.0238.

References

[1] Sadowsky A I 1987 *Zhurn. Russ. Fiz.-Khim. Ob.-va*, Fiz. I, Vyp. **2** 82 (in Russian) see also 1899 *Acta et Comment. Imp. Univ. Jurien* 7, No. 1-2 (in French)
[2] Beth B A 1935 *Phys. Rev.* **48** 471; 1936 **50** 115
[3] Holborn A N S 1936 *Nature* **137** 31
[4] Akhmanov S A and Nikitin S Yu 1998 *Fiz. Optica* (MSU, Moskow) (in Russian)
[5] Ohanian H C 1986 *Amer. J. Phys.* **54** 500
[6] Wulfson K S 1987 *Usp. Fiz. Nauk* **152** 668
[7] Sokolov I V 1991 *Usp. Fiz. Nauk* **161** 175
[8] Barabanov A L 1993 *Usp. Fiz. Nauk* **163** 75
[9] Rosenberg G V 1950 *Usp. Fiz. Nauk* **40** 328 (in Russian)
[10] Ivanenko D D and Sokolov A A 1949 *Classical Field Theory* (GITTL, Moskow-Leningrad), in Russian; see also 1953 *Klassische Feld theorie* (Akademie-Verlag, Berlin), in German
[11] Teitelboim C and Villarroel D 1980 *Nuovo Cim.* **3** 1
[12] Ch. G. van Weert 1973 *Physica* **65** 452
[13] Bordovitsyn V A, Ternov I M and Bagrov V G 1995 *Usp. Fiz. Nauk* **165** 1083
[14] Ternov I M, Bagrov V G and Khapaev A M 1965 *JETP* **48** 961
[15] Lobanov A E and Studenikin A I 2003 *Phys. Lett* **564B** 27
[16] Bondar A E and Saldin E L 1982 *Nuel. Instrum. Meth* **195** 577
[17] Sokolov I V and Ternov I M 1963 *Dokl. Akad. Nauk* **153** 1052
[18] Ternov I M 1995 *Usp. Fiz. Nauk* **38** 1037
[19] *Synchrotron Radiation Theory and its Development* 1999 In Memory of I. M. Ternov. ed. V. A. Bordovitsyn (World Scientific, Singapore)
[20] Kulipanov G N, Bondar A E, Bordovitsyn V A et al 1995 *Nuel. Instrum. Meth* **359A** 34
[21] Bordovitsyn V and Epp V Ya 1998 *Nuel. Instrum. Meth* **405A** 220
[22] Lobanov A E 2006 *J Phys.* **39A** 7517; 2005 *Phys. Lett* **619B** 136
[23] Grigoriev A, Shinkevich S and Studenikin A 2006 *Particle Physics at the year of 250th Lomonosov Conf. on Element. Particle Physics* ed. by A. Studenikin, (World Scientific, Singapore)
[24] Dvornikov M, Grigoriev A and Studenikin A 2005 *Int. J. Mod. Phys.* **14D** 309

New approach to the theory of a moving charge's radiation in dispersive medium and its application to the case of left-handed materials

S.N. Galyamin, A.V. Tyukhtin

Radiophysics Department, Saint Petersburg State University, Saint Petersburg, 198504, Russia

galiaminsn@yandex.ru, tyukhtin@bk

Abstract. We present a new approach to the analysis of the field of a uniformly moving charge in medium with frequency dispersion. This is based on methods of complex function theory. First, we give the deduction of expressions for the field components directly from the causality principle. Second, the properties of integrands and integrals for the field components in the complex plane are discussed. Further, we use our approach for the case of a "left-handed" medium. The charge's field is presented as the sum of different components having clear physical interpretation: the "quasi-Coulomb" field, the wave field, and the "plasma trace". An efficient method for computing the field is developed. Additionally, the case of a boundary between an ordinary medium and a left-handed one is considered. The phenomenon of reversed Cherenkov-transition radiation is described.

1. On using the causality principle for the problem of a point charge moving in a medium

It is well known that solving the field equations calls for the use of one of the radiation principles [1]. The solution of Maxwell equations for the case of the uniformly moving point charge has been obtained primarily by Tamm and Frank [2] with the use of the so-called Sommerfeld's radiation condition (i.e., requirement that the phase velocity is directed away from the source). Later the absorption principle was used by Fermi for a similar problem [3]. This condition demands the field be exponentially decreased with increasing distance from the source (in lossy medium). However, the Sommerfeld's condition is false for the "left-handed medium", and the absorption principle is not true for active medium. Therefore, if we deal with a medium having unusual ("exotic") properties it is essential to obtain a solution from the most fundamental condition, which is the causality principle [1].

We can use any initial-value problem implying uniform linear motion as a limit case (one of examples is the Tamm's problem, but this is commonly solved with the use of the Sommerfeld's condition [4]). In our opinion, the following problem is more convenient for the indicated purpose. Two point particles with charges q and $-q$ are born at a moment t_1 at a point $x = y = 0$, $z = z_1$ and start to move with constant velocities $\vec{V} = V\vec{e}_z$ and $-\vec{V} = -V\vec{e}_z$, correspondingly. In this case the charge density ρ and the current density $\vec{j} = j\vec{e}_z$ have the form

$$\{\rho, j\} = q\{1, V\}\,\delta(x)\delta(y)[\delta(z - Vt) \mp \delta(z + V(t - 2t_1))]\Theta(t - t_1)\,, \qquad (1)$$

where $\Theta(\xi)$ is the Heaviside step function. For source (1), the causality principle means that the field equals 0 for $t < t_1$.

It is assumed that the medium described by permittivity $\varepsilon(\omega)$ and permeability $\mu(\omega)$ is passive, isotropic, and does not have spatial dispersion. We use potentials $\vec{A} = A\vec{e}_z$ and Φ with the Lorenz gauge condition. The solution of the initial-value problem can be obtained by the Fourier method. The result of calculations can be written in the following form:

$$\begin{Bmatrix} A \\ \Phi \end{Bmatrix} = \frac{-\mathrm{i}q}{4\pi^2 V}\int_{-\infty}^{+\infty}\mathrm{d}k_\rho \int_{-\infty}^{+\infty}\mathrm{d}k_z \int_{-\infty}^{+\infty}\mathrm{d}\omega \begin{Bmatrix} \beta\mu(\omega+\mathrm{i}0) \\ \varepsilon^{-1}(\omega+\mathrm{i}0) \end{Bmatrix} \frac{k_\rho H_0^{(1)}(k_\rho\rho)}{k_z^2 + k_\rho^2 - (\omega+\mathrm{i}0)^2 n^2(\omega+\mathrm{i}0)c^{-2}} \times$$
$$\times[(k_z - \omega V^{-1} - \mathrm{i}0)^{-1} \mp (k_z + \omega V^{-1} + \mathrm{i}0)^{-1}]\exp[\mathrm{i}k_z(z - Vt_1) + \mathrm{i}\omega(t_1 - t)], \tag{2}$$

where $n^2(\omega) = \varepsilon(\omega)\mu(\omega)$, $\rho = \sqrt{x^2 + y^2}$, and $H_0^{(1)}(k_\rho\rho)$ is the Hankel function. Note that the addend $+\mathrm{i}0$ in the square brackets has emerged owing to the Fourier-transform of $\Theta(\xi)$, but the term $+\mathrm{i}0$ in the denominator and figure brackets has been added to provide the causality. One can see that this addend shifts the corresponding poles from the real axis into the lower half-plane ω. At the same time, if $t < t_1$ then the integration path (real axis) can be enclosed in the upper half-plane ω. Since there are no poles inside it, $A = \Phi = 0$ for $t < t_1$. Note that if losses are taken into account, the poles lie below the integration path and causality is provided automatically.

To obtain the solution for the infinite particle motion we should let t_1 approach $-\infty$ at finite t, ρ, and z. Integration over k_z can be calculated using the residue theorem. As $z - Vt_1 > 0$, only the poles in the upper half plane k_z are of importance. One can obtain that $A = A^{\mathrm{I}} + A^{\mathrm{II}}$, $\Phi = \Phi^{\mathrm{I}} + \Phi^{\mathrm{II}}$, where

$$\begin{Bmatrix} A^{\mathrm{I}} \\ \Phi^{\mathrm{I}} \end{Bmatrix} = \frac{q}{2\pi V}\int_{-\infty}^{\infty}\mathrm{d}\omega \begin{Bmatrix} \beta\mu(\omega+\mathrm{i}0) \\ \varepsilon^{-1}(\omega+\mathrm{i}0) \end{Bmatrix}\int_{-\infty}^{\infty}\mathrm{d}k_\rho \frac{k_\rho H_0^{(1)}(k_\rho\rho)}{k_\rho^2 - s^2(\omega+\mathrm{i}0)}\exp\left[\mathrm{i}\omega\left(\frac{z}{V} - t\right)\right], \tag{3}$$

$$\{A^{\mathrm{II}}, \Phi^{\mathrm{II}}\} = q(4\pi V)^{-1}\int_{-\infty}^{+\infty}\mathrm{d}\omega\{\beta\mu(\omega+\mathrm{i}0), \varepsilon^{-1}(\omega+\mathrm{i}0)\}\int_{-\infty}^{+\infty}\mathrm{d}k_\rho k_\rho H_0^{(1)}(k_\rho\rho)\times$$
$$\times[k_{z0}^{-1}(k_{z0} - \omega V^{-1} - \mathrm{i}0)^{-1} \mp k_{z0}^{-1}(k_{z0} + \omega V^{-1} + \mathrm{i}0)^{-1}]\exp[\mathrm{i}k_{z0}(z - Vt_1) + \mathrm{i}\omega(t_1 - t)], \tag{4}$$

$$s^2(\omega+\mathrm{i}0) = (\omega+\mathrm{i}0)^2 V^{-2}(n^2(\omega+\mathrm{i}0)\beta^2 - 1), \quad k_{z0} = \sqrt{(\omega+\mathrm{i}0)^2 n^2(\omega+\mathrm{i}0)c^{-2} - k_\rho^2}, \quad \mathrm{Im}\,k_{z0} > 0. \tag{5}$$

One can integrate in (3) over k_ρ utilising the residue theorem once again. Integral (4) is convenient for the asymptotic evaluation with the steepest-descent technique. As a result, A^{II} and Φ^{II} contain the saddle point contribution ($\sim[\rho^2 + (z - Vt_1)^2]^{-1/2}$) and possibly the contribution of poles ($\sim\rho^{-1/2}$). Omitting calculations, we write the result for the limit $t_1 \to -\infty$ at finite t, ρ, and z:

$$\begin{Bmatrix} A \\ \Phi \end{Bmatrix} = \frac{\mathrm{i}q}{2V}\int_{-\infty}^{+\infty}\mathrm{d}\omega \begin{Bmatrix} \beta\mu(\omega+\mathrm{i}0) \\ \varepsilon^{-1}(\omega+\mathrm{i}0) \end{Bmatrix} H_0^{(1)}(s(\omega+\mathrm{i}0)\rho)\exp\left[\mathrm{i}\omega\left(\frac{z}{V} - t\right)\right], \quad s(\omega+\mathrm{i}0) = \sqrt{s^2(\omega+\mathrm{i}0)}, \tag{6}$$

where $\mathrm{Im}\,s(\omega+\mathrm{i}0) > 0$. Formula (6) represents the field of a uniformly moving charge for any isotropic homogeneous passive medium. The analogous result is obtained with the help of the absorption principle. However, the Sommerfield's condition can give false results for some situations (for example, for left-handed media).

2. The approach to the analysis of the integrals for the charge field components

Using (6), one can write expressions for the field components in the following form:

$$\{E_\rho, E_z, H_\varphi\} = q/(2c)\int_{-\infty}^{+\infty}\mathrm{d}\omega\{e_\rho(\omega+\mathrm{i}0), e_z(\omega+\mathrm{i}0), h_\varphi(\omega+\mathrm{i}0)\}\exp(\mathrm{i}\omega\zeta V^{-1}), \tag{7}$$

$$e_\rho(\omega) = \frac{\mathrm{i}s(\omega)}{\beta\varepsilon(\omega)}H_1^{(1)}(s(\omega)\rho), \quad e_z(\omega) = \frac{-c\omega}{\varepsilon(\omega)}s^2(\omega)H_0^{(1)}(s(\omega)\rho), \quad h_\varphi(\omega) = \mathrm{i}s(\omega)H_1^{(1)}(s(\omega)\rho), \tag{8}$$

where $\zeta = z - Vt$. First of all, let us extract consequences from the requirement $\operatorname{Im} s(\omega + \mathrm{i}0) > 0$ in the case of a lossless medium. Expanding $s^2(\omega + \mathrm{i}0)$ into a Taylor series in the vicinity of ω we get

$$s(\omega + i0) = \begin{cases} |s| \operatorname{sgn} \mathrm{d}s^2/\mathrm{d}\omega & \text{for } s^2(\omega) > 0 \\ \mathrm{i}|s| & \text{for } s^2(\omega) < 0 \end{cases}, \quad \operatorname{sgn}\frac{\mathrm{d}s^2}{\mathrm{d}\omega} = \operatorname{sgn}\left[\omega\left(1 + \frac{\beta^2}{2(n^2\beta^2 - 1)}\omega\frac{\mathrm{d}n^2}{\mathrm{d}\omega}\right)\right]. \quad (9)$$

For the normal dispersion (where $\omega \mathrm{d}n^2/\mathrm{d}\omega > 0$), one obtains $\operatorname{sgn} \mathrm{d}s^2/\mathrm{d}\omega = \operatorname{sgn}\omega$ and Sommerfeld's condition is fulfilled, but for anomalous dispersion ($\omega \mathrm{d}n^2/\mathrm{d}\omega < 0$) $\operatorname{sgn} \mathrm{d}s^2/\mathrm{d}\omega$ can equal both $\operatorname{sgn}\omega$ and $-\operatorname{sgn}\omega$. In the latter case Sommerfeld's condition fails. Note that expression (9) for s satisfies the so-called Mandelshtam's radiation condition [1] requiring the group velocity $\vec{V}_{\mathrm{g}}$ (and the energy flow density) be directed away from the source: $V_{\mathrm{g}\rho} = \mathrm{d}\omega/\mathrm{d}s = 2s(\mathrm{d}s^2/\mathrm{d}\omega)^{-1} > 0$.

The essence of our approach is the following. We determine the function s in the complex plane ω in such a way that (9) is fulfilled on the real axis. We draw the cuts in segments where $\operatorname{Im} s = 0$, fixing the "physical" sheet of the Riemann surface by the rule $\operatorname{Im} s > 0$. Next, the following helpful properties take place (for more detail see [5–7]). If the integration path Γ consists of two parts, with one of them (Γ_+) lying in the domain $\operatorname{Re}\omega > 0$, another ($\Gamma_-$) lying in the domain $\operatorname{Re}\omega < 0$, and the total Γ being symmetrical with respect to the imaginary axis, then we get for integrals (7):

$$\int_{\Gamma} f(\omega)\exp(\mathrm{i}\omega\zeta V^{-1})\mathrm{d}\omega = 2\int_{\Gamma_+} \operatorname{Re}[f(\omega)\exp(\mathrm{i}\omega\zeta V^{-1})\mathrm{d}\omega], \quad f(\omega) = \{e_\rho(\omega), e_z(\omega), h_\varphi(\omega)\}. \quad (10)$$

As we will see later, all our integration paths possess the required symmetry. Thus, we can consider the domain $\operatorname{Re}\omega > 0$ only. Next, the asymptote of the steepest-descent path (SDP) for integrands in (7) is determined by the equation

$$\rho\sqrt{1-\beta^2}\, \operatorname{Im}\omega = \zeta\,|\operatorname{Re}\omega| + \text{const}. \quad (11)$$

Further manipulations are based on applying methods of complex function theory together with properties (9) – (11) to the calculation of (7).

3. The case of a left-handed medium

Now we apply our approach to the case of a left-handed medium (LHM). LHM has simultaneously negative permittivity $\varepsilon(\omega)$ and permeability $\mu(\omega)$. Contrary to the ordinary "right-handed" medium (RHM), vectors $\vec{E}$, $\vec{H}$, and $\vec{k}$ form a left-handed orthogonal set in LHM. Therefore, the Poyting vector $\vec{S} = c(4\pi)^{-1}\vec{E}\times\vec{H}$ is opposite to the phase velocity. These peculiarities result in the unusual properties of wave processes in LHM [8,9].

Note that the "left-handed properties" can be realised only for a limited frequency range [8,9]. Therefore, it would be more correct to refer to a "left-handed frequency band" (LHFB) as opposed to a "right-handed frequency band" (RHFB). Nevertheless, we will use the term LHM with the understanding that LHM is a medium possessing both LHFB and RHFB, whereas RHM is a medium with RHFB only.

LHM were realised recently in the gigahertz frequency band by means of artificial metamaterials [10–13]. Cherenkov radiation (CR) in an infinite or semi-infinite LHM was investigated in [14–16]. However, the main attention was given to the analysis of energetic characteristics of the wave field while the field structure was practically not studied.

We use the following typical model for the left-handed metamaterial:

$$\varepsilon(\omega) = 1 - \omega_{\mathrm{pe}}^2(\omega^2 + 2\mathrm{i}\omega_{\mathrm{de}}\omega)^{-1}, \qquad \mu(\omega) = 1 + \omega_{\mathrm{pm}}^2(\omega_{\mathrm{rm}}^2 - 2\mathrm{i}\omega_{\mathrm{dm}}\omega - \omega^2)^{-1}, \quad (12)$$

where ω_{rm}, ω_{pe}, ω_{pm}, ω_{de}, and ω_{dm} are real positive parameters. Sometimes the function $F\omega^2$ ($0 < F < 1$) is used instead of ω_{pm}^2 in the formula for μ, but it is not important for the description of principal properties. Further analysis is performed for the lossless medium ($\omega_{\mathrm{de}} = \omega_{\mathrm{dm}} = 0$).

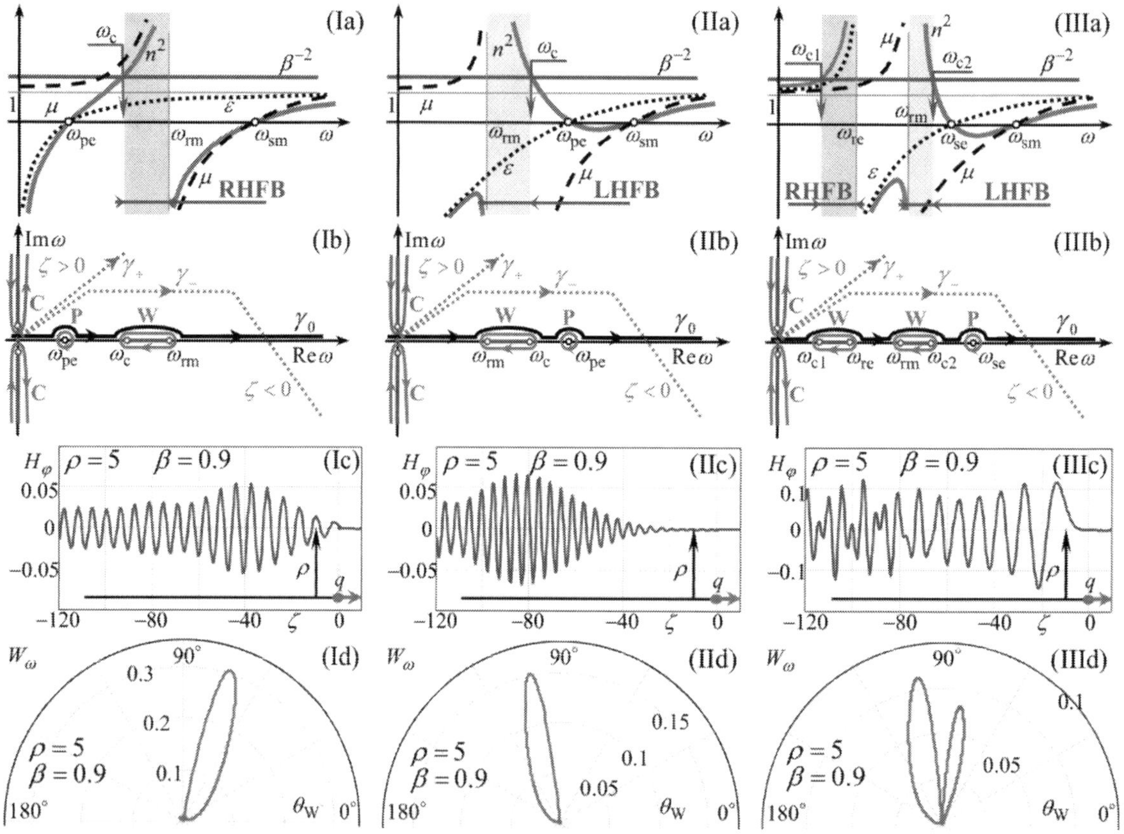

Figure 1. Typical view of dispersion laws (a), complex plane ω (b), magnetic field plots (c) and energy patterns (d) for the case of RHM (I), LHM (II), and a medium with both RHFB and LHFB (III). For cases (I), (II) the media are described by (12) with the following parameters (in units of ω_{rm}): (I) – $\omega_{\mathrm{pm}}=1$, $\omega_{\mathrm{pe}}=0.6$; (II) – $\omega_{\mathrm{pm}}=1$, $\omega_{\mathrm{pe}}=2$. Case (III) deals with both ε and μ of resonant type: $\omega_{\mathrm{pm}}=1$, $\omega_{\mathrm{pe}}=1.7$, $\omega_{\mathrm{re}}=0.9$. H_{φ} is in units $q^2\omega_{\mathrm{rm}}^2 c^{-2}$; W_{ω} is in units $q^2\omega_{\mathrm{rm}}^3 c^{-3}$; ζ, ρ are in units $c\omega_{\mathrm{rm}}^{-1}$.

The frequency range of radiation is determined by the condition $s^2(\omega)>0$. If $\omega_{\mathrm{pe}}<\omega_{\mathrm{rm}}$ then the medium is RHM with the radiation frequency range $\omega_{\mathrm{c}}<\omega<\omega_{\mathrm{rm}}$ (figure 1 (Ia)). In the case of $\omega_{\mathrm{pe}}>\omega_{\mathrm{rm}}$, the medium is LHM with the radiation frequency range $\omega_{\mathrm{rm}}<\omega<\omega_{\mathrm{c}}$ (figure 1 (IIa)). The parameter ω_{c} is determined by $s(\omega_{\mathrm{c}})=0$:

$$\omega_{\mathrm{c}}^2=\frac{\omega_{\mathrm{rm}}^2(1-\beta^2)-(\omega_{\mathrm{pe}}^2+\omega_{\mathrm{pm}}^2)+\{[\omega_{\mathrm{rm}}^2(1-\beta^2)-(\omega_{\mathrm{pe}}^2+\omega_{\mathrm{pm}}^2)]^2+4\beta^2(1-\beta^2)\omega_{\mathrm{pe}}^2(\omega_{\mathrm{rm}}^2+\omega_{\mathrm{pm}}^2)\}^{1/2}}{2(1-\beta^2)}. \quad (13)$$

Figure 1 (Ib), (IIb) shows the cuts (determined by equation $\mathrm{Im}\,s(\omega)=0$) and the integration path in the complex plane ω. It is important that the integration path γ_0 goes on the upper bank of cuts, where $\mathrm{sgn}(s)=\mathrm{sgn}(\omega)$ for RHM and $\mathrm{sgn}(s)=-\mathrm{sgn}(\omega)$ for LHM. This means that the projection of the phase velocity in the radial direction $\vec{\rho}$ is positive for RHM and negative for LHM (see (9)).

One of the possible analytical manipulations consists of the formation of the closed integration contour and the calculation of contributions of cuts and poles. The contours surrounding these singularities are shown in figure 1 (Ib), (IIb). As a result, we obtain

$$E_\rho = E_{\rho \mathrm{C}} + E_{\rho \mathrm{W}} + E_{\rho \mathrm{P}}\ , \quad E_z = E_{z\mathrm{C}} + E_{z\mathrm{W}} + E_{z\mathrm{P}}\,, \quad H_\varphi = H_{\varphi \mathrm{C}} + H_{\varphi \mathrm{W}}\,, \tag{14}$$

$$\begin{Bmatrix} E_{\rho \mathrm{C}} \\ E_{z\mathrm{C}} \\ H_{\varphi \mathrm{C}} \end{Bmatrix} = \frac{q}{c} \int_0^\infty \begin{Bmatrix} \varepsilon^{-1}(\mathrm{i}\tilde{\omega})\beta^{-1} s(\mathrm{i}\tilde{\omega}) J_1(\rho \,|\, s(\mathrm{i}\tilde{\omega})\,|) \\ c^{-1}\mu(\mathrm{i}\tilde{\omega})[n^{-2}(\mathrm{i}\tilde{\omega})\beta^{-2} - 1]\tilde{\omega}\, J_0(\rho\, s(\mathrm{i}\tilde{\omega})) \\ s(\mathrm{i}\tilde{\omega}) J_1(\rho\, s(\mathrm{i}\tilde{\omega}) \end{Bmatrix} \exp\left(-\tilde{\omega}\frac{|\zeta|}{V} \right) \mathrm{d}\tilde{\omega}\,, \tag{15}$$

$$\begin{Bmatrix} E_{\rho \mathrm{W}} \\ E_{z\mathrm{W}} \\ H_{\phi \mathrm{W}} \end{Bmatrix} = \frac{2q}{c}\Theta(-\zeta) \int_{\min(\omega_c,\omega_r)}^{\max(\omega_c,\omega_r)} \begin{Bmatrix} \varepsilon^{-1}(\omega)\beta^{-1} s(\omega) J_1(\rho\, s(\omega))\sin(\omega\zeta V^{-1}) \\ c^{-1}\mu(\omega)[1 - n^{-2}(\omega)\beta^{-2}]\,\omega\, J_0(\rho\, s(\omega))\cos(\omega\zeta V^{-1}) \\ s(\omega) J_1(\rho\, s(\omega))\sin(\omega\zeta V^{-1}) \end{Bmatrix} \mathrm{d}\omega\,, \tag{16}$$

$$\begin{Bmatrix} E_{\rho \mathrm{P}} \\ E_{z\mathrm{P}} \end{Bmatrix} = \frac{2q\omega_{\mathrm{pe}}^2}{c^2\beta^2}\Theta(-\zeta) \begin{Bmatrix} K_1(\rho\,\omega_{\mathrm{pe}} V^{-1})\,\sin(\omega_{\mathrm{pe}}\zeta V^{-1}) \\ K_0(\rho\,\omega_{\mathrm{pe}} V^{-1})\,\cos(\omega_{\mathrm{pe}}\zeta V^{-1}) \end{Bmatrix}, \tag{17}$$

where $J_k(\xi)$ and $K_k(\xi)$ are the Bessel and modified Hankel functions, correspondingly. Here the index "C" is assigned to the "quasi-Coulomb" part of the field (contributions of cuts along the imaginary axis), the index "W" is assigned to the wave part (contributions of cuts along the real axis), and the index "P" is assigned to the so-called "plasma trace" (contributions of poles $\omega = \pm\omega_{\mathrm{pe}}$) (see figure 1 (Ib), (IIb)). The "quasi-Coulomb" field exists both behind and in front of the moving charge and quickly decreases with distance from it. The wave field (CR) exists only behind the charge and oscillates with distance from it. Unlike the radiation field, the "plasma trace" is concentrated close to the charge movement trajectory and exponentially decreases with increasing ρ. Note that analogous expressions for a resonant RHM are given in our papers [5,6]. As one can see, representation (14) has a clear physical meaning. Additionally, it possesses certain advantages for computing. One of them is that integrands in (15), (16) are free from Hankel functions, contrary to the initial formulae (7), (8). Therefore, (15), (16) are convenient for computation of the wave and quasi-Coulomb fields for small values of ρ including $\rho = 0$.

Another possible way of investigation is based on certain transformation of the integration path [5–7]. For relatively small values of ω, it is convenient to transform contour γ_0 in such a way that it bypasses all branch points and poles. For the best convergence at large values of ω we can employ the contour being parallel to the SDP asymptote (11). So, for $\zeta > 0$, we use the semi-infinite rays parallel to the SDP as the new integration path, while for $\zeta < 0$ the trapezoidal contour with the half-infinite part parallel to the SDP asymptote is utilised (contours $\gamma_\pm$ in figure 1 (Ib), (IIb)). This transformation allows the avoidance of the intersection of the integrands' singularities. The essential advantage of such an approach is that it is possible to choose the most convenient parameters of the new contour for the concrete parameters of the problem.

Some examples of computations of the H_φ component are shown in figure 1 (Ic), (IIc). Figure 1 (Id), (IId) shows corresponding energetic patterns as well. The latter are defined by following the method in [14]. The total energy passing through the unity square is equal to $W_\rho = \int_{-\infty}^{+\infty} S_\rho \mathrm{d}t$ (for the square orthogonal to the z axis) and $W_z = \int_{-\infty}^{+\infty} S_z \mathrm{d}t$ (for the square parallel to the z axis). These values can be represented as integrals from the spectral density of energy: $W_\rho = \int_0^\infty W_{\rho\omega}\mathrm{d}\omega, W_z = \int_0^\infty W_{z\omega}\mathrm{d}\omega$. For $|\,s\,|\,\rho >> 1$ the following approximations take place:

$$W_{\rho\omega} \approx q^2(2\pi\rho)^{-1}\mathrm{Re}[\,|\,s\,|\,s\omega^{-1}\varepsilon^{-1}\exp(-2\rho\,\mathrm{Im}s)]\,, \quad W_{z\omega} \approx q^2(2\pi V\rho)^{-1}\mathrm{Re}[\,|\,s\,|\,\varepsilon^{-1}\exp(-2\rho\,\mathrm{Im}s)]\,. \tag{18}$$

The dependence $W_\omega = \sqrt{W_{\rho\omega}^2 + W_{z\omega}^2}$ on the angle $\theta_{\mathrm{W}} = \mathrm{atan}(W_{\rho\omega}/W_{z\omega})$ is the energetic pattern of radiation.

Figure 1 (Ic,d) is related to the case of a medium with RHFB only, while figure 1 (IIc,d) is related to the case of a medium with LHFB only. One can see that the pattern in the LHFB case is "reversed". Additionally, the increase in the field amplitude with $|\zeta|$ in LHM is slower compared with that in RHM. Obviously, this is due to the fact that the wave field in LHM lags more behind the charge in comparison with the RHM case.

Figure 1 (IIIa–d) is related to a more complex case. Here we use the model of resonant permeability (as in (14)) and resonant permittivity as well: $\varepsilon(\omega)=1+\omega_{\mathrm{pe}}^2(\omega_{\mathrm{re}}^2-2\mathrm{i}\omega_{\mathrm{dm}}\omega-\omega^2)^{-1}$. Parameters for the figure 1 (IIIa–d) are selected so that the medium has both RHFB and LHFB. Therefore, the pattern has two lobes corresponding to ordinary and reversed CR (figure 1 (IIId)). At the magnetic field plot (figure 1 (IIIc)) one can see that, for relatively small distance $|\zeta|$ the field is close to sinusoidal, however for $|\zeta|>70$ the field behaviour is more complicated. This is explained by the addition of the "left-handed" CR for large enough distances.

4. The case of a boundary between RHM and LHM

Here we analyse the electromagnetic field generated by the charge q intersecting the interface (located at $z=0$) between two passive, homogeneous, isotropic, frequency-dispersive media described by permittivity and permeability: $\varepsilon_1(\omega)$, $\mu_1(\omega)$ for $z<0$ and $\varepsilon_2(\omega)$, $\mu_2(\omega)$ for $z>0$. The medium filling the region $z<0$ is RHM, while the volume $z>0$ is filled with LHM. The charge motion rule is $z=Vt=c\beta t$, $x=y=0$.

The electromagnetic field in the present geometry can be described by formulas similar to those from [17]. If the index 1 refers to the area $z<0$ and the index 2 refers to the area $z>0$, then

$$\{A^{1,2},\Phi^{1,2}\}=\{A^{\mathrm{q}1,2}+A^{\mathrm{b}1,2},\Phi^{\mathrm{q}1,2}+\Phi^{\mathrm{b}1,2}\}=\int_{-\infty}^{+\infty}\mathrm{d}\omega\{A_\omega^{\mathrm{q}1,2}+A_\omega^{\mathrm{b}1,2},\Phi_\omega^{\mathrm{q}1,2}+\Phi_\omega^{\mathrm{b}1,2}\}\exp(-\mathrm{i}\omega t). \qquad (19)$$

Index "q" is assigned to the so-called "forced" field as the charge field in the corresponding infinite medium. Thus, $A^{\mathrm{q}1,2}$ and $\Phi^{\mathrm{q}1,2}$ are given by (6). Index "b" is assigned to the so-called "free" field arising because of the presence of the boundary:

$$\{A_\omega^{\mathrm{b}1,2},\Phi_\omega^{\mathrm{b}1,2}\}=\int_{-\infty}^{+\infty}\mathrm{d}k_\rho\{A_{\omega k_\rho}^{\mathrm{b}1,2},\Phi_{\omega k_\rho}^{\mathrm{b}1,2}\}\exp[\mathrm{i}(k_z^{(1,2)}|z|-\omega t)], \qquad (20)$$

$$\begin{Bmatrix}A_{\omega k_\rho}^{\mathrm{b}1,2}(\omega,k_\rho)\\ \Phi_{\omega k_\rho}^{\mathrm{b}1,2}(\omega,k_\rho)\end{Bmatrix}=\frac{-q}{2\pi V}\begin{Bmatrix}\dfrac{\mu_{1,2}\,\mathrm{sgn}(z)}{k_z^{(1,2)}}\\ c(\omega\varepsilon_{1,2})^{-1}\end{Bmatrix}\frac{H_0^{(1)}(\rho k_\rho)k_z^{(1,2)}}{k_\rho^{-1}g_3}\left[\frac{\omega\varepsilon_{2,1}/c\mp\beta\varepsilon_{1,2}k_z^{(2,1)}}{k_\rho^2-s_{1,2}^2}-\frac{\beta^2\varepsilon_{2,1}}{\omega/c\pm\beta k_z^{(2,1)}}\right], \qquad (21)$$

where $k_z^{(1,2)}(\omega,k_\rho)=\sqrt{\omega^2n_{1,2}^2c^{-2}-k_\rho^2}$, $g_3(\omega,k_\rho)=\varepsilon_1k_z^{(2)}+\varepsilon_2k_z^{(1)}$. In accordance with the absorption condition we must define that $\mathrm{Im}k_z^{(1)}>0$, $\mathrm{Im}k_z^{(2)}>0$. One can obtain that $\mathrm{Re}k_z^{(1)}>0$ for all positive frequencies, but $\mathrm{Re}k_z^{(2)}>0$ for RHFB and $\mathrm{Re}k_z^{(2)}<0$ for LHFB (the same results follow from the Mandelshtam's condition as well).

For analysis of the free field two methods were utilised. In the first way, we analysed the free field Fourier harmonics A_ω^{b}, Φ_ω^{b} using the steepest-descent technique (for more detail see [18]). As usual, we introduce new spatial coordinates R and θ, where R denotes the radial distance from the origin of coordinates to the observation point, while θ is the angle between R and the negative z direction. Turning from k_ρ to a new complex variable ψ ($k_\rho=\kappa_{1,2}\sin\psi$, $\kappa_{1,2}=\sqrt{\omega^2c^{-2}n_{1,2}^2}$, $\mathrm{Im}\kappa_{1,2}>0$), we find the SDP passing through the saddle point θ and transform the initial contour to it.

Let us look more closely at the situation where the frequency lies in LHFB for medium 2. We have obtained the uniform asymptotic approximation [19] of the integral along SDP valid for $R|\kappa_{1,2}|>>1$. This takes into account the possibility of poles $\pm s_2$ situated near the saddle point. In this case, θ is close to the reflected or refracted Cherenkov angle, respectively. Provided that θ is far from any of these angles, the aforementioned approximation represents the spherical wave of transition radiation (TR).

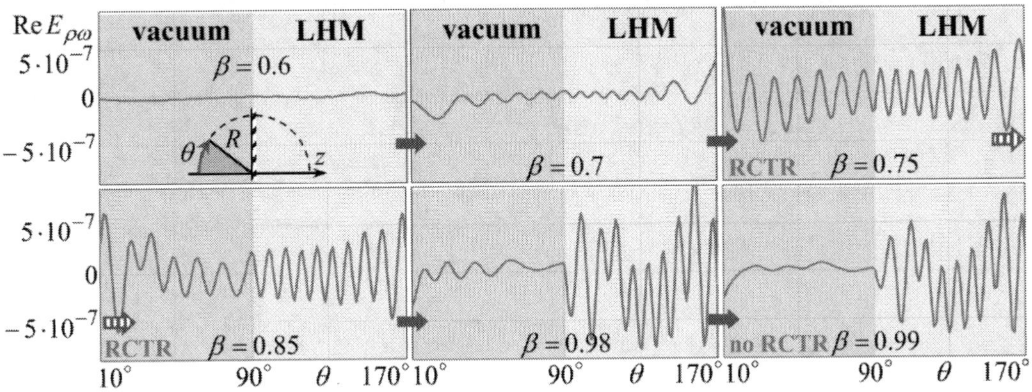

Figure 2. Modification of the dependence of $\mathrm{Re}\,E_{\rho\omega}$ ($\mathrm{Vm^{-1}s}$) on θ with increase in β. The LHM parameters are: $\varepsilon=-1.7$, $\mu=-1.2$, $\beta_{CR}=0.7$, $\beta_{TIR}=0.98$; $q=-1\mathrm{nC}$, $\nu=10\mathrm{GHz}$, $R=15\mathrm{cm}$.

Additionally, the poles $\pm s_2$ contribute to the free field at certain values of θ. These contributions represent the cylindrical waves of reflected or refracted CR. We call these waves the "reversed Cherenkov-transition radiation" (RCTR) because they possess certain features of both CR and TR [20]. As was shown, the refracted RCTR arises if the charge velocity is enclosed between the lower and upper thresholds (see also [21,22]):

$$\beta_{CR} < \beta < \beta_{TIR}, \qquad \beta_{CR} = |n_2|^{-1}, \qquad \beta_{TIR} = (n_2^2 - n_1^2)^{-1/2}, \tag{22}$$

where β_{CR} is the Cherenkov threshold, while β_{TIR} corresponds to the total internal reflection.

In the second way we produce an effective algorithm for numerical computation of both A_ω^b, Φ_ω^b and A^b, Φ^b. In the LHFB case, the poles $\pm s_2$ are located near the integration path, leading to drastic changes of (21). Our numerical algorithm is adapted for overcoming this difficulty. The free field for the graphical representations shown here was calculated with the use of the second method.

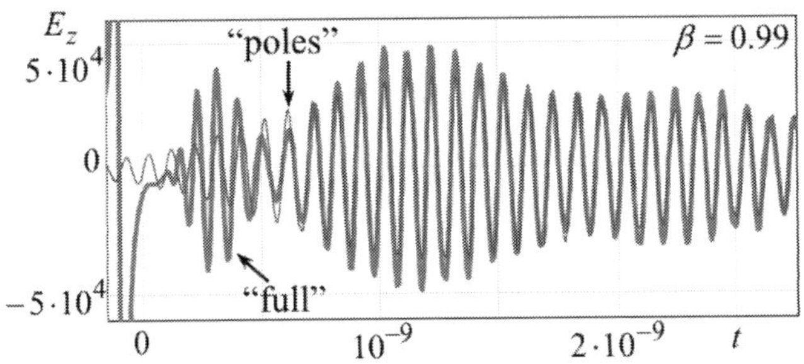

Figure 3. The dependence of E_z ($\mathrm{Vm^{-1}}$) component on time (sec) at fixed spatial point in vacuum: $z=-3.2\mathrm{cm}$, $\rho=0.03\mathrm{cm}$. Medium is described by (12) with the following parameters:
$\omega_{rm}=0$, $\omega_{pm}=\omega_{pe}=2\pi\cdot10^{10}\mathrm{s^{-1}}$,
$\omega_{de}=\omega_{dm}=10^{-4}\omega_{pm}$, $q=-1\mathrm{nC}$.

Figure 2 presents the modification in the dependence $E_{\rho\omega}(\theta)$ ($R=\mathrm{const}$) with increase in β. It is supposed that the medium 1 is a vacuum and the frequency is in LHFB for medium 2 (for more detail see [20]). This result confirms that RCTR in vacuum has the double threshold effect, in accordance with (22).

The results for the total field (figure 3) demonstrate that RCTR can be the dominant radiation. The medium 1 is supposed to be vacuum, the medium 2 is described by (12). The line marked "full" in figure 3 corresponds to the full field, while the line marked "poles" corresponds to the poles contribution (determining RCTR) as if it were alone. One can see that, for $t > 0.7\,\text{ns}$, the total field is very well approximated by the pole contribution. This means that, in this case, RCTR gives the main contribution to the spatial radiation. For $t < 0.7\,\text{ns}$ the dominant radiation is TR.

5. Conclusion

The presented approach to the theory of uniformly moving charge radiation has some advantages. One of them consists in obtaining a convenient field decomposition having clear physical meaning. An efficient numerical algorithm for computing the total field is another benefit. We have used the left-handed medium to demonstrate the possibilities of this technique. Additionally, the effect of the reversed Cherenkov-transition radiation occurring at the interface between the ordinary medium and the left-handed one has been analysed both analytically and numerically.

Acknowledgments

This research was supported by the Education Agency of Russian Federation, the Saint Petersburg Government, US Dept. of Energy SBIR (DE-FG02-08ER85031), and the Russian Foundation for Basic Research (09-02-00921).

References

[1] Bolotovskiy B M and Stolyarov S N 1972 *Problems of Theoretical Physics. Collection of I.E.Tamm's Memory* (Moscow: Nauka) p 267 (in Russian)
[2] Tamm I E and Frank I M 1937 *Compt. Rend. Acad. Sci. USSR* **14** 107
[3] Fermi E 1940 *Phys. Rev.* **57** 485
[4] Afanasiev G N, Kartavenko V G and Ruzicka J 2000 *J. Phys.* A: *Math. Gen.* **33** 7585
[5] Tyukhtin A V and Galyamin S N 2008 *Phys. Rev.* E **77** 066606
[6] Galyamin S N and Tyukhtin A V 2006 *Vestnik SPbGU, Series 4* **1** 21–30 (in Russian)
[7] Tyukhtin A V and Galyamin S N 2007 *Tech. Phys. Lett.* **33** 632
[8] Veselago V G 1968 *Sov. Phys. Usp.* **10** 509
[9] Veselago V G 2003 *Sov. Phys. Usp.* **46** 764
[10] Pendry J B, Holden A J, Stewart W J and Youngs I 1996 *Phys. Rev. Lett.* **76** 4773
[11] Pendry J B, Holden A J, Robins D J and Stewart W J 1999 *IEEE Trans. Microwave Theory Tech.* **47** 2075
[12] Smith D R, Padilla W, Vier D C, Nemat-Nasser S C and Shultz S 2000 *Phys. Rev. Lett.* **84** 4184
[13] Ziolkowski R W 2003 IEEE Trans. Antennas and Propag. **51** 1516
[14] Lu J, Grzegorczyk T M, Zhang Y, Jr J P, Wu B-I and Kong J A 2003 *Opt. Express* **11** 723
[15] Duan Z Y, Wu B-I, Xi S, Chen H S and Chen M 2009 *Progress in Electromagnetic Research, PIER* **90** 75
[16] Averkov Yu O, Kats A V and Yakovenko V M 2009 *Phys. Rev.* B **79** 193402
[17] Ginzburg V L and Tsytovich V N 1990 *Transition Radiation and Transition Scattering* (London: Hilger)
[18] Galyamin S N, Tykhtin A V, Alekhina T Yu and Doil'nitsina E G 2009 *Proc. Int. Conf. "Days on Diffraction" (Saint Petersburg)* (in press)
[19] Felsen L B and Marcuvitz N 2003 *Radiation and Scaterring of Waves* (New Jersey: Wiley Interscience) p 399
[20] Galyamin S N, Tyukhtin A V, Kanareykin A D and Shoessow P 2009 *Phys. Rev. Lett.* **103** 194802
[21] Pafomov V E 1959 *Sov. Phys. JETP* **9** 1321
[22] Averkov Yu O 2005 *Telecommunications and Radio Engineering* **63** 419

"Shadowing" of the electromagnetic field of relativistic charged particles

G Naumenko[1], X Artru[2], A Potylitsyn[3], Yu Popov[3], L Sukhikh[3] and M Shevelev[1]

[1]Nuclear Physics Institute of Tomsk Polytechnic University, Lenina str. 2a Tomsk, 634050, Russia

[2]Institut de Physique Nucl.eaire de Lyon, Université de Lyon,CNRS- IN2P3 and Université Lyon 1, 69622 Villeurbanne, France

[3]Tomsk Polytechnic University, Lenina str. 2, Tomsk, 634050, Russia

x.artru@ipnl.in2p3.fr

Abstract. In radiation processes such as a transition radiation, diffraction radiation, etc. based on relativistic electrons passing through or near an opaque screen, the electron self-field is partly shadowed after the screen over a distance of the order of the formation length $\gamma^2\lambda$. This effect has been investigated on coherent diffraction radiation (DR) by electron bunches. Absorbing and conductive half-plane screens were placed at various distances L before a standard DR source (inclined half-plane mirror). The radiation intensity was reduced when the screen was at small L and on the same side as the mirror. No reduction was observed when the screen was on the opposite side. It is worth noting that absorbing and conductive half-plane screens produce the same shadowing effect. The shadowing effect is responsible for a bound on the intensity of Smith-Purcell radiation.

1. Introduction

Macroscopic Maxwell equations may be used (in principle) for calculating the radiation emitted by relativistic electrons passing through or near material targets. Nevertheless phenomenological concepts like *equivalent photons* and *formation zone* are useful for an intuitive understanding of the main features. In transition radiation (TR), the *formation zone* is a region of length $l_f \sim \gamma^2\lambda$ where the Coulomb field of the fast electron and the forward TR from the target cannot be measured separately (in this paper we assume that the electron is ultra-relativistic: $\gamma \gg 1$). The total electromagnetic field in this region is reduced by a destructive interference between the two fields. This effect can also be interpreted with the *equivalent photons* method [1] (the terms *virtual quanta*, *quasi-real photons* or *pseudo-photons* are also used). In this picture the target intercepts the beam of equivalent photons, leaving a *shadow* downstream. It takes a length $\sim l_f$ for the electron to re-form its pseudo-photon cloud. This explains in a simple way the suppression of Optical TR in a Wartski interferometer when the distance between the two foils is small compared to l_f.

The similar concept of *half-naked* electron was introduced in [2,3] in the physics of electron scattering on atoms or nuclei. These authors consider the Landau-Pomeranchuk effect [4] and the density effect of Ter-Mikaelian [5] as experimental evidences for the half-naked electron effect.
In this paper it will be shown, both experimentally and theoretically, that a similar shadow region exists behind a Diffraction Radiation (DR) target (either conductive or absorbing). However, unlike for TR, the shadow is only on one side of the particle trajectory (the same side as the target), as expected from a naive application of optical geometry to the pseudo-photons. In this region radiation by a second DR target is inhibited.

2. Shadowing effect from different points of view

Here we present different points of view about the shadow effect. Let us consider a fast electron moving near two targets like in figure 1.

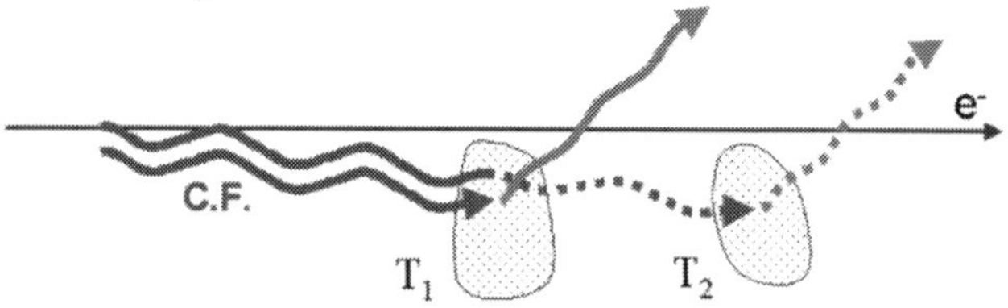

Figure 1. Shadowing of the target by another target.

In the first point of view the Coulomb field (C.F.) is considered as a beam of quasi-real photon. Scattering of this field by the first target gives Diffraction Radiation. The second target is in the shadow of the first one, therefore emits almost no radiation. The Coulomb field is gradually "repaired" during the formation zone of length $l_f \sim \gamma^2\lambda$.
A precise space-time description of a "half-naked electron", in the context of the Landau-Pomeranchuk effect mentioned above, is given in [3], where figures 1.1 and 2.4 depict the restoration of the Coulomb field. In figure 2 we adapted this picture to the case of a particle passing through a narrow hole.

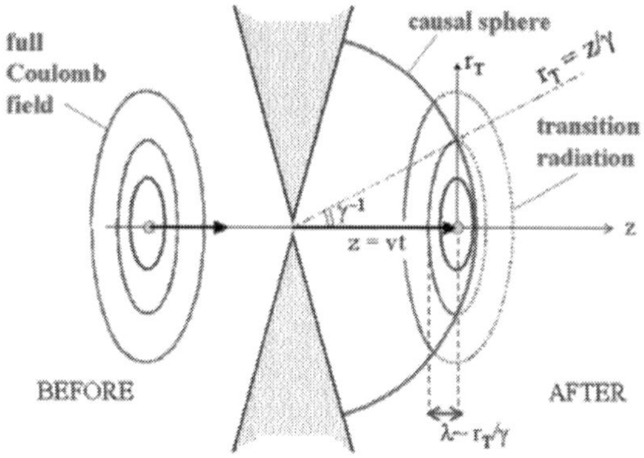

Figure 2. Field of a particle passing through a hole (analogous to figure 1.1 of [3]).

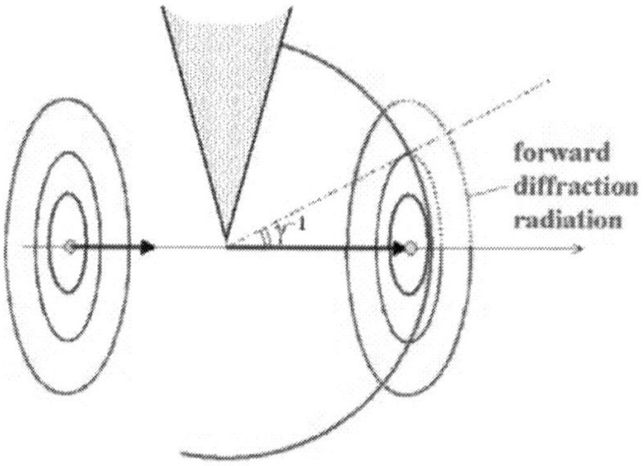

Figure 3. Shadowing by a half-screen.

On the right-hand side, according to the relativistic causality principle, no field exists outside the expanding sphere $|\mathbf{r}| \le t$. Inside, the field is fully restored. Outside the cone $|\mathbf{r}_T| = z/(\gamma v) \simeq z/\gamma$, there is no field before the electron arrives at the same z, therefore the Coulomb field is more than half-amputed. Taking the temporal Fourier decomposition of the Coulomb field, we have $\omega \le \gamma v/|\mathbf{r}_T|$, therefore the condition for a good restoration of the frequency ω is $z \ge \gamma|\mathbf{r}_T|$ for all $|\mathbf{r}_T| \in [0, \gamma/\omega]$. Up to a factor 2π, we recover the condition $z \ge l_f$.

Figure 3 depicts an analogous situation where the particle passes near the border of a screen. In this case only one side of the Coulomb field is removed. The goal of the experiment presented below is to observe this one-sided shadow region.

The second point of view has been illustrated by B.M. Bolotovskii in figure 4 [6]. The traveling Coulomb field (represented by ellipses centered on the successive positions of the particle) induces current in the half-plane screen, which in turn emit Diffraction Radiation (DR), represented by small pieces of ellipses. The radiation field is such that close to the screen it kills part of the particle field. In this point of view, shadowing appears as a destructive interference effect between the Coulomb field, CF, and the forward diffraction radiation field, FDR.

The interference gradually disappears (positions 3, 4, 5 of the figure) due to the different velocities, $v \simeq 1-\gamma^{-2}/2$ and $c=1$ (in our units), of the Coulomb and radiation fields. These fields get out of phase after a time $t_f \sim \lambda/(c-v) \sim l_f$.

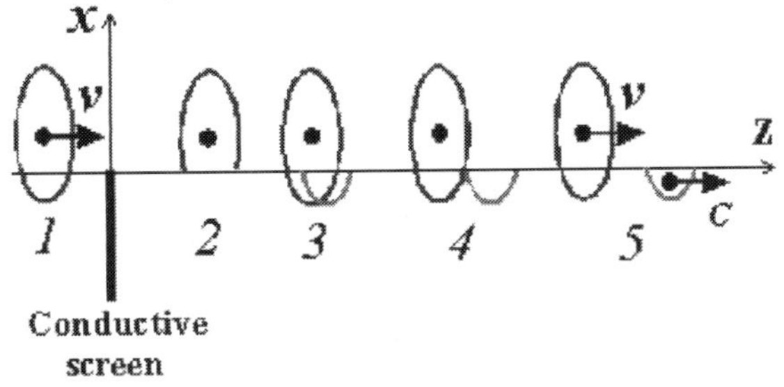

Figure 4. Illustration of the radiation formation length effect by B.M. Bolotovskil in [6].

In this second point of view, the target T_2 of Fig.1 is submitted to the usual Coulomb field of the particle plus the field of the forward diffraction radiation from T_1, denoted $FDR^{(1)}$. Each of these fields induces a current in T_2 and this current generates a new radiation. Outgoing from T_2, we have then the superposition of

- the ordinary diffraction radiation from T_2, denoted $DR^{(2)}$,
- a part of FDR from T_1 rescattered by T_2, denoted $R_2\{FDR^{(1)}\}$.

If T_2 is at distance less than l_f from T_1, the total inducing field, the total induced current and the sum $DR^{(2)} + R_2\{FDR^{(1)}\}$ are suppressed. In this viewpoint, illustrated in Fig. 5, the inhibition of radiation from the second target is a combined result of rescattering and destructive interference.

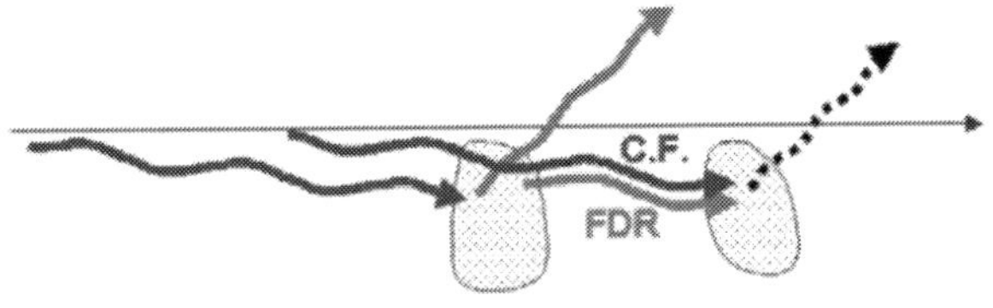

Figure 5. Shadow effect as a re-scattering effect.

3. The Tomsk experiment

In the Landau-Pomeranchuk effect the formation zone is not directly observed. In optical transition radiation, it has a macroscopic longitudinal size. However its transverse size $\sim \gamma\lambda$ is usually small compared to the width of the electron beam (this is the effective transverse size of the Coulomb field at observed wavelength λ). Therefore a direct investigation of its transverse properties needs higher values of the product $\gamma\lambda$.

3.1. Principle of the experiment

The Tomsk microtron delivers an electron beam with a Lorenz-factor $\gamma \simeq 12$. The radiation is observed in the microwave domain ($\lambda \sim 1$ cm). Thus the transverse scale $\gamma\lambda$ is about 12 cm, which is quite comfortable. The formation length $\gamma^2\lambda$ is about 1.4 m.

The energy of the radiation coming from the interaction of one electron with the targets, in the bandwidth $\Delta\omega$, is typically of the order of $\alpha\,\hbar\Delta\,\omega$. In the microwave region, it is too small to be measurable, even if we multiply it by the number N_e of electrons in a bunch. However, if the bunch length σ_z is smaller than λ the radiation has a coherent character which makes it proportional to N_e^2 rather than N_e. More precisely, the radiation intensity from the bunch is given by

$$I_{\text{bunch}} = \left[N_e(N_e-1)\,f^2(\lambda/\sigma_z)+N_e\right]I_e$$

where $f(\lambda/\sigma_z)$ is the longitudinal form-factor of the bunch. In this experiment, $N_e \sim 10^8$, therefore the bunch coherence amplifies the radiation by 8 orders of magnitude, making it easily observable.

Shadowing is achieved by the reflection of the pseudo-photons by a half-plane mirror or their absorption in a half-plane screen. We can indeed assume that pseudo-photons are reflected in the mirror or absorbed in absorber almost like real photons. In particular, it does not matter whether we use an absorber or a mirror for shadowing. This is supported by the analysis of diffraction radiation from a "black body" absorber in [7], although this reference does not analyze the electromagnetic field inside the radiation formation zone.

The electron field in the shadowing region is probed with a second diffraction radiation target, from which one measures the intensity and angular distribution of backward diffraction radiation (BDR), according the scheme shown in figure 6. BDR is easier to measure than FDR, since the separation from the Coulomb field occurs very close to the target.

For the case where targets are on opposite sides of the beam we keep the same T_2 target but over-turn T_1 to the opposite side (see figure 7).

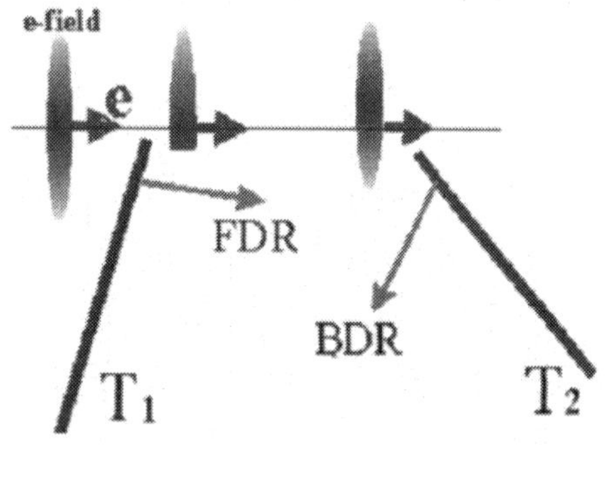

Figure 6. Experimental scheme for the investigation of shadowing. T_1 is an absorber or conductive target, T_2 is a conductive mirror.

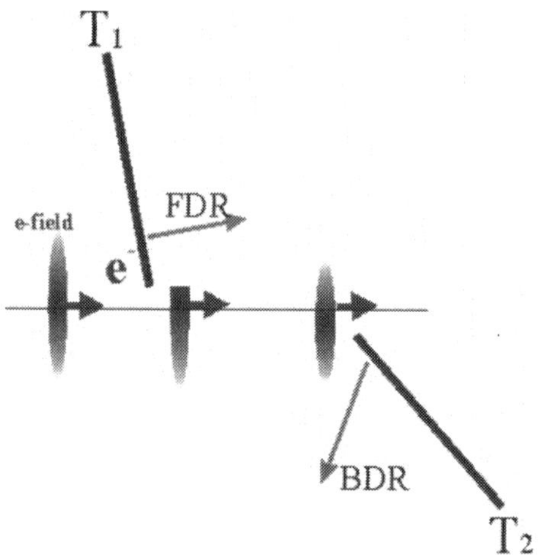

Figure 7. Scheme with the targets on opposite beam sides

3.2. Experimental setup

The experiment was performed on the extracted electron beam of the microtron of the Tomsk Nuclear Physics Institute. The beam is extracted from the vacuum chamber through a 20 μm thick beryllium foil. Its parameters are listed in table 1.

Table 1. Electron beam parameters.

Electron energy	6.1 MeV ($\gamma = 12$).	Bunch period	380 psec
Train duration	$\tau \approx 4$ μsec	Bunch population	N_e=6·10^8
Bunches in a train	n_b≈1.6·10^4	Bunch length	σ≈1.3~1.6mm

The window caused a beam divergence ($\simeq 0.08$ radian). For this reason the distance L between targets T_1 and T_2 was limited to L_{max} = 220 mm. At larger distance, it would be necessary to increase the impact parameter b_2 of T_2, to prevent direct interactions with the side of the beam. This would strongly reduce the diffraction radiation yield. L was changed by moving the upstream target T_1, keeping the diffraction radiation mirror T_2 fixed. In this way, the influence of a transversal bunch form-factor on the DR yield from was T_2 avoided.

The beryllium window with diameter of 34 mm may be also considered as a source of a transition radiation. A simple estimation showed that the intensity of this radiation is negligible downstream T_1.

The coherent radiation intensity for $\lambda > 9$ mm is by 8 orders larger than the incoherent one and has a power level =1 Watt per steradian. It means that one can investigate coherent radiation in this wavelength range without difficulty.

For the radiation measurements we used the room temperature detector DP20M, with parameters described in [8]. Its main elements are a low-threshold diode, a broadband antenna and a preamplifier. The detector efficiency in the wavelength region λ=3~16 mm is estimated to be constant to a ±15% accuracy. The detector sensitivity is 0.3 V/mWatt. A wave-guide with a cutoff λ_{cut}=17 mm was used to cut the long-wave background of the accelerator RF system. The high frequency limit of the wavelength interval is defined by the bunch form-factor. This limit (λ_{min}=9 mm) was measured using discrete wave filters [9] and a grating spectrometer.

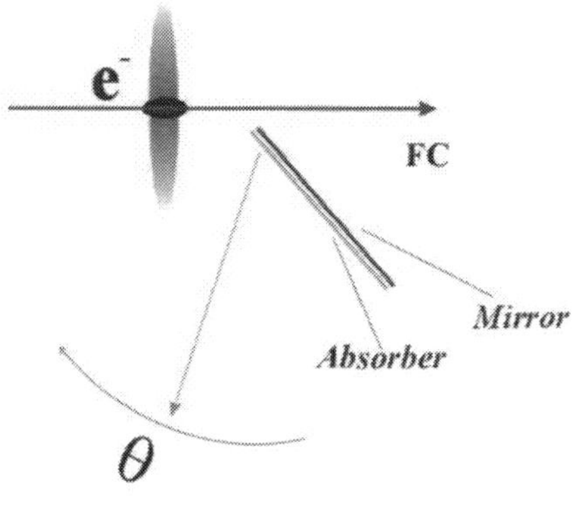

Figure 8. Scheme of the absorber test.

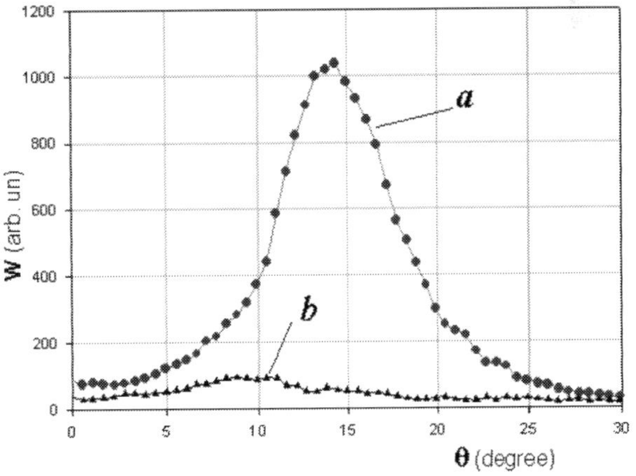

Figure 9. Angular distribution of the radiation intensity, measured in the absorber test (figure 8). *a* – BDR from conductive target without absorber, b – radiation from the same target covered by absorber.

To exclude the prewave zone effect (see [10]) a parabolic telescope was used to investigate the angular distribution. This method was suggested and tested in [11] and gives the same angular distribution as in the far field zone ($R \gg \gamma^2\lambda$).
The absorber properties were checked both with a real photon beam from a source of wavelength λ=6 mm and with the measurement of reflected pseudo-photons of beam electrons, according to the scheme shown in figure 8. In figure 9 curve *a* is the BDR angular distribution from a conductive target without absorber and curve *b* is for the same target covered by the absorber. We can see that almost no pseudo-photon reflection is registered. The test with the real photon beam showed that, within $\simeq 3\%$ experimental accuracy, no real photons passed through the absorber and no photons were reflected by it.

3.3. Experimental results

Using the described method, the angular distributions of backward radiation from the conductive target T_2 were measured for inter-target distances L=0 to 220 mm with steps of 20 mm. Two configurations of the experiment were used :

- Configuration *a* (figure 10*a*) is intended to demonstrate the shadowing of the electron field. In order not to restrict the measured angular range of the radiation from T_2 at small values of L, the absorber was inclined at 45° as shown in figure 10.
- In configuration *b* (figure 10b) the absorber is placed on the side opposite to T_2 with respect to the electron beam. In this geometry the shadow effect is expected not to be observed.

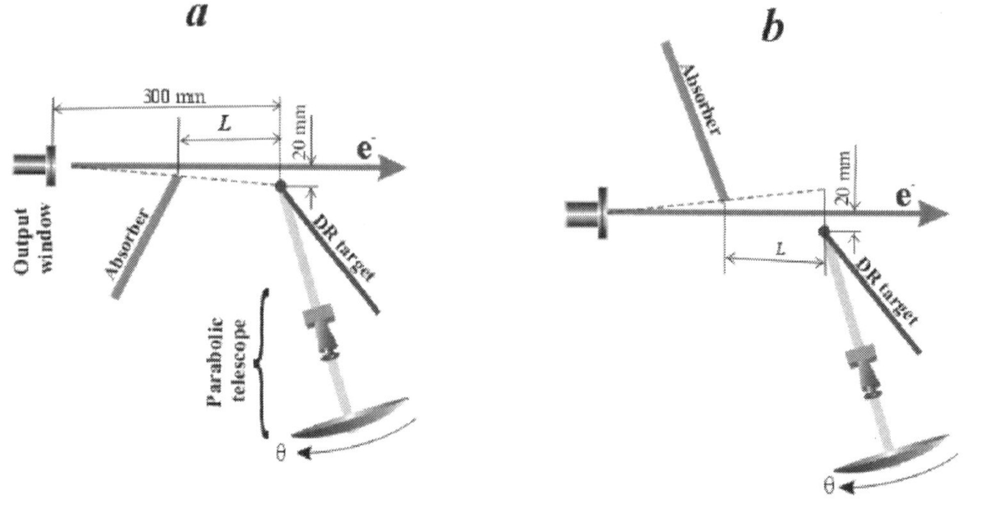

Figure 10. Scheme of experiment on the shadow effect observation. *a* – the targets are on the same side of the electron beam. *b* – the targets are on the opposite sides of the electron beam.

Samples of the measured distributions using configuration ***a*** are shown in figure 11 and the simultaneous dependence on the observation angle θ and on the distance L is shown in figure 13*a*. The corresponding results for the configuration ***b*** (opposite-side screens) are presented in figures 12 and 13*b*. Figures 11 and 12 have the same units of radiation intensity.

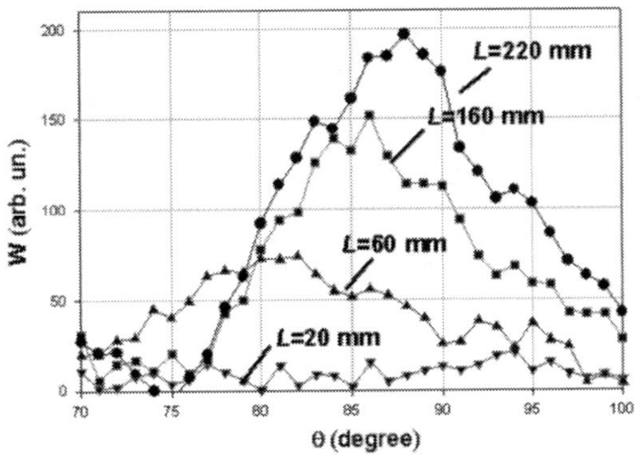

Figure 11. Samples of the measured angular distributions for different distance to the absorber using scheme ***a*** (figure 10*a*).

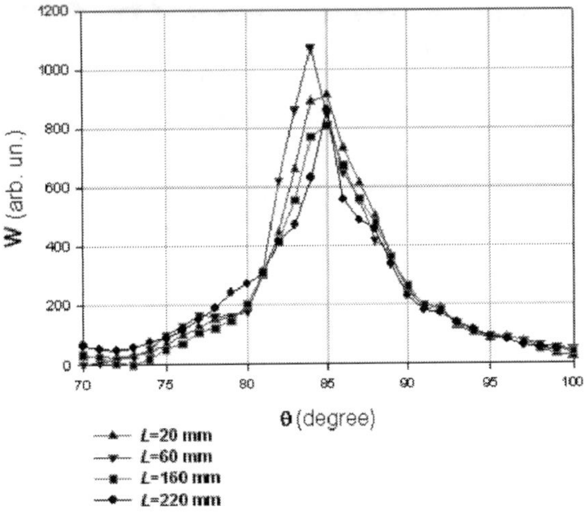

Figure 12. Samples of the measured angular distributions for different distance to the absorber using scheme ***b*** (figure 10*b*).

Figures 13*a* and 13*b* have the same intensity scale to emphasize the shadowing effect. These figures exhibit the qualitative properties of shadowing :

- the intensity in scheme *a* is much mower than in scheme *b*,
- it decreases together with *L*. By contrast, no systematic dependence on *L* is seen in scheme *b*.

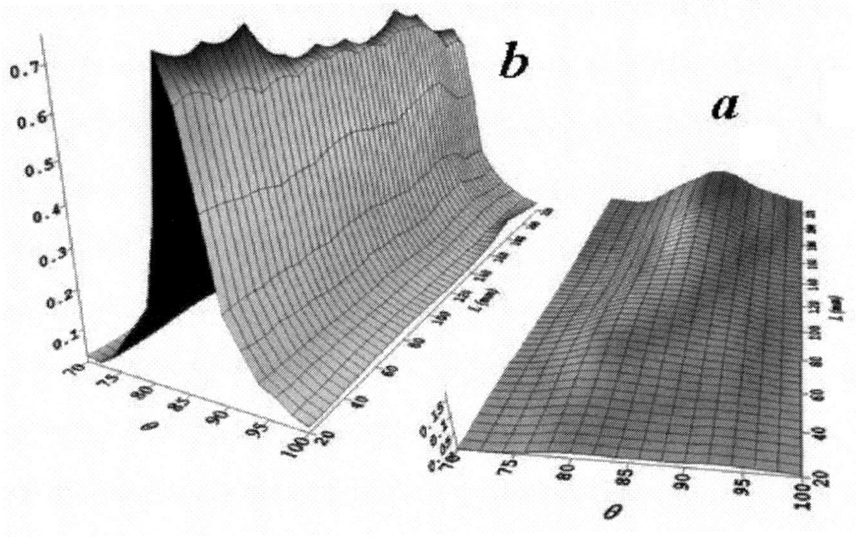

Figure 13. Measured dependence of the radiation intensity on the observation angle θ and the distance L to the absorber. Plots a and b correspond to the configurations a and b of figure 10.

To show that the shadow effect does not depend on the type of the target, the experiment was repeated replacing the absorber target T_1 by a conducting screen. The double dependence on θ and L is shown in figure 14. Comparison with figure 13 confirms the independence of the shadow effect on the target type.

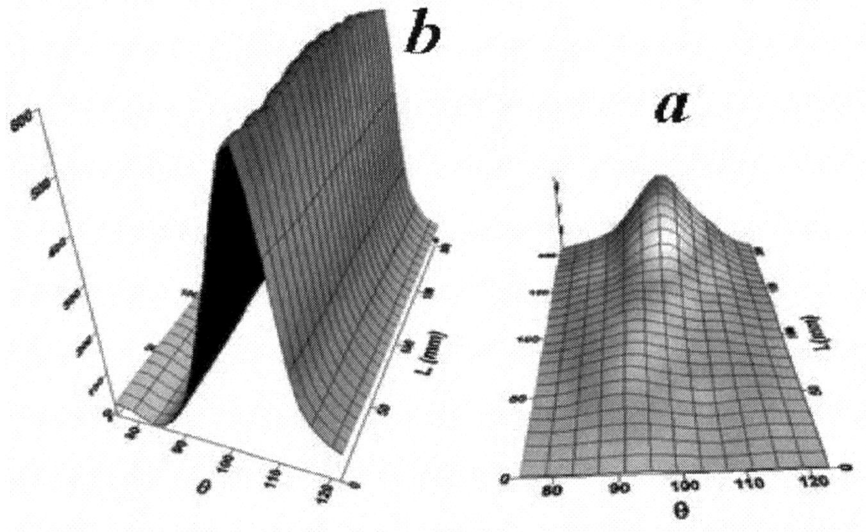

Figure 14. Measured dependence similar to the one shown in figure 13, using a conducting instead of an absorbing target T_1.

4. Theoretical calculation and comparison with experiment

The calculation of the expected angular distribution of the radiation from the conductive target T_2 (see figures 6 and 7), for one electron, can be done with the method outlined at the end of section 2, *i.e.*, express the amplitude as $DR^{(2)} + R_2\{FDR^{(1)}\}$. The results presented below are derived in the ultra-relativistic approximation for the electron ($\gamma \gg 1$) and the paraxial approximation for the photon ($k_T \ll k_z$).

4.1. Free field evolution in z

In the paraxial approximation of the Huyghens-Fresnel formula, the evolution of a free field from a transverse plane at z_0 to a transverse plane at z can be written as

$$\mathbf{E}_{\text{free}}(z) = G(z-z_0)\,\mathbf{E}_{\text{free}}(z_0), \tag{1}$$

In this symbolic formula, $\mathbf{E}(z)$ represents the field state in the transverse plane of abscissa z at fixed frequency ω. In the (x,y) representation, it stands for $\mathbf{E}(\omega,x,y,z)$ considered as function of the bi-vector (x,y). ω is considered as a parameter and z plays the role similar to time in quantum mechanics. By a 2-dimensional Fourier transform, one passes to the $\mathbf{k}_T$-representation $\mathbf{E}(\omega,k_x,k_y,z)$. The *propagator* G is an operator. $G\mathbf{E}$ is an ordinary product in the $\mathbf{k}_T$ representation and a convolution product in the $\mathbf{r}_T$ representation. In the $\mathbf{k}_T$ representation,

$$G(\mathbf{k}_T;z) = \exp\left\{ i\left(\omega - \frac{k_T^2}{2\omega}\right) z \right\}. \tag{2}$$

In the mixed (x,k_y) representation,

$$G(x,k_y;z) = (iz\lambda)^{-1/2} \exp\left\{ i\omega z + i\omega\frac{x^2}{2z} - i\frac{k_y^2}{2\omega} z \right\}. \tag{3}$$

Here $G\mathbf{E}$ is an ordinary product in k_y and a convolution product in x.

In the limit $z=+\infty$, Eq.(1) gives

$$\mathbf{E}_{\text{free}}(\omega,\mathbf{r}_T,z_0+Z)\simeq\frac{\omega}{2\pi Z}\exp\left\{i\omega\left(Z+\frac{r_T^2}{2Z}\right)\right\}\mathbf{E}_{\text{free}}(\omega,\mathbf{k}_T,z_0)\,|_{\mathbf{k}_T=\omega\mathbf{r}_T/Z} \tag{4}$$

with $Z=z-z_0$. This formula relates the asymptotic angular distribution to the $\mathbf{k}_T$ representation at finite z_0.

4.2. Coulomb field

The Coulomb field of an electron fast moving along the z-axis is[1]

$$\mathbf{E}_{\text{C}}(t,x,y,z)\simeq\frac{-e}{4\pi}\gamma\left[\mathbf{r}_T^2+\gamma^2(z-vt)^2\right]^{-3/2}\mathbf{r}_T \tag{5}$$

We have neglected the longitudinal component. Using partial Fourier transformation, it can also be represented in energy, transverse momentum and z space,

$$\mathbf{E}_{\text{C}}(\omega,\mathbf{k}_T,z)=\frac{ie}{v}\frac{\mathbf{k}_T}{\mathbf{k}_T^2+q_0^2}e^{i\omega z/v}, \tag{6}$$

or in energy, x,k_y and z space,

$$\mathbf{E}_{\text{C}}(\omega,x,k_y,z)=\frac{ie}{2v}e^{-\mu|x|}\,e^{i\omega z/v}\,(i\,\text{sign}(x),\tau), \tag{7}$$

with $q_0=\omega/(\gamma v)$, $\mu=\sqrt{k_y^2+q_0^2}$, $\tau=k_y/\mu$.

The Coulomb field is made of virtual, or *bound* photons, of phase velocity $\omega/k_z=v$, whereas diffraction radiation is made of real photons, *i.e.*, $\omega/|\mathbf{k}|=1$. If the electron is suddenly stopped, or scattered at large angle, at $z=z_0$ the field (5-7) is "released" and propagates as a free field downstream the z_0 point.

4.3. Diffraction Radiation from one target

The forward diffraction radiation field $\mathbf{E}_{\text{FDR}}$ from a target T can be considered as the negative of the intercepted Coulomb field, propagating freely in the forward half-space:

$$\mathbf{E}_{\text{FDR}}(z)=-G(z-z_t)\,A\,\mathbf{E}_{\text{C}}(z_t), \tag{8}$$

where z_t is the point where the z axis crosses the target plane and $A(x,y)$ is the supporting function of the transverse area of the target. We put the screen in the $x\geq b$ half-space, where b is the impact parameter, therefore $A(x,y)=A(x)=\theta(x-b)$ ($A\mathbf{E}$ is an ordinary product in (x,y) space). Forward transition radiation is obtained in the limit $b=-\infty$. $\mathbf{E}_{\text{FDR}}$ is independent on the target material (once it is fully opaque), on the surface roughness and is unchanged when the screen is tilted. The field after the screen is the sum of FDR and the full Coulomb field $\mathbf{E}_{\text{C}}$ of (5-7).

For a mirror target, the backward diffraction radiation field $\mathbf{E}_{\text{BDR}}$ is symmetrical of $\mathbf{E}_{\text{FDR}}$ with respect to the target plane.

$$\mathbf{E}_{\text{back}}=-S\left\{G(z-z_t)\,A\,\mathbf{E}_{\text{C}}(z_t)\right\} \tag{9}$$

where S is symmetry operator about the target plane.

---

[1]We use relativistic units $(c=1)$ and rational definitions of fields and charge, for instance $\nabla\mathbf{E}=\rho$, $e^2/(4\pi\hbar)=\alpha=1/137$.

The FDR amplitude at $z = z_t$ is most conveniently calculated in the (x, k_y) representation (7) :

$$\mathbf{E}_{\mathrm{FDR}}(\omega, x, k_y, z_t) = -\frac{ie}{2}\theta(x-b)\, e^{-\mu|x|}\, e^{i\omega z_t/v}\,(i\,\mathrm{sign}(x), \tau). \tag{10}$$

We have approximated v by 1, except in the exponentials. Taking the Fourier transform in k_x, we obtain the full $\mathbf{k}_T$ representation which, for $b > 0$ (non-intercepting screen), reads

$$\mathbf{E}_{\mathrm{FDR}}(\omega, \mathbf{k}_T, z_t) = -\frac{ie}{2(\mu + ik_x)} e^{-(\mu+ik_x)b}\, e^{i\omega z_t/v}\,(i, \tau). \tag{11}$$

Using (4) one obtains the well-known energy-angle spectrum of Diffraction Radiation (here $\mathbf{k}_T = \omega\vec{\theta}$):

$$\frac{dW}{\hbar\, d\omega\, d\Omega} = \frac{R^2}{\pi}|\mathbf{E}_{\mathrm{FDR}}(\omega, \mathbf{R})|^2 = \frac{\alpha\omega^2}{4\pi^2\mu^2}\frac{q_0^2 + 2k_y^2}{q_0^2 + \mathbf{k}_T^2} e^{-2\mu b}. \tag{12}$$

4.4. Generalization to two targets

Let us now consider two successive targets T_1 (absorbing or reflecting) and T_2 (reflecting), located at z_1 and z_2 and of supporting functions A_1 and A_2. The field between T_1 and T_2 is

$$\mathbf{E}_{\mathrm{betw}}(z) = \mathbf{E}_{\mathrm{FDR}}^{(1)}(z) + \mathbf{E}_{\mathrm{C}}(z), \tag{13}$$

with $\mathbf{E}_{\mathrm{FDR}}^{(1)}(z)$ given by (8) with $z_t = z_1$. In the backward direction from the mirror T_2 one observes the reflection of $\mathbf{E}_{\mathrm{betw}}$ by T_2 :

$$\mathbf{E}_{\mathrm{back}\,2} = -S_2\{G(z - z_2)\, A_2 \mathbf{E}_{\mathrm{betw}}(z_2)\} \tag{14}$$

where S_2 is symmetry operator about the T_2 plane. Leaving aside the symmetry about the T_2 plane, the energy-angle distribution of $\mathbf{E}_{\mathrm{back}\,2}$ is determined by the $\mathbf{k}_T$ representation of the field

$$\mathbf{F} = A_2\mathbf{E}_{\mathrm{betw}}(z_2) = A_2\, G(z_2 - z_1)\, \mathbf{E}_{\mathrm{FDR}}^{(1)}(z_1) - \mathbf{E}_{\mathrm{FDR}}^{(2)}(z_2), \tag{15}$$

where (8) has been used for target T_2 in the last term.

4.5. Analytical results

The analytical calculation starting from (6) are most easily done in the (x, k_y) representation. Here we only give the final results for the same-side and opposite-side cases. The details of the derivation can be found in [12]. We take $z_1 = 0$, $z_2 = L$.

4.5.1. Targets on the same side.

Their supporting functions are $A_1(x, y) = \theta(x - b_1)$ and $A_2(x, y) = \theta(x - b_2)$ with $b_1, b_2 > 0$. The amplitude in the $\mathbf{k}_T$ representation is

$$\mathbf{F}(\omega, \mathbf{k}_T) = \frac{ie}{2}(i, \tau)\frac{e^{-(\mu+ik_x)b_2}}{\mu + ik_x}\{e^{i\omega L/v} - e^{i\omega L}\,[$$

$$e^{(\mu+ik_x)(b_2-b_1)} \exp\left(-iL\frac{k_T^2}{2\omega}\right)\frac{1}{2}\mathrm{Erfc}\left[(2iL/\omega)^{-1/2}(b_2 - b_1 - k_x L/\omega)\right]$$

$$+\exp\left(iL\frac{\mu^2-k_y^2}{2\omega}\right)\frac{1}{2}\mathrm{Erfc}\left[(2iL/\omega)^{-1/2}(b_1-b_2+i\mu L/\omega)\right]\}. \quad (16)$$

The function $\mathrm{Erfc}(z)=2\pi^{-1/2}\int_z^{+\infty}dt\,e^{-t^2}=1-erf(z)$ is the complementary error function for complex argument. One can check that $\mathbf{F}(\omega,\mathbf{k}_T)$ is zero for $L\simeq 0$ and $b_1<b_2$ (full shadowing) and is equal to $\mathbf{E}_{\mathrm{FDR}}(b_2)-\mathbf{E}_{\mathrm{FDR}}(b_1)$ (partial shadowing) for $L\simeq 0$ and $b_1>b_2$.

4.5.2. Targets on opposite sides.
We have still $A_1(x,y)=\theta(x-b_1)$ with $b_1>0$ but $A_2(x,y)=\theta(b_2-x)$ with $b_2<0$. We obtain

$$\frac{2}{ie}\mathbf{F}_{\mathrm{opp}}(\omega,\mathbf{k}_T)=\frac{(-i,\tau)}{\mu-ik_x}e^{(\mu-ik_x)b_2}\exp(i\omega L/v)-\frac{(i,\tau)}{\mu+ik_x}e^{i\omega L}\{$$
$$e^{-(\mu+ik_x)b_1}\exp\left(-iL\frac{k_T^2}{2\omega}\right)\frac{1}{2}\mathrm{Erfc}\left[(2iL/\omega)^{-1/2}(b_2-b_1-k_xL/\omega)\right]$$
$$+e^{-(\mu+ik_x)b_2}\exp\left(iL\frac{\mu^2-k_y^2}{2\omega}\right)\frac{1}{2}\mathrm{Erfc}\left[(2iL/\omega)^{-1/2}(b_1-b_2+i\mu L/\omega)\right]\}. \quad (17)$$

4.6. Numerical results
Numerical calculations of the quantity $W=|F/e|^2$, which is proportional to the spectral-angular distribution and given by (16) or (17), were performed for the electron beam parameters indicated above for impact-parameter $b_1=b_2=10$ mm and wavelength $\lambda=12$ mm. First we consider the calculations for the case where targets are on the same side of the electron beam (see figure 6). In figure 15 the radiation angular distribution calculated with (16) for different values of the distance L between absorber and conductive target for value interval of $L=0\sim 220$ mm is shown.
We can see the manifestation of the shadowing effect and the effect of the recovering of the electron field. Figure 16 shows the numerical results, calculated with (17), for the case where the targets are on opposite sides of the beam (figure 7). The intensity is much higher than in the same-side case and practically independent of L, which indicates that in this case the influence of shadowing is negligible. The theoretical plots of figures 15-16 are quite similar to the experimental ones of figures 13-14, in spite of the fact that concern only one wavelength. θ is the angle with respect to the direction of specular reflection on T_2. It corresponds to 85-θ (in degree) of figure 13 or 100-θ of figure 14.

5. Discussion

The present experiment has demonstrated the feasibility of a direct observation of the shadowing of an electron electromagnetic field in a macroscopic mode. The used experimental method allowed us to make a basic analysis of this phenomenon. Unfortunately the experimental conditions in this experiment could not allow us to measure the L-dependence of the Coulomb field recovering up to $\gamma^2\lambda$ (L is the distance from the absorber). Nevertheless, in the observed L range, the expected properties of shadowing are clearly seen and the dependence of the radiation intensity on L and on the observation angle is in a good agreement with the theoretical one. A complementary confirmation of the pseudo-photons viewpoint is the identity of the shadow effects with absorbing and conductive screens.

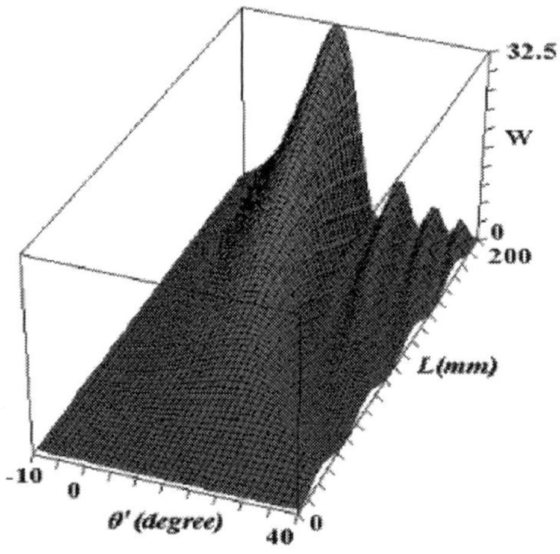

Figure 15. Calculated dependence of the radiation intensity on the observation angle θ' and distance L between the targets for targets on the same side.

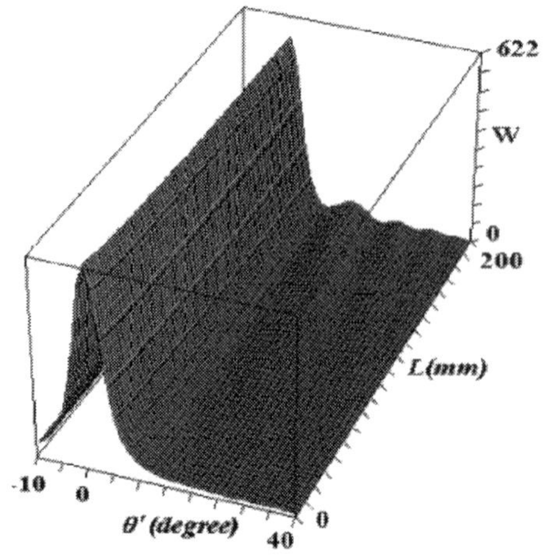

Figure 16. Same quantity as in Fig. 15, but for targets on opposite sides

We have presented the shadow effect from two points of view:

A) For the phenomenology, the electron field properties at $\gamma^2 \gg 1$ are close to those of an electromagnetic field in free space, therefore pseudo-photons are reflected by the mirror or absorbed in the absorber almost like real photons.

B) For the calculation, we have considered the shadowed electron field as a sum of the electron field in free space and the forward diffraction radiation (FDR) emitted from the target (or, if the electron crosses the target, forward transition radiation FTR).

The qualification "emitted from the target" applied to FDR or FTR may be confusing. In a traditional interpretation (for example in [6]), DR and TR from perfectly conducting targets are indeed considered as radiations emitted by surface currents. For non-perfectly conducting targets, surface currents are replaced by volume polarization currents. These surface or polarization currents are induced by the electromagnetic field of the relativistic electron. If we now take the viewpoint A and apply it to the absorber case, the electron electromagnetic field is evanescing in the target and no surface or polarization current may occur on the downstream side. One could then erroneously conclude (in this traditional interpretation) that no FDR or FTR is generated from an absorber.

The solution to this paradox is: either one uses Maxwell equations in matter (MEM), which include absorption, but do not take the polarizations currents as field sources, or one uses Maxwell equations in vacuum (MEV), which take the polarizations currents as sources but do not include absorption. It is therefore not correct to "screen" the field of the polarization currents. Let us show how these two formalisms are dual and give the same prediction in the case of forward optical transition radiation.

The Coulomb field can be decomposed as $F_C=F_u+F_d$, where F_u is the retarded field of the upstream electron trajectory (suddenly stopped electron) and F_d the retarded field of the downstream trajectory (suddenly accelerated electron). In the MEM formalism, FTR is generated by the current of the traveling electron after traversal of the screen [13]}. This current also re-creates the Coulomb field. The total field in the forward region is then $F_d = F_{FTR} + F_C$. Far enough from the trajectory, $F_{FTR} \approx F_d$. We can then say that FTR is generated during the transition of electron from an unstable "naked" state to the stable "dressed" state (here the term "transition radiation" seems particularly appropriate). A similar interpretation of this phenomenon can be found in [2]. Similarly, forward diffraction radiation can be viewed as a consequence of the transit from a "half-naked" state to the dressed state.

In the MEV formalism, F_{FTR} is equal in the forward region to the field F_m generated by the currents circulating on or under the upstream surface of the target. These currents are induced by F_C, which is nearly equal to F_u in the target, and screen it, therefore $F_m \sim - F_u$ and the total forward field is $F_C + F_m = F_d$, as in the MEM formalism. Nevertheless the MEM approach is more intuitive for FTR.

Another example where the shadow effect is at work is Smith-Purcell (SP) radiation. One can consider the SP effect as a reflection of the pseudo-photons on the ridges of the grating. Then one expects that each ridge makes a shadow on the next one. The grating period, $\Lambda = \lambda/(1 - v\cos\theta)$ is indeed smaller than $l_f = \lambda\gamma^2$ for a typical emission angle θ. The shadow effect suggests a limit of the form $W \leq C \cdot \hbar L/(137 \cdot b^2)$ on the energy of Smith-Purcell radiation per electron [14,12]. C is a numerical constant, L the length of the electron path above the grating and b the impact parameter, *i.e.*, the distance between the electron trajectory and the tops of the ridges. In fact, such a bound with the precise value $C=1/(2\pi)$ can be derived from unitarity and analyticity of the reflection coefficients and detailed bounds are obtained for fixed ω and fixed component k_y of the photon momentum [15].

Acknowledgment

This work was partly supported by the warrant-order 1.226.08 of the Ministry of Education and Science of the Russian Federation and by the Federal agency for science and innovation, contract 02.740.11.0245

References

[1] Jackson J D *Classical Electrodynamics, 3rd ed.* (J. Willey&Sons, New-York, 1998)

[2] Feĭnberg E L. *SOV PHYS USPEKHI*, 1979, **22** (6), 479-479

[3] Shul'ga N F and Syshchenko V V. *Journal Physics of Atomic Nuclei*, **63**, 11, (2000), 2018

[4] Anihuity P L, Bfickef-Susfldy R, Bosled P R, et.al. *Phys. Rev. Let.* **75** 10(1995) 1949

[5] Ter-Mikaelian N L High-Energy Electromagnetic Processes in Condensed Media (Wiley, New York, 1972).

[6] BolotovskiI B N. Preprints of Lebedev Institute of Physics, Russian Academy of Sciences,Vol **140** p. 95

[7] Bolotvskiy B M, Serov A V. *Pis'ma v Zhurnal Eksperimental'noy i Teoreticheskoy Fiziki*, 2006, Vol. **90**, No. 6, pp. 486–488.

[8] Kalinin B N, Naumenko G A, Potylitsyn A P et al. *JETP Letters*, 2006, Vol. **84**, No. 3, pp. 110–114.

[9] Hanke K. *CLIC note* 298, 19. 04.1996

[10] Verzilov V A. *Phys. Lett.* A **273** (2000) 135.

[11] Kalinin B N, Naumenko G A, Potylitsyn A P et al. *JETP Letters*, **84**, 3, (2006), p. 110.

[12] Naumenko G, Artru X, Potylitsyn A et al. Shadowing of the electromagnetic field of a relativistic electron. *The 51st Workshop Channeling 2008.* Erice, Arxiv.org/pdf/0901.2630v1

[13] Artru X, Chehab R, Honkavaara K, Variola A. *Nucl. Inst. Meth.* B**145** (1998) 160.

[14] Artru X., Ray C. Interferences between light emitted by charged particles crossing or skimming past optical fibers. *International Symposium RREPS'07*, Prague, 2007

[15] Artru X, in preparation.

The synchrotron radiation of first excited state electrons

V. G. Bagrov [1,2], **A. N. Bourimova** [1]

[1] Department of Physics, Tomsk State University, Lenin Avenue, Tomsk, Russia
[2] Institute of High Current Electronics of the Siberian Branch of the Russian Academy of Sciences, Tomsk, Russia

E-mail: `bagrov@phys.tsu.ru`

Abstract. We give the precise analysis of first excited state electron synchrotron radiation (SR) power using quantum theory methods.

1. Introduction

The precise analytical expressions for the basic synchrotron radiation (SR) characteristics such as spectral-angular distribution, spectral and angular distribution, total output, radiation polarization are entirely formulated in terms of classical theory. These expressions could be found in monographic and educational literature [1 - 5].

As for the quantum theory the only expressions which could be represented analytically at any quantum factor value are the expressions for the spectral-angular and spectral distributions and for the total SR polarization components output in ultrarelativistic limit [1-5] when we can successfully use the quasi-classical approximation method modification. However even in ultrarelativistic limit the precise analytical expression for the angular distribution at the arbitrary quantum factor value is unknown. Still we can obtain the precise analytical expressions for SR characteristics in terms of quantum theory e. g. in case of initial state charge being in low energy level. This kind of analysis in non-relativistic limit was taken in [6, 7]. The results of the precise analytical and numerical analysis of the first excited state electron SR characteristics are represented below.

2. Basic theoretical expressions

We study the synchrotron radiation (SR) angular distribution structure of an electron in the first excited state (charge $-e, \quad e > 0$).

In this case the only possible quantum transition is the transition from the first excited state to the ground state. So, the electron radiates one harmonic with the frequency

$$\omega = \frac{mc^2}{\hbar}\frac{\gamma^2 - 1}{\gamma + \sqrt{\gamma^2\cos^2\theta + \sin^2\theta}} = \frac{E}{\hbar}\frac{\beta^2}{1 + \sqrt{1 - \beta^2\sin^2\theta}}, \quad 0 \leqslant \theta \leqslant \pi. \tag{1}$$

Here m - is the rest mass of an electron, c is the velocity of light, $\hbar$ is the Planck constant, $\gamma = E/mc^2 = (1 - \beta^2)^{-1/2}$ is the relativistic factor, E is the electron energy, $\beta = v/c$, v is the particle velocity, θ is the angle between the photon's impulse direction and the magnetic field. For the first excited state we have the relation

$$\gamma = \sqrt{1 + 2\frac{H}{H_0}}, \quad H_0 = \frac{m^2c^3}{e\hbar} \text{ – the Schwinger field,} \tag{2}$$

where $H > 0$ - is the magnetic field strength.

Let the quantum number $\zeta = \pm 1$ specify the initial state electron spin orientation with respect to the magnetic field direction, ($\zeta = 1$ means that electron spin and the magnetic field do have the same direction, $\zeta = -1$ signifies that they have opposite directions). The final (ground) state of the electron is distinguished by that the electron being in it has the spin opposite to the field. So, the jump from the initial state $\zeta = 1$ to the ground state occurs with a spin reorientation (spin - flip) only , the transition from the initial state $\zeta = -1$ to the ground state happens without the spin-flip.

In the case under consideration the SR power W and its polarization components angular distribution can be represented as follows

$$\begin{gathered}
\frac{dW_i}{d\Omega} = QMG_i, \quad Q = \frac{e^2m^2c^2}{\hbar^2}, \quad d\Omega = \sin\theta d\theta, \\
M = \frac{\beta^6(1+\sqrt{1-\beta^2})}{64(1-\beta^2)} = \frac{2}{(1-x_0^2)^2}\left(\frac{x_0}{1+x_0}\right)^3 = \frac{(\gamma+1)(\gamma^2-1)^3}{64\gamma^5}, \\
x_0 = \frac{1-\sqrt{1-\beta^2}}{1+\sqrt{1-\beta^2}} = \frac{\gamma-1}{\gamma+1}, \quad 0 \leqslant x_0 < 1.
\end{gathered} \tag{3}$$

where the index i denotes the radiation polarization: $i = 2$ corresponds to the radiation power of the "σ" - linear polarization component, $i = 3$ is for the "π" - linear polarization component, $i = 0$ is for the total (summed over polarizations) output, $i = l$ ($l = \pm 1$) is for the right ($l = 1$) and left ($l = -1$) circular polarizations. The angular distribution structure is completely defined by the functions $G_i = G_i(\zeta, \beta, \theta)$, given below

$$\begin{gathered}
G_2(\zeta, \beta, \theta) = (1-\zeta)R_1(\beta, \theta) + (1+\zeta)x_0R_2(\beta, \theta), \\
G_3(\zeta, \beta, \theta) = (\gamma-\zeta)(\gamma+\zeta)^{-1}\, G_2(-\zeta, \beta, \theta), \\
G_0(\zeta, \beta, \theta) = 2[1 + x_0 - \zeta(1-x_0)]R_0(\beta, \theta), \\
2R_0(\beta, \theta) = R_1(\beta, \theta) + R_2(\beta, \theta), \\
G_l(\zeta, \beta, \theta) = [1 + x_0 - \zeta(1-x_0)]\, S_l(\beta, \theta), \\
S_l(\beta, \theta) = R_0(\beta, \theta) + l\, S(\beta, \theta)\cos\theta.
\end{gathered} \tag{4}$$

Here we introduce the notations

$$R_1(\beta, \theta) = \frac{(1-x_0x)(1+x)^3}{1-x}e^{-x}, \quad R_2(\beta, \theta) = \frac{(x_0-x)(1+x)^3}{x_0(1-x)}e^{-x},$$

$$R_0(\beta, \theta) = \frac{[2\,x_0 - (1+x_0^2)x](1+x)^3}{2\,x_0(1-x)}e^{-x}, \quad S(\beta, \theta) = \frac{(1+x)^4}{1-x}e^{-x}, \tag{5}$$

$$x = \frac{1-\sqrt{1-\beta^2\sin^2\theta}}{1+\sqrt{1-\beta^2\sin^2\theta}} = \frac{\gamma - \sqrt{\gamma^2\cos^2\theta + \sin^2\theta}}{\gamma + \sqrt{\gamma^2\cos^2\theta + \sin^2\theta}}, \quad 0 \leqslant x \leqslant x_0 < 1\,.$$

Thus we can see that in our case the synchrotron radiation angular distribution structure is completely defined by the functions $R_s(\beta,\, \theta),\quad s = 0,\, 1,\, 2;\quad S_l(\beta,\, \theta)$. The following properties are easy to check

$$G_k(\zeta,\, \beta,\, \theta) = G_k(\zeta,\, \beta,\, \pi - \theta),\quad k = 0,\, 2,\, 3;\quad G_l(\zeta,\, \beta,\, \theta) = G_{-l}(\zeta,\, \beta,\, \pi - \theta),$$
$$R_s(\beta,\, \theta) = R_s(\beta,\, \pi - \theta),\quad s = 0,\, 1,\, 2;\quad S_l(\beta,\, \theta) = S_{-l}(\beta,\, \pi - \theta). \tag{6}$$

It follows from (6) that it is quite enough to consider the $R_s(\beta,\, \theta)$ function behavior on the segment $0 \leqslant \theta \leqslant \pi/2$ .Besides it is sufficient to investigate one of the functions $S_l(\beta,\, \theta)$ (say, $S_1(\beta,\, \theta)$).

3. SR polarization components angular distribution structure of the first excited state electron

At first we mention that the functions $R_s(\beta,\, \theta),\quad S_l(\beta,\, \theta)$ are finite at any argument value $0 \leqslant \beta \leqslant 1$ (including the point $\beta = 1$). In particular, at the boundary points of the segment $0 \leqslant \beta \leqslant 1$ segment we have

$$R_1(0,\, \theta) = 1,\quad R_2(0,\, \theta) = \cos^2\theta,\quad R_0(0,\, \theta) = \frac{1}{2}(1 + \cos^2\theta),\quad S_l(0,\, \theta) = \frac{1}{2}(1 + l\cos\theta)^2. \tag{7}$$

$$R_s(1,\, \theta) = (1 + p)^3 e^{-p},\quad S_l(1,\, \theta) = \left(1 + l\frac{\cos\theta}{|\cos\theta|}\right)(1 + p)^3 e^{-p};\quad p = \frac{1 - |\cos\theta|}{1 + |\cos\theta|}. \tag{8}$$

It is not difficult to obtain the relations

$$R_s(\beta,\, 0) = R_s(\beta,\, \pi) = 1,\quad S_l(\beta,\, 0) = 1 + l,\quad S_l(\beta,\, \pi) = 1 - l;$$
$$R_1(\beta,\, \pi/2) = 2R_0(\beta,\, \pi/2) = 2S_l(\beta,\, \pi/2) = (1 + x_0)^4 e^{-x_0},\quad R_2(\beta,\, \pi/2) = 0. \tag{9}$$

It follows from (4) and (8) that in the ultrarelativistic limit ($\gamma \to \infty$)the functions $G_i = G_i(\zeta,\, \beta,\, \theta)$ do not longer depend on the spin orientation ζ. It means that in ultrarelativistic limit the spin-flip quantum jumps and the jumps without spin reorientation become equiprobable.

The second expression in (4) implies that the angular distribution structure of the "π" -linear polarization component for an electron with spin ζ and the "σ"-linear polarization component for an electron with spin $-\zeta$ are the same (the corresponding expressions differs only by the factors which do not depend on θ).

The "π" -linear polarization component does not vanish at $\theta = \pi/2$ for an electron with spin $\zeta = 1$ and this is an essentially quantum effect.

As it ensues from (8) in the ultrarelativistic case ($\gamma \to \infty$)the only right circular polarization is radiated in the upper half plane ($\theta < \pi/2$),while the lower half plane($\theta > \pi/2$) contains a pure left circular polarization. It is a characteristic feature of the transitions to the ground state.

The analysis of the functions $R_s(\beta,\, \theta),\ S_1(\beta,\, \theta)$ behavior shows up that the function $R_1(\beta,\, \theta)$ is a monotone increasing function of θ on $0 < \theta < \pi/2$ at any β.

The function $R_2(\beta,\, \theta)$ is monotone decreasing function of θ on $0 < \theta < \pi/2$ and while $\beta^2 \leqslant 3/4$ it has its maximum at some inner point $\theta = \theta_2^{(max)}(\beta)$ of this interval.

The functions $R_0(\beta,\, \theta)$ on $0 < \theta < \pi/2$ and $S_1(\beta,\, \theta)$ on $0 < \theta < \pi$ at $\beta^2 \leqslant 1/2$ are monotone decreasing functions of θ. At $\beta^2 > 1/2$ they reach maxima on $0 < \theta < \pi/2$ at the points $\theta = \theta_0^{(max)}(\beta)$ and $\theta = \theta_1^{(max)}(\beta)$, respectively.

The figures illustrate the graphs of the functions $R_s(\beta,\, \theta),\quad S_1(\beta,\, \theta)$ for different β and the graphs of $\theta_n^{(max)}(\beta)$, ($n = 0,\, 2$)

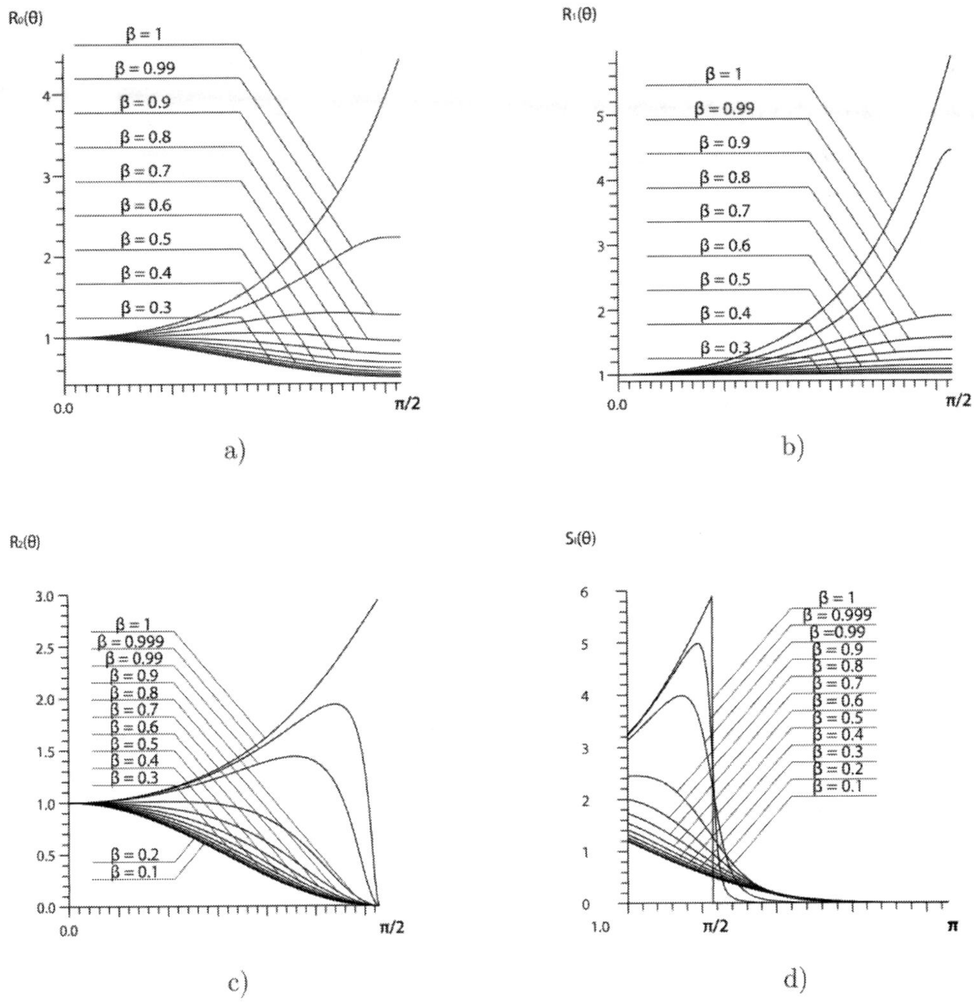

Figure 1. This functions define the character of the polarization components angular distribution $R_0(\theta)$ (a), $R_1(\theta)$ (b), $R_2(\theta)$ (c) and $S_l(\theta)$ $(l = 1)$ (d).

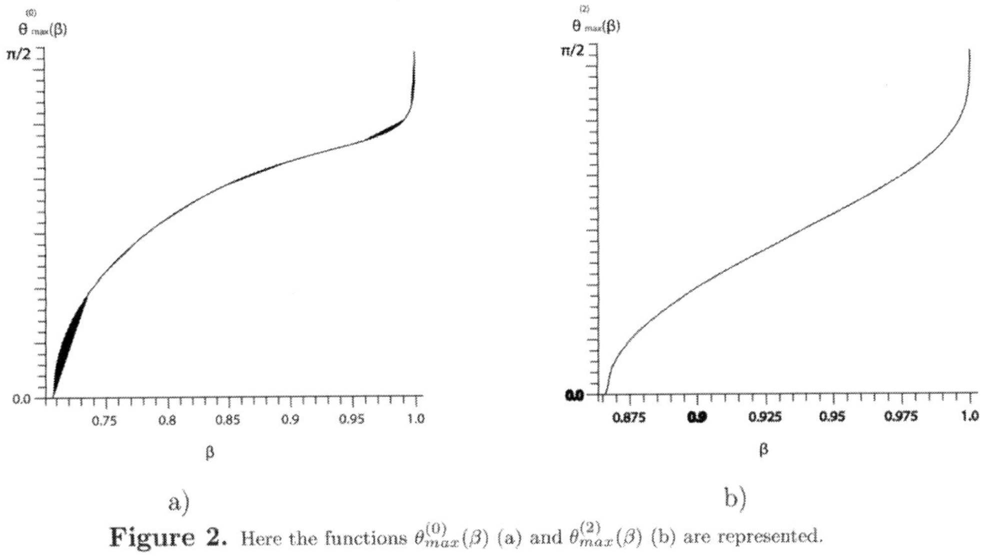

Figure 2. Here the functions $\theta^{(0)}_{max}(\beta)$ (a) and $\theta^{(2)}_{max}(\beta)$ (b) are represented.

4. The total output

For the total output radiated in the upper half plane (integrated over θ from $0 \leqslant \theta \leqslant \pi/2$) we obtain

$$W_i = Q\overline{M}P_i(\zeta,\,\beta),\quad \overline{M} = \frac{\beta^6}{32(1-\beta^2)(1+\sqrt{1-\beta^2})} = \frac{1}{(1-x_0)^2}\left(\frac{x_0}{1+x_0}\right)^3 = \frac{(\gamma^2-1)^3}{32\gamma^3(\gamma+1)},$$
$$P_2(\zeta,\,\beta) = (1-\zeta)f_1(\beta) + (1+\zeta)x_0f_2(\beta),$$
$$P_3(\zeta,\,\beta) = (\gamma-\zeta)(\gamma+\zeta)^{-1}\,P_2(-\zeta,\,\beta), \tag{10}$$
$$P_0(\zeta,\,\beta) = [1+x_0-\zeta(1-x_0)]f_0(\beta),\quad f_0(\beta) = f_1(\beta)+f_2(\beta),$$
$$2P_l(\zeta,\,\beta) = [1+x_0-\zeta(1-x_0)][f_0(\beta)+2l\phi(\beta)],$$

In (10) the following notations are used

$$f_1(\beta) = \int_0^1 (1+x_0y)\exp(-x_0y)\sqrt{\tfrac{1-x_0^2y}{1-y}}\,dy,$$
$$f_2(\beta) = \int_0^1 (1+x_0y)\exp(-x_0y)\sqrt{\tfrac{1-y}{1-x_0^2y}}\,dy,\quad \phi(\beta) = \frac{2-(2+x_0)\exp(-x_0)}{x_0}. \tag{11}$$

Substituting $l \longrightarrow -l$ in (10) we get the total output radiated in the lower half plane (integrated over θ from $\pi/2 \leqslant \theta \leqslant \pi$).

The functions defined by (11) are positive and bounded, the functions $f_0(\beta)$, $f_1(\beta)$ and $\phi(\beta)$ are monotone decreasing and $f_2(\beta)$ is monotone increasing on the segment $0 \leqslant \beta \leqslant 1$. It is easy to calculate the values of the function (11) at the segment boundary points

$$f_1(0) = 3f_2(0) = 2\phi(0) = 2,\quad f_1(1) = f_2(1) = \phi(1) = 2 - \frac{3}{e}. \tag{12}$$

At $\beta \ll 1$, $(x_0 \ll 1)$ we find for the functions $f_n(\beta)$

$$f_1(\beta) \approx 2\left(1 - \frac{13}{15}x_0^2\right),\quad f_2(\beta) \approx \frac{2}{3}\left(1 + \frac{11}{35}x_0^2\right),\quad f_0(\beta) \approx \frac{8}{3}\left(1 - \frac{4}{7}x_0^2\right), \tag{13}$$

and for $\phi(\beta)$ we have the convergent at any x_0 series expansion

$$\phi(\beta) = \sum_{n=0}^{\infty}\frac{(-1)^n(1-n)}{(n+1)!}x_0^n \approx 1 - \frac{x_0^2}{6}. \tag{14}$$

For the functions $f_s(\beta)$ $(s = 0,\,1,\,2)$ and $\phi(\beta)$ at $\beta \approx 1\,(\gamma \gg 1)$ we find

$$f_0(\beta) \approx 2\left(2 - \tfrac{3}{e} + \tfrac{2}{e}\tfrac{\ln\gamma}{\gamma}\right) + O\left(\tfrac{1}{\gamma}\right),\quad f_1(\beta) \approx 2 - \tfrac{3}{e} + \tfrac{8}{e}\tfrac{\ln\gamma}{\gamma} + O\left(\tfrac{1}{\gamma}\right),$$
$$f_2(\beta) \approx 2 - \tfrac{3}{e} - \tfrac{4}{e}\tfrac{\ln\gamma}{\gamma} + O\left(\tfrac{1}{\gamma}\right),\quad \phi(\beta) \approx 2 - \tfrac{3}{e} + \left(2 - \tfrac{5}{e}\right)\tfrac{1}{\gamma} + O\left(\tfrac{1}{\gamma^2}\right). \tag{15}$$

The precise analytical expression (G.A. Schott formula) for the total SR output in terms of classical theory is known [1-5] and we can rewrite it using our own designation

$$W^{cl} = \frac{2}{3}Q\left(\frac{H}{H_0}\right)^2(\gamma^2-1) = \frac{2}{3}Q\left(\frac{H}{H_0}\right)^2(\beta\gamma)^2. \tag{16}$$

In the case under consideration in accordance with (2) there is a relation between the field tension H and Lorentz - factor γ,

$$\frac{H}{H_0} = \frac{1}{2}(\gamma^2 - 1), \tag{17}$$

which let us rewrite(16) as follows

$$W^{cl} = \frac{1}{6}Q(\gamma^2 - 1)^3 = \frac{1}{6}Q(\beta\gamma)^6, \tag{18}$$

and therefore

$$W^{cl} \approx \frac{1}{6}Q\beta^6 \text{ at } \beta \ll 1; W^{cl} \approx \frac{1}{6}Q\gamma^6 \text{ at } \gamma \gg 1 .$$

For the upper half plane radiation $W^{cl}_{(+)}$ we finally obtain

$$\begin{gathered} W^{cl}_{(+)} = \tfrac{1}{2}W^{cl} = \tfrac{1}{12}Q(\gamma^2 - 1)^3 = \tfrac{1}{12}Q(\beta\gamma)^6 ; \\ W^{cl}_{(+)} \approx \tfrac{1}{12}Q\beta^6 \text{ at } \beta \ll 1; \quad W^{cl}_{(+)} \approx \tfrac{1}{12}Q\gamma^6 \text{ at } \gamma \gg 1 . \end{gathered} \tag{19}$$

Comparing the classical expressions (19) with the results obtained at $\beta \ll 1$ for an electron with spin $\zeta = -1$ (the transitions without spin-flip)and taking (10) and (12) into account we find

$$W_i(\zeta = -1) = W^{cl}_{(+)} q_i ; \quad q_0 = 1 , \quad q_2 = \frac{3}{4} , \quad q_3 = \frac{1}{4} , \quad q_l = \frac{(4 + 3l)}{16} \quad (l = \pm 1) . \tag{20}$$

Thus, here we have the total coincidence with the classical theory (the value of q_l was for the first time obtained in [8]).

If the electron spin $\zeta = 1$ (the spin-flip transitions)then we gain

$$W_i(\zeta = 1) = \frac{\beta^2}{4} W^{cl}_{(+)} p_i ; \quad p_0 = 1 , \quad p_2 = \frac{1}{4} , \quad p_3 = \frac{3}{4} , \quad p_l = \frac{(4 + 3l)}{16} . \tag{21}$$

Thus, here the radiated output is small ($\approx \beta^2/4$) with respect to the case $\zeta = -1$ and the linear polarization components switch their places.

Averaging on the initial spin we gain

$$\overline{W}_i = \frac{1}{2} W_i(\zeta = -1) \tag{22}$$

- nonpolarized nonrelativistic electron radiates the half of classical output (that coincides with [6,7]).

At $\beta \approx 1\,(\gamma \gg 1)$ from (10) and (15) we gain

$$\begin{gathered} W_i = Q\,\gamma^2\, \tfrac{2e-3}{8e}\, r_i, \quad r_0 = 1 + \tfrac{2}{2e-3}\, \tfrac{\ln\gamma}{\gamma} , \quad r_2 = \tfrac{1}{2}\left(1 + \tfrac{2-6\zeta}{2e-3}\, \tfrac{\ln\gamma}{\gamma}\right) , \\ r_3 = \tfrac{1}{2}\left(1 + \tfrac{2+6\zeta}{2e-3}\, \tfrac{\ln\gamma}{\gamma}\right) , r_l = \tfrac{1}{2}\left(1 + l + \tfrac{2}{2e-3}\, \tfrac{\ln\gamma}{\gamma}\right) . \end{gathered} \tag{23}$$

Here with γ rising the quantum output increases $\approx \gamma^2$ that is much slower than the classical output growth $\approx \gamma^6$. The left circular polarization ($l = -1$) in the upper half plane is strongly suppressed ($\approx \frac{\ln \gamma}{\gamma}$) in comparison with the right one ($l = 1$). And there is no prior linear polarization: the difference between "σ" and "π" linear polarization components is small ($\approx \frac{\ln \gamma}{\gamma}$) with respect to the primary terms. The SR power dependance on spin disappears. The spin dependance shows up in the amendments $\approx \frac{\ln \gamma}{\gamma}$.

Acknowledgments

The work is partially supported by Russian Science and Innovations Federal Agency under contract No 02.740.11.0238 and Russia President grant SS-871.2008.2.

References

[1] A. A. Sokolov and I. M. Ternov.
Synchrotron Radiation.
Akademie – Verlag, Berlin; Pergamon Press, New York, 1968.

[2] A. A. Sokolov, I. M. Ternov
Radiation from relativistic electrons.
American Institute of Physics, New York, 1986.

[3] I. M. Ternov, V. V. Mikhailin.
Synchrotron Radiation. Theory and Experiment.
Energoatomizdat, Moskow, 1986. (in Russian).

[4] V. G. Bagrov, G. S. Bisnovatyi – Kogan, V. A. Bordovitsyn et all.
Synhrotron Radiation Theory and Its Development. Editor V. A. Bordovitsyn.
World Scientific. Singapore • New Jersey • London • Hong Kong. 1999.

[5] V. G. Bagrov, G. S. Bisnovaty – Kogan, V. A. Bordovitsyn et all.
Radiation Theory of Relativistic Particles. Editor: V. A. Bordovitsyn.
Moscow, Fizmatlit. 2002. (in Russian).

[6] V. G. Bagrov, O. F. Dorofeev.
Radiation from Polarized Electrons Occupying Low Energy Levels in a Magnetic Field.
Vestnik Moskovskogo Universiteta, Seriya 3. Fizika. Astronomiya. (Moscow University Bulletin, Ser. 3. Physics. Astronomy.) 1966, v. 7, No 2, p. 97–101. (in Russian).

[7] I. M. Ternov, V. G. Bagrov, O. F. Dorofeev.
Characteristic of Electrons Moving in a Magnetic Field at Low Energy Levels.
Soviet Physics Journal. 1968, v. 11, No 10, p. 50-55.

[8] V. G. Bagrov, M. V. Dolzhin, V. B. Tlyachev, A. T. Jarovoi
Evolution of the Angular Distribution of Circularly Polarized
Synchrotron Radiation with Charge Energy.
Russian Physics Journal. 2004, v. 47, No 4, p. 414–423.

Radiation of X-ray Entangled Photons by High Energy Electrons

K.A. Ispirian and M.K. Ispiryan
Yerevan Physics Institute, Brothers Alikhanians 2, Yerevan, 0036, Armenia

E-mail: karo@mail.yerphi.am

Abstract. Without developing a correct theory the rate of the production of entangled photon pairs in the X-ray region by high energy electrons is estimated for experimental parameters used in a few experiments on X-ray parametric down conversion. Taking into account the need for X-ray entangled X-ray photons it is proposed to begin the experimental study of such mechanism of nonlinear X-ray optics before the developing of the theory and the launch of the X-ray FELs.

Keywords: Parametric down conversion, non-linear X-ray optics, parametric X-ray radiation

1. Introduction

QED predicts four nonlinear (NL) effects: scattering of light by light, Delbruck scattering, photon splitting and coalescence of photons. These processes are due to nonlinearity of vacuum and are described by Euler-Heisenberg Lagrange function and electron-positron box diagrams. For instance, the photon splitting due to virtual pair production is the splitting of a photon into two photons, $\omega_0 \rightarrow \omega_1 + \omega_2$.

There are more than 10 optical NL effects due to interactions of photons in NL media which are described by NL susceptibilities. For instance, the optical parametric down conversion (OPDC) is the splitting of a "pump" photon into two entangled photons, "signal" and "idler" photons, $\omega_p \rightarrow \omega_s + \omega_i$ in a optically NL media. OPDC semiclasically is interpreted as mixing of the pump photons with zero point fluctuation photons. OPDC has been theoretically and experimentally studied in many works. At present OPDC has found application in quantum cryptography, teleportation, metrology, for Einstein-Podolsky-Rosen experiments, testing Bell inequalities, etc.

The NL effects in X-ray region have been studied much less then in UV, optical, IR and microwave regions. Among the proposed processes are the X-ray parametric down conversion (XPDC) [1], X-ray modulation by optical field [2], X-ray nonlinear absorption [3], X-ray second harmonic generation (SHG) [4]. A beautiful review on XPDC is written by B. Adams (see 5-th Chapter of [5]). Although XPDC has been observed for the first time in [6], nevertheless, it is necessary to carry out more theoretical and experimental studies of X-ray NL effects because they also can find applications.

Since the cross section of XPDC is small it has been further experimentally studied only in a few works [7-10] without satisfactory development and comparison between the theory and experiment. The highest event rate observed for the third generation SR sources [8-10] is ~ 0.1/s, while the expected one for TESLA FELs is ~10000 times higher [11]. Similar results are expected at LCLS [12]. As OPDC, XPDC semiclasically is interpreted as mixing of the «pump» photon with a zero point fluctuation photon (ZPF), and the very small X-ray nonlinear suspectibilities are somewhat compensated by higher ZPF fields. XPDC takes place as non-linear Bragg diffraction in crystals with energy and momentum conservation laws

$$\hbar\omega_p = \hbar\omega_s + \hbar\omega_i, \quad \vec{k}_p + \vec{H}_{hkl} = \vec{k}_s + \vec{k}_i, \tag{1}$$

where $\vec{k}_j\,(j = p, s, i)$ are the momenta of the photons and $\vec{H}(h,k,l)$ is the crystal reciprocal lattice vector.

The demand of intense source of entangled X-ray photon pairs is so high that even such exotic processes as Unruh radiation [13] and inverse double Compton scattering [14] are proposed for production of entangled X-ray photon pairs. Postponing the development of the correct theory of photon entangled pair production by electrons (PEPPE) and using the results obtained for XPDC [1,8] in this short note we shall make simple estimates of the corresponding PEPPE yields for arrangement parameters coinciding with those of the works [1,8] in order to accelerate the development of the correct theory and experiment.

2. Formulae for Yield Estimates

According to X-ray PDC theory [1] the number N_S or the probability P_S of a PEPPE signal photon produced by a single pump photon and emitted into a solid angle $\Delta\Omega_S$ is equal to

$$\Delta N_S(\omega_P,\omega_S) = P_S(\omega_P,\omega_S) = A\Delta\Omega_S(\hbar\omega_P), \tag{1}$$

where

$$A = \frac{8\pi\omega_S^3(\omega_P-\omega_S)}{c^4[1-\cos\theta_{Si}]}\left|\frac{G(hkl)}{V}\right|^2 \vec{\theta}_{spi}^2(hkl)L_{eff}, \tag{2}$$

$$G(hkl) = \sum_n g_n(hkl)\exp[-2\pi i(hu_n + kv_n + lw_n) \tag{3}$$

is the nonlinear structure factor, $g_n(hkl)$ are coefficients of the Fourier expansion of the linear susceptibility $\chi(\omega_P)$, V is the volume of unit cell, $\vec{\theta}(spi)$ is a vector determining the polarization and depends on the experiment geometry (see eq. (11) of [1]), L_{eff} is the crystal effective length, which for Bragg geometry is equal to

$$L_{eff} \approx (\alpha_P + \alpha_S/\cos\varphi)^{-1}, \tag{4}$$

α_P, α_S are the linear extinction coefficients of the crystal and φ is the angle between the signal photon direction and the back-normal to the slab, $\Delta\Omega_s = 2\pi\sin\theta\Delta\theta$ is the detector acceptance around $\vec{k}_S$, θ is approximately the deviation of one of PEPPE photon emission direction from the specular or from diffraction direction and θ_{Si} is the angle between the signal and idler photons or the opening angle between the signal and idler detectors.

The incidence angle onto the crystal of the pump photon must differ (see [9]) slightly by $\Delta\theta_{Br}$ from the Bragg angle θ_{Br}

$$\Delta\theta_{Br} = (\theta_S^2/2 + 3\chi_0)/\sin(2\theta_{Br}) + \chi_0/\tan\theta_{Br}, \tag{5}$$

where χ_0 is the linear susceptibility of the crystal. In [8,9] $\Delta\theta_{Br}$ is negative and of the order of (0.01-0.02) degrees.

The expression (1) differs from the corresponding formula (13) of [1] by the fact that in (13) it is given $dN_S/dt = AP_P\Delta\Omega_S$ where P_P is the power of the pump photon beam. In [1] it is also numerically calculated the dependence of $dN_S/dt = AP_P\Delta\Omega$ upon $\hbar\omega_P$ (see Fig.2 of [1]) in the assumption $\omega_S \approx \omega_i \approx \omega_P/2$ for the (004) reflection of diamond crystal with effective thickness and $\Delta\Omega_S = 10^{-3}$ sr, when P_P=10 mW. In particular, when $\hbar\omega_P = 18.9$ keV or $\lambda_P = 0.65\,A^\circ$ (this is the pump photon wavelength of the experiment [8]) $dN_S/dt = AP_P\Delta\Omega$=0.78 s^{-1}, and therefore, $A \approx 7.8\text{x}10^4$ $W^{-1}.s^{-1}sr^{-1}$ or $A \approx$ $1.25\text{x}10^{-14}$ $eV^{-1}sr^{-1}$. Note that at such wavelength P_P=10 mW is obtained if $3.3\text{x}10^{12}$ pump photons per second hit the radiator, and this means that the probability $P_S^{1PP} = \Delta N_S(\hbar\omega_P,\hbar\omega_S)$ of the detection of one entangled signal photon produced by one pump photon for the experimental parameters [1,8] is equal to P_S^{1PP} = $2.4\text{x}10^{-13}$. Taking into account the differences between the experimental parameters ($\theta_{Si}, \Delta\Omega_S$, detection efficiency, etc) this probability P_S^{1PP} is in more or less satisfactory agreement with the results [8,9] of the coincidence experiments on 3-rd generation SR source beams. Very close value for P_S^{1PP} is obtained in the assumption $\omega_S \approx \omega_i \approx \omega_P/2$ using the derived cross section per electron $d\sigma/d\Omega \approx (1/4\pi\alpha)r_e^4 k_P^2\Delta x$, in which $r_e = e^2/mc^2$ is the classical electron radius, $\alpha = e^2/\hbar c$ is the fine structure constant, and it is assumed $\Delta x = \Delta(\hbar\omega_P)/(\hbar\omega_P) \approx 10^{-2}$ [9]. Let us note that using the Fig. 2 of [1] without complicated calculations of $G(hkl)$, $\vec{\theta}(spi)$ and other magnitudes one can find $P_S^{1PP}(\omega_S)$ or $P_S^{1PP}(\omega_P)$, of course, in the assumption $\omega_S \approx \omega_i \approx \omega_P / 2$ as function of λ_P and obtain curve with its maxima and zeros as the Fig. 2 of [1]. Let us also remind that for measuring or calculating such dependence as it is given in Fig.2 of [1] for each λ_P the values of θ_{Br}, $\Delta\theta_{Br}$, etc are different.

It is evident that such pump photons can be provided also by the Williams-Weizsaker pseudophotons of high energy electrons the spectral distribution of which for our approximation and for $\gamma = E/mc^2 = 1/(1-\beta)^2 >> 1$ and $\hbar\omega_{WW} << \gamma mc^2$ according to [15] is equal to

$$\frac{dN_{WW}(\gamma, \hbar\omega_{WW})}{d(\hbar\omega_{WW})} \approx \frac{2\alpha}{\pi(\hbar\omega_{WW})} \ln\left(\frac{\gamma mc^2}{\hbar\omega_{WW}}\right). \qquad (6)$$

Using the Williams-Weizsaker method the probability or the number $\Delta N_{PEPPE}(\omega_P)$ of PEPPE photons produced in a crystal by a single pseudophoton with frequency ω_P will be

$$\Delta N_{PEPPE}(\hbar\omega_P) = \frac{dN_{WW}(\gamma, \hbar\omega_P)}{d(\hbar\omega_P)} \Delta N_S(\hbar\omega_P). \qquad (7)$$

Instead of integration of (7) over $(\hbar\omega_{WW})$ in order to derive the number of PEPPE photons produced by the pseudophotons of a passing charged particle into the solid angle $\Delta\Omega_s = 2\pi \sin\theta\Delta\theta$ we shall assume that in a narrow bandwidth $\Delta(\hbar\omega_P)$ the spectrum of pseudophotons is not varied. Therefore

$$\Delta N_{PEPPE}^{1electron}(\gamma, \hbar\omega_P) = \int_{\omega_{WW}} \frac{dN_{WW}(\gamma, \hbar\omega_{WW})}{d(\hbar\omega_{WW})} \Delta N_S(\hbar\omega_{WW}) d\hbar\omega_{WW} \approx \frac{dN_{WW}(\gamma, \hbar\omega_P)}{d(\hbar\omega_P)} \Delta N_S(\hbar\omega_P)\Delta(\hbar\omega_P). \qquad (8)$$

After all the substitutions one obtains

$$\Delta N_{PEPPE}^{1electron}(\gamma, \hbar\omega_P) = \frac{2\alpha A}{\pi} \ln\left(\frac{\gamma mc^2}{\hbar\omega_P}\right)(h\omega_P)\Delta\Omega\Delta x. \qquad (9)$$

Just as the Darwin width in X-ray diffraction, $\Delta\omega_P$ and therefore Δx must be small. However this width is not determined only by the rocking width, but also by the size of the detector entrance (one can use also gradient crystal to increase this width). For further estimate we take $\Delta x \approx 10^{-3}$, though in [9] it is taken $\Delta x \approx 10^{-2}$. Let us remind that as in [1,8] the spectral distributions of the photons are not considered: it is only assumed that $|\vec{k}_S| \approx |\vec{k}_i|$.

3. Numerical Estimates and Experimental arrangement

Before making estimates let us consider briefly the PEPPE arrangement shown in figure1. In order to escape the complicated calculations of the necessary parameters, especially, the nonlinear X-ray susceptibility, we shall consider the same experimental parameters as in [1,8], i.e., we assume a 0.5 mm thick, symmetrically cut (004) diamond, photon energies $\hbar\omega_P \approx 19$ keV and $\hbar\omega_{S,i} \approx \hbar\omega_P / 2 \approx 9.5$ keV, $\Delta\Omega_{S,i} \approx 10^{-3}$ sr, θ_{Br}, θ, θ_{Si} etc. Therefore, we can use the value of A

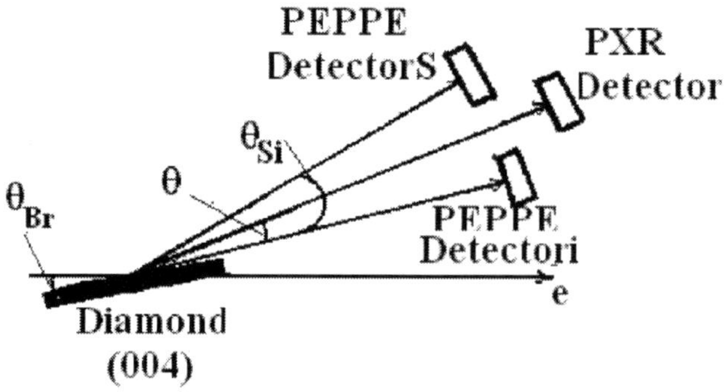

Figure 1. The experimental arrangement

from [1,8]. The 1 GeV electron beam with small cross section passes through a symmetrically cut diamond under incident angle which as it has been mentioned above has slightly smaller, $\Delta\theta_{Br}$ ~1.5 degrees. The crystal with thickness ~0.5 mm is slightly thicker than the necessary effective thickness [1,8] and has transversal sizes larger than the electron beam spot. The PXR detector placed at the specular direction for X-ray diffraction serves for monitoring purposes and detects the PXR as well as the diffracted transition radiation (DTR) photons produced with a probability ~10^{-6} per electron. The PEPPE detectors S and I, as in [1] and [8], are in a plane under angle θ with respect to the specular direction in or outside of the scattering plane. Despite to the small yield of PEPPE the coincidence of PEPPE signal and idler photons helps to separate the PEPPE events from the single and double Compton as well as other radiation background emitted under small angles ~1/γ.

For the Mainz cw electron accelerator MAMI 850 MeV electron beam with average current 0.1 mA or electron number per second ~$6.2\text{x}10^{14}$, taking before the development of the correct theory conservatively $\Delta x = 10^{-3}$ (in [9] it is assumed $\Delta x = 10^{-2}$) with the help of (9) one expects $\Delta N_{PEPPE}^{1electron}(E_e = 850MeV, \hbar\omega_p = 19keV) \approx 1.1\text{x}10^{-14}$ per electron, either 7.9 PEPPE photons per second or 28440 h^{-1}. This last yield is much better than it can be obtained on SR sources and is close to that expected for future X-ray FELs. The use a system of 2D-segmented position-sensitive detectors with much larger acceptance angle as in [13] can significantly increase the number of detected PEPPE photons.

4. Conclusion

The building of the correct PEPPE theory requires to use the non-linear part of the susceptibilities as it is has been done in the Chapter 11 of [5] for XPDC, and to solve the necessary Maxwell equations.

It is necessary to note: Firstly, as it is considered in [13] the X-ray microbunching of the electrons after the long SASE undulators [11], which can be measured before being spoiled by the method considered in [16], will make the proposed scheme of PEPPE a FEL for PEPPE due to the ~ N_{mb}^2 gain of the PEPPE yield, where N_{mb} is the number of electrons in the microbunches. Secondly, as we have seen PEPPE is produced accompanying PXR. Therefore, when an electron passes a plate of NL substance, simultaneously with the production of transition (TR) and Cherenkov (CR) photons PEPPE can be produced [17]. Due to the proportionality of the TR photon yield to Z^2 the rate will be enhanced if instead of electrons one uses relativistic heavy ions with high Z values, i.e, beams at RHIC or LHC as in the proposal [18] for ring TRD. As in the above considered case of crystal radiator in order to separate the production of PEPPE from the background in the case of TR one can measure the polarization of the both of entangled photons in directions different from the TR (CR) polarization plane made of the direction of the primary particle and emission direction of TR (CR). Finally one can use charged (especially multiply charged) particle beams for the study of PEPPE.

References

[1] Freund I and Levine B F 1969 *Phys. Rev. Lett.* **23** 854
[2] Eisenberg P and McCall S L 1971 *Phys. Rev.* **A 3** 1145
[3] Shkolnikov P I and A.E. Kaplan A E 1991 *Phys. Rev.* **A 44** 6951
[4] Nazarkin A et al. 2003 *Phys. Rev.* **A 67** 041804(R)
[5] *Nonlinear Optics, Quantum Optics and Ultrafast Phenomena with X-Rays* 2003 Ed.Adams B New York Kluwer Academic Publications
[6] Eisenberg P and McCall S L 1971 *Phys. Rev. Lett.* **26** 684
[7] Danino H and Freund I 1981 *Phys. Rev. Lett.* **46** 1127
[8] Yoda Y et al. 1998 *J. Synchrotron Rad.* **5** 980
[9] Adams B et al. 2000 *J. Synchrotron Rad.* **7** 81
[10] Tamazaku K and Ishikawa T 2007 *J. Synchrotron Rad.* **98** 244801
[11] *TESLA Technical Design Report* 2001 Part V Eds. Materlik B DESY-011
[12] Arthur J et al 2002 Preprint SLAC-R-593; Emma P *First Lasing of the LCLS, X-Ray FEL at 1.5 A^0* 2009 to be published in Proc. PAC09
[13] Schutzhold R, Schaller G and Habs D *Phys. Rev. Lett.* 2006 **97**, 121302; 2008 **100**,091301
[14] Bell F *On Generation of X-Ray Pairs; A Verification of UE* 2008 ArXiv:0809.1505
[15] Akhiezer A I and Berestetsky B V 1969 *Kvantovaya Elektrodinamika* Nauka Moscow (In Russian).
[16] Gazazian E et al. *Nucl. Instr. and Meth.* **2001 B173** 160
[17] In a letter to K.A.I. from 2003 F. Piuz from CERN asked: *"Can entangled photon pairs be produced simultaneously with transition or Cherenkov radiation?"*, and since then the above considerations are in process of maturation
[18] Aginian M A et al. 2004 *Nucl. Instr. and Meth.* **A522** 112

Some features of electromagnetic field of charged particle revolving about dielectric ball

S R Arzumanyan, L Sh Grigoryan[1], H F Khachatryan and M L Grigoryan
Institute of Applied Problems in Physics, 25 Hr. Nersessian Str., 0014 Yerevan, Armenia

E-mail: levonshg@mail.ru

Abstract. A relativistic electron uniformly rotating along an equatorial orbit around a dielectric ball may generate Cherenkov radiation tens of times more intense as that in case of revolution of a particle in a continuous, infinite and transparent medium. The root-mean-square values of electric and magnetic field strengths of particle are practically not localized in the central part of the equatorial plane of ball and close to the poles of ball.

Keywords: Cherenkov radiation, relativistic particle, dielectric ball.

1. Introduction and background

The presence of matter may essentially influence the characteristics of high energy electromagnetic processes giving rise to the production of Cherenkov radiation (CR), transition radiation etc. [1-9]. The operation of a number of devices assigned to production of electromagnetic radiation is based on the interaction of relativistic electrons with matter (see e.g. [8,9]).

The effects of interest arise in stratified media of different configurations (refer, e.g., [10-12] and references therein). The interfaces of media are widely used to control the radiation flow emitted by various systems. In the present work the case of one-layer medium of spherical configuration has been investigated.

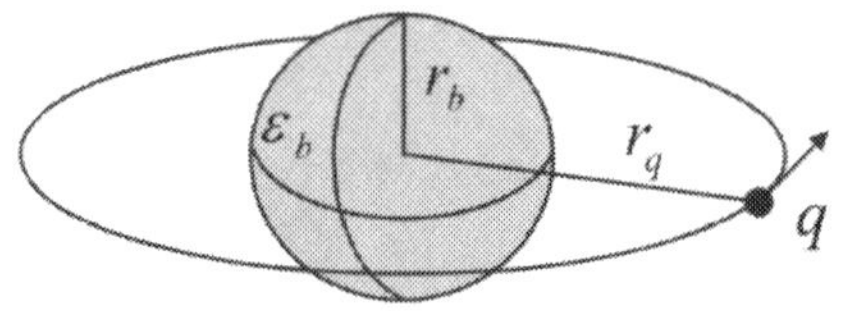

Figure 1. A relativistic particle rotating about a dielectric ball in it's equatorial plane.

Now consider a relativistic particle uniformly rotating in the magnetic field in vacuum about a dielectric ball in its equatorial plane. In spherical coordinates r, θ, φ with origin in the center of ball, the permittivity $\varepsilon(r)$ of the substance is a step function of radial coordinate:

$$\varepsilon(r) = \varepsilon_b \quad \text{for } r \le r_b \qquad \text{and} \qquad \varepsilon(r) = 1 \quad \text{for } r > r_b, \tag{1}$$

[1] To whom any correspondence should be addressed.

where $\varepsilon_b = \varepsilon_b' + i\varepsilon_b''$ is the complex valued permittivity of ball with radius r_b. The magnetic permeability is taken to be 1. The current density may be written as

$$\vec{j}(\vec{r},t) = \frac{qv\vec{e}_\varphi}{r_q^2}\delta(r - r_q)\delta(\theta - \pi/2)\delta(\varphi - \omega_q t), \quad (2)$$

where q, $v = r_q\omega_q$ and ω_q are the charge, linear velocity and cyclic frequency of particle rotation, and r_q is the radius of particle orbit.

The rotation of particle entails radiation at some discrete frequencies (harmonics) $\omega_k = k\omega_q$ with $k = 1;2;3...$ We assumed that an exterior force would make up for the braking of particle due to the radiation, by forcing the particle to uniformly rotate about the ball.

At large distances from the ball the radiation intensity I_k after averaging over the period $T = 2\pi/\omega_q$ of revolution is determined by the expression [13]

$$I_k = \frac{c}{2\pi}\lim_{r\to\infty} r^2 \oint \left|\text{rot}\vec{A}_k(\vec{r})\right|^2 d\Omega, \quad (3)$$

where Ω is the solid angle, and $\vec{A}_k(\vec{r})$ is the Fourier component of vector-potential of the electromagnetic field that satisfies the equation

$$(\Delta + \frac{\omega_k^2}{c^2}\varepsilon)\vec{A}_k(\vec{r}) - \frac{1}{\varepsilon}(\vec{\nabla}\varepsilon)\text{div}\vec{A}_k(\vec{r}) = -\frac{4\pi}{c}\vec{j}_k(\vec{r}). \quad (4)$$

It is convenient to introduce a non-dimensional quantity

$$TI_k/\hbar\omega_k \equiv n_k, \quad (5)$$

where TI_k is the energy emitted at ω_k frequency during one period of particle revolution, and $\hbar\omega_k$ is the energy of quantum of corresponding electromagnetic wave. So, n_k is the "number of electromagnetic field quanta" emitted during one revolution period of particle.

The dependence of n_k on v, r_b/r_q and ε_b was studied in [14]. It was shown that at some harmonics, in case of weak absorption of radiation in the ball material ($\varepsilon_b'' << \varepsilon_b'$), the particle may generate $n_k \sim 1$ radiation field quanta exceeding in several dozens of times those generated by particle rotating in a continuous, infinite and transparent medium having the same permittivity as the ball material ($\varepsilon = \varepsilon_b'$). In the present paper we have studied an angular distribution of electric and magnetic field strengths of particle inside the ball and on its surface.

2. Basic formulae

Due to the azimuthal symmetry of system one may write the following Fourier expansions

$$\vec{E}(r,\theta,\varphi,t) = \sum_{s=-\infty}^{\infty}[E_{sr}(r,\theta)\vec{e}_r + E_{s\theta}(r,\theta)\vec{e}_\theta + E_{s\varphi}(r,\theta)\vec{e}_\varphi]\exp[is(\varphi - \omega_q t)]$$

$$\vec{H}(r,\theta,\varphi,t) = \sum_{s=-\infty}^{\infty}[H_{sr}(r,\theta)\vec{e}_r + H_{s\theta}(r,\theta)\vec{e}_\theta + H_{s\varphi}(r,\theta)\vec{e}_\varphi]\exp[is(\varphi - \omega_q t)] \quad (6)$$

for the strengths of electric and magnetic fields of particle ($\vec{e}_r$, $\vec{e}_\theta$ and $\vec{e}_\varphi$ are the unit vectors of the spherical system of coordinates). For this reason the root-mean-square values of electromagnetic field strengths may be written in the following form

$$\frac{1}{T}\int_0^T \vec{F}^2(\vec{r},t)dt = F_0^2(r,\theta)+\sum_{k=1}^{\infty} F_k^2(r,\theta), \tag{7}$$

where

$$F_0^2 = |F_{0r}|^2 + |F_{0\theta}|^2 + |F_{0\varphi}|^2, \quad F_k^2 = 2(|F_{kr}|^2 + |F_{k\theta}|^2 + |F_{k\varphi}|^2), \qquad k = 1,2,3,... \tag{8}$$

and $F = E, H$. The problem is reduced to the calculation of $E_{kr}, E_{k\theta}, E_{k\varphi}$ and $H_{kr}, H_{k\theta}, H_{k\varphi}$.

In our calculations we took advantage of the method for obtaining appropriate exact solutions of equation (4) proposed in [15]. Final expressions for $F_{kr}, F_{k\theta}, F_{k\varphi}$ inside a ball are given below:

$$E_{kr} = \frac{ic}{\omega_k \varepsilon_b r}\sum_{l=k}^{\infty} p_{lk}^{(2)}(r) Y_{lk}(\theta,0), \qquad E_{k\theta} = \sum_{l=k}^{\infty}[p_{lk}^{(1)}(r) g_{lk}(\theta)\frac{ir\omega_k}{cl(l+1)} + p_{lk}^{(4)}(r) d_{lk}(\theta)],$$

$$E_{k\varphi} = \sum_{l=k}^{\infty}[p_{lk}^{(4)}(r) g_{lk}(\theta) - p_{lk}^{(1)}(r) d_{lk}(\theta)\frac{ir\omega_k}{cl(l+1)}] \tag{9}$$

for an electric field and

$$H_{kr} = \sum_{l=k}^{\infty} p_{lk}^{(1)}(r) Y_{lk}(\theta,0), \quad H_{k\theta} = \sum_{l=k}^{\infty}[p_{lk}^{(2)}(r)\frac{g_{lk}(\theta)}{l(l+1)} + p_{lk}^{(3)}(r) d_{lk}(\theta)],$$

$$H_{k\varphi} = \sum_{l=k}^{\infty}[p_{lk}^{(3)}(r) g_{lk}(\theta) - p_{lk}^{(2)}(r)\frac{d_{lk}(\theta)}{l(l+1)}] \tag{10}$$

for a magnetic field. Here we introduce the following notations

$$p_{lk}^{(1)}(r) = 4\pi q v d_{lk}(\pi/2) h_{\underline{l}}(\sigma u)\frac{j_l(\tau u_b)}{c r_b^2 \tau}, \qquad p_{lk}^{(2)}(r) = j_l(\tau u_b)\frac{4\pi q v u_b c_{lk} g_{lk}(\pi/2)}{c r_b^2 (2l+1)},$$

$$p_{lk}^{(3)}(r) = [(l+1) j_{l-1}(\tau u_b) - l j_{l+1}(\tau u_b)]\frac{4\pi q v u_b d_{lk}(\pi/2) h_{\underline{l}}(\sigma u)}{c r_b^2 l(l+1)(2l+1)}, \tag{11}$$

$$p_{lk}^{(4)}(r) = [(l+1) j_{l-1}(\tau u_b) - l j_{l+1}(\tau u_b)]\frac{4\pi i q v \omega_k c_{lk} g_{lk}(\pi/2)}{r_b c^2 l(l+1)(2l+1)^2},$$

where $h_l(y) = j_l(y) + i n_l(y)$ (j_l and n_l are spherical Bessel and Neumann functions respectively), $Y_{lm}(\theta,\varphi)$ is the spherical function,

$$\sigma = \frac{r_q}{r_b}, \quad u = \frac{\omega_k r_b}{c}, \quad \tau = r/r_b \le 1, \quad u_b = \frac{\omega_k r_b}{c}\sqrt{\varepsilon_b},$$

$$g_{lk}(\theta) = ikY_{lk}(\theta,0), \qquad d_{lk}(\theta) = \frac{\partial Y_{lk}(\theta,0)}{\partial\theta}, \tag{12}$$

$$f_{\underline{l}}(\tau) = f_l(\tau)/[u_b j_{l+1}(u_b) h_l(u) - j_l(u_b) u h_{l+1}(u)],$$

and

$$c_{lk} = (l+1) h_{\underline{l-1}}(\sigma u) - l h_{\underline{l+1}}(\sigma u) +$$
$$+ (\varepsilon_b - 1)[h_{\underline{l-1}}(u) + h_{\underline{l+1}}(u)][h_{\underline{l-1}}(\sigma u) + h_{\underline{l+1}}(\sigma u)]\frac{l(l+1) u_b j_l(u_b)}{l\gamma_{l-1}^l + (l+1)\gamma_{l+1}^l}. \tag{13}$$

In the last expression

$$\gamma_\nu^l = \frac{\varepsilon_b j_\nu(u_b) u h_l(u) - u_b j_l(u_b) h_\nu(u)}{j_\nu(u_b) u h_l(u) - u_b j_l(u_b) h_\nu(u)}. \tag{14}$$

3. Numerical results and discussion

Let us consider an electron uniformly revolving about a dielectric ball. The electromagnetic field of electron is determined by its energy E_q and parameters r_q, r_b and ε_b. It is also convenient to introduce a dimensionless parameter

$$\nu = \mathrm{v} r_b / c r_q \qquad (15)$$

and assume for certainty in what follows that $\varepsilon_b = 3.78$ (fused quartz in the range of microwaves), $r_b = 36.2mm$, $r_q \cong 36.9mm$, $k = 8$ and $E_q = 2MeV$.

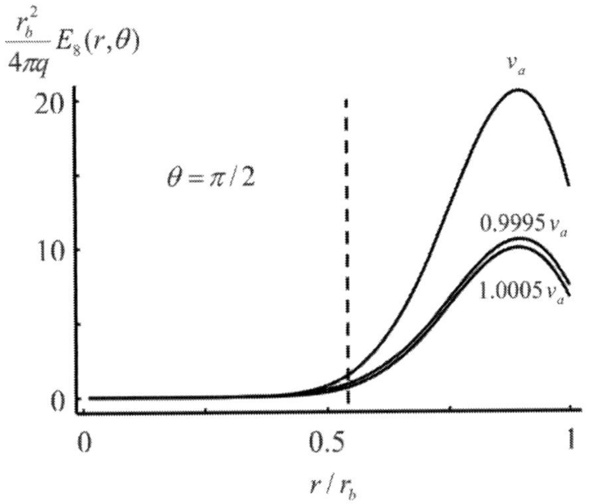

Figure 2. r-dependence of the root-mean-square strength $E_k(r,\theta)$ of the k-th harmonic of electric field inside the ball at $\theta = \pi/2$ (the equatorial plane of the ball) for $2MeV$ electron. The values of parameter (15) are shown beside the curves, $\nu_a = 1.2507Ghz$ [14], $\varepsilon_b = 3.78$, $k = 8$.

In figure 2 the root-mean-square strength $E_k(r,\theta)$ of the electric field inside a dielectric ball at the equatorial plane ($\theta = \pi/2$) is plotted versus the ratio r/r_b. Here we give three curves of r-dependence of E_k for $\nu = \nu_a$; $0.9995\nu_a$ and $1.0005\nu_a$. The material of ball is assumed to be transparent ($\varepsilon_b'' = 0$). It is seen that large values of $E_k(r)$ are possible on the curves with $\nu \cong \nu_a$. The comparison of curves with ν/ν_a=1; 0.9995 and 1.0005 shows how rapidly the function $E_k(r)$ tends to the maximum when $\nu \to \nu_a$. A slight departure of ν in either direction from ν_a would suffice for the point describing the state of system to notably fall. For this reason one may term ν_a as *the resonance value of* ν *parameter*.

As it is seen from the plots in figure 2, the function $E_k(r)$ steeply increases rightwards of the vertical dashed line. Corresponding to this line is the value

$$r/r_b = 0.54 \cong \cos\theta_C, \qquad \text{where } \theta_C = \arccos(c/\mathrm{v}\sqrt{\varepsilon_b}) \qquad (16)$$

is the Cherenkov radiation emission angle for an electron at rectilinear motion in a continuous medium with $\varepsilon = \varepsilon_b$. This fact testifies to the relation of localization of the electromagnetic field in the range of $\cos\theta_C \le r/r_b \le 1$ to the generation of CR inside the dielectric ball.

In figure 3 the electric field strength E_k is plotted versus θ at the surface of the ball ($r = r_b$). As in figure 2 here we give three curves of θ- dependence of E_k for $v = v_a$; $0.9995 v_a$ and $1.0005 v_a$. It is seen again that large values of $E_k(\theta)$ are possible on the curves with $v \cong v_a$.

As it is seen from the plots in figure 3, the function $E_k(\theta)$ increases sufficiently fast rightwards of the vertical dashed line. Corresponding to this line is the value

$$\theta = \arccos(\sin^2\theta_C) \cong 44.5^o. \qquad (17)$$

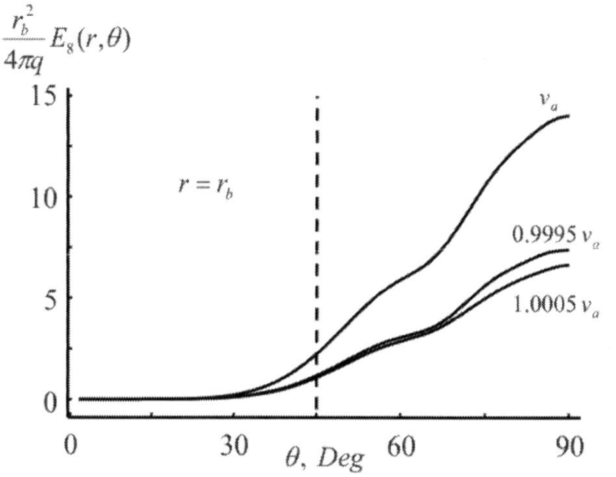

Figure 3. The root-mean-square strength $E_8(r,\theta)$ of the 8-th harmonic of electric field on the surface of ball ($r = r_b$) as the function of polar angle θ. The notations and values of parameters are the same as those in figure 2.

This fact again testifies to the relation of localization of the electromagnetic field in the range of

$$\arccos(\sin^2\theta_C) \le \theta \le 90° \qquad (18)$$

to the generation of CR inside the dielectric ball.

The results of our numerical calculations indicate that close to the maxima in figures 2,3 the value of E_k are tens of times as large as that in the case of revolution of a particle in a continuous, infinite and transparent medium having the same permittivity as the ball material. The situation with the root-mean-square strength $H_k(r,\theta)$ of the magnetic field is simalar. Sure, the electromagnetic oscillations of CR induced by the particle along the trajectory on the whole are partially locked inside the ball by superimposing each other mostly in the destructive way. However, in some specific cases ($v \cong v_a$ in figures 2,3) the oscillations may superimpose nondestructively, as a result of which the electromagnetic field of particle inside the ball is amplified.

The maxima of functions $E_k(r,\theta)$ and $H_k(r,\theta)$ at resonance are highly sensitive to the values of permittivity of the ball and its radius. This fact may be used in applications.

References

[1] Jelley J V 1958 *Cherenkov Radiation and its Applications* (London: Pergamon Press)

[2] Bolotovskii B M 1961 *Physics-Uspekhi* 75 295

[3] Zrelov V I 1968 *Vavilov-Cherenkov Radiation (and its Applications in High Energy Physics)* (Moscow: Atomizdat) (in Russian)

[4] Garibian G M and Yang C 1983 *X-Ray Transition Radiation* (Yerevan: AN Arm. SSR Press) (in Russian)

[5] Ginzburg V L and Tsytovich V N 1990 *Transition Radiation and Transition Scattering* (Bristol: Adam Hilger)

[6] Frank I M 1988 *Vavilov-Cherenkov Radiation. Theoretical Aspects* (Moscow: Nauka) (in Russian)

[7] Berman B L and Datz S 1985 in: *Coherent Radiation Sources* eds A W Saenz and H Uberall (Berlin: Springer-Verlag)

[8] Rullhusen P, Artru X and Dhez P 1998 *Novel Radiation Sources Using Relativistic Electrons* (Singapore: World Scientific)

[9] Potylitsyn A P 2009 *Radiation from Electrons in Periodic Structures* (Tomsk: NTL Press) (in Russian)

[10] Saharian A A and Kotanjyan A S 2009 *J. Phys. A: Math. Theor.* **42** 135402

[11] Arzumanyan S R, Grigoryan L Sh, Khachatryan H F and Grigoryan M L 2008 *Nucl. Instr. and Meth.* B **266** 3715–3720

[12] Arzumanyan S R, Grigoryan L Sh, Khachatryan H F, Kotanjyan A S and Saharian A A *2008 Nucl. Instr. and Meth.* B **266** 3703–3707

[13] Landau L D and Lifshitz E M 1980 *The Classical Theory of Field* (Oxford: Pergamon Press)

[14] Grigoryan L Sh, Khachatryan H F, Arzumanyan S R and Grigoryan M L 2006 *Nucl. Instr. and Meth.* B **252** 50–56

[15] Arzumanyan S R, Grigoryan L Sh, Saharian A A, Kotanjian Kh V *Izv. Nats. Akad. Nauk Arm Fiz.* (Engl. Transl.: *J. Contemp. Phys.*) **30** 106-113

Acknowledgements

L Sh Grigoryan is thankful to K V Lekomtsev for useful discussion. This work was supported in part by the Grant No. 077 from Ministry of Education and Science of RA.

A novel method for sub-micrometer transverse electron beam size measurements using optical transition radiation

A. Aryshev*[,1], S. T. Boogert[2], D. Howell[3], P. Karataev[2], N. Terunuma[1], and J. Urakawa[1]

[1] KEK, 1-1 Oho, Tsukuba, Ibaraki 305-0801, Japan

[2] John Adams Institute at Royal Holloway, Egham, Surrey, TW20 0EX, UK

[3] John Adams Institute at Oxford University, Denys Wilkinson Building, Keble Road, Oxford OX1 3RH UK

E-mail: alar@post.kek.jp

Abstract. Optical Transition Radiation (OTR) appearing when a charged particle crosses a boundary between two media with different dielectric properties has widely been used as a tool for transverse profile measurements of charged particle beams in various facilities worldwide. The resolution of the monitor is defined by so-called Point Spread Function (PSF), source distribution generated by a single electron and projected by an optical system onto a screen. In this paper we represent the development of a novel sub-micrometre electron beam profile monitor based on the measurements of the PSF structure. The first experimental results are presented and future plans on the optimization of the monitor are discussed

1. Introduction

Optical Transition Radiation (OTR) appearing when a charged particle crosses a boundary between two media with different dielectric properties has widely been used as a tool for transverse profile measurements of charged particle beams in various facilities worldwide. OTR monitors are simple, robust, and give a direct image of a two-dimensional beam profile. The resolution of the monitor is defined by so-called Point Spread Function (PSF) [1]. The PSF is the source distribution generated by a single electron and projected by an optical system onto a screen. The root-mean-square (RMS) size of the PSF is considered as a resolution quantity. All existing theories predict that the PSF is not a uniform function, but it is structured and has a minimum in the centre. Moreover, when considering such a small dimensions, the PSF form significantly depends on various parameters of the optical system like diffraction of the OTR tails, spherical and chromatic aberrations, etc. Recently the first observation of the PSF of OTR was made [2]. One of the biggest challenges for experimental verification is a difficulty in obtaining such a small electron beam size, i.e. much smaller than the PSF structure. The experimental results have clearly shown that the PSF structure with a minimum in the centre gives an opportunity to measure the beam size with micrometer resolution. And with proper optimization of the optical system sub-micrometer resolution can also be achieved.

2. Experimental setup

Last year the KEK Accelerator test facility (ATF) extraction line (see Fig.1) has been significantly upgraded. The ATF-II [3] is a final-focus test beam line that aims to focus the low-emittance beam from the ATF damping ring to a vertical beam size of about 37 nm, and at the same time to demonstrate nm beam stability, using numerous advanced beam diagnostics and feedback tools. The laser-wire (LW) system [4, 5] has also been upgraded in order to allow the measurement of micron-scale transverse electron beam size. In the course of this upgrade an OTR monitor has been integrated into the LW interaction chamber. The aims of the monitor were to simplify the tuning procedure for the micrometer electron beam optics to be used for the LW experiment and to investigate a possibility to cross check the LW micron-scale beam size measurements.

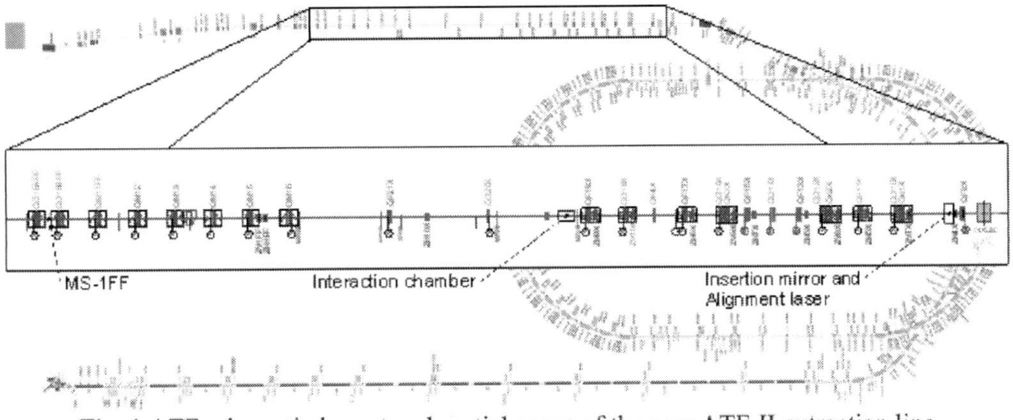

Fig. 1 ATF schematic layout and partial zoom of the new ATF-II extraction line.

To observe the PSF of the OTR an optical system with large magnification factor was constructed. It consists of a 100mm focal length lens, two aluminum mirrors forming a periscope, an optical polarizer and a set of the optical filters which can be inserted onto the optical path right in front of a CCD camera (see Fig.2).

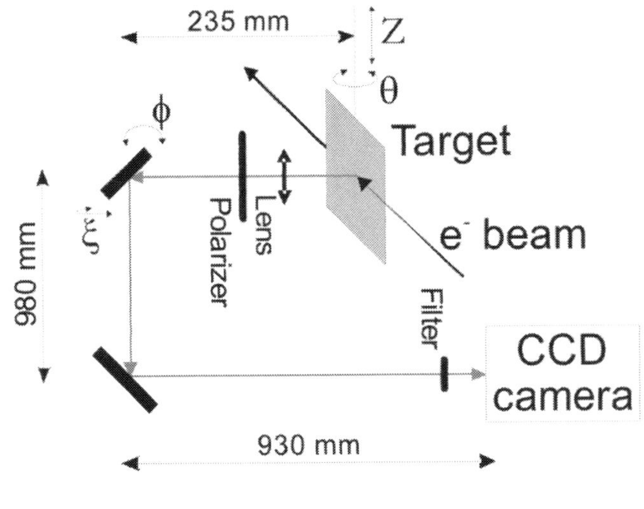

Fig. 2 Experimental layout.

Aluminum-coated silicon target 30x30x0.3 mm.
Lens: Sigma-koki SLB-50.8-100PY2, 50mm diameter.
Linear Polarizer, 75mm diameter.
Top aluminum mirror, 50mm diameter.
Bottom aluminum mirror, 75mm diameter.
Set of optical filters: 459±18nm, 558±20nm, and 609±18nm center band pass wavelength, 25mm diameter.
CCD camera Alta E4000, 2048 X 2048 pixel array @ 7.4 microns and 16 bit resolution, input flange - 25mm diameter.

Table. 1 Specification of the main components of the experimental setup

The vacuum chamber with attached 2 inch travel motorized vacuum translator with rotatable axis was used to position the OTR target in the center of the vacuum chamber and at 45 degrees with respect to the electron beam trajectory. The 300 μm thick, square (size 30 mm) silicon wafer, coated with aluminum was used as the target. Transition radiation generated by the target was extracted out of the vacuum chamber through the fused silica vacuum window and then transported into the cooled CCD camera with peak QE 55% at 500 nm. Two axes of the 50 millimeter diameter top periscope mirror were remotely controlled by the ThorLabs Z706 DC-servo motors with minimal step of the order of 3 μrad allowing the adjustment of the OTR spot position at the CCD during the experiment.

To align the optical system a special vacuum mirror was installed downstream the ZH4X steering magnet. That mirror allows us to send the alignment laser along the electron beam in the ATF-II extraction line. The laser system consists of a laser stage, a Helium-Neon (He-Ne) continuous wave (CW) laser (Melles Griot 05-LHP-121 laser head with oscillation wavelength of 632.8 nm), a spatial filter, and a focusing lens. Changing the distance between the spatial filter and the lens, one may focus the laser at any point of the experimental setup (screen monitors, target, and detector). The MS1FF screen located about 38 meters downstream of the vacuum mirror (Fig. 1) was used as a second reference point. The laser direction could be corrected using precise adjustment screws of the laser stage. When the laser alignment system was optimized, the OTR target placed at 45 degrees with respect to the beam line reflected the laser light away from the accelerator chamber, and the entire optical system was aligned.

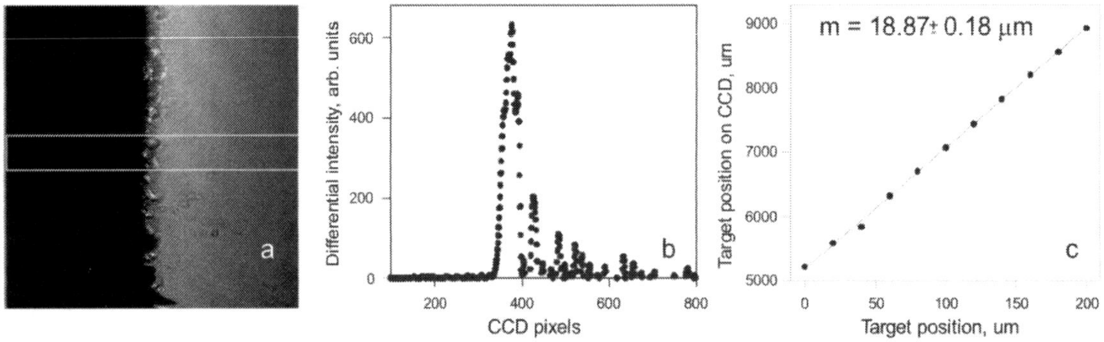

Fig. 3 The CCD image of the OTR target edge (a), its differential intensity (b) and calibration curve with the linear fit (c).

The laser alignment tests showed that the alignment system accuracy was smaller than 0.05 mrad, which is quite enough for our purposes. The optical system has a large magnification factor which was measured using laser alignment system. The target was gradually moved out of the chamber until its edge image was observed on the CCD camera. At this point the lens longitudinal position was adjusted to provide a clear target edge image (see Fig. 3a). The lens height position and the angular corrections were applied in order to have no image position shifts while changing longitudinal lens position (i.e. reflected alignment laser beam passed through the center of the lens at a right angle of incidence to the tangent of the lens surface). The image was digitized and a vertical projection in a narrow horizontal box shown in Fig. 3a with white lines was differentiated (see Fig. 3b). The maximum of the differential intensity curve was defined as the position of the target edge. The relative position of the target was also obtained from the manipulator motor controller encoder system. The resulting dependence of the target position measured at the CCD camera versus manipulator readout was used as a calibration curve (see, for instance, Fig. 3c). The magnification factor extracted from the linear fit was $18.87 \pm 0.18\ \mu m$.

Special electron beam optics was simulated using Strategic Accelerator Design (SAD) code [6] to generate a micron-scale transverse beam size at the OTR target location resulting in beam dimensions to be 1.5x20 μm. Figure 4 represents the SAD simulations of the ATF extraction line optics performed for initial electron beam parameters summarized in Table 2. The target chamber is located after two quadrupole magnets QM19X and QD18X used to change the horizontal and the vertical beam sizes respectively. The ATF-II lattice was designed so that the first order dispersion could be corrected in the diagnostics section of the extraction line. The calculated beam sizes were confirmed by the wire-scanner (MW2X) measurements located at about 0.8 m downstream of the target chamber.

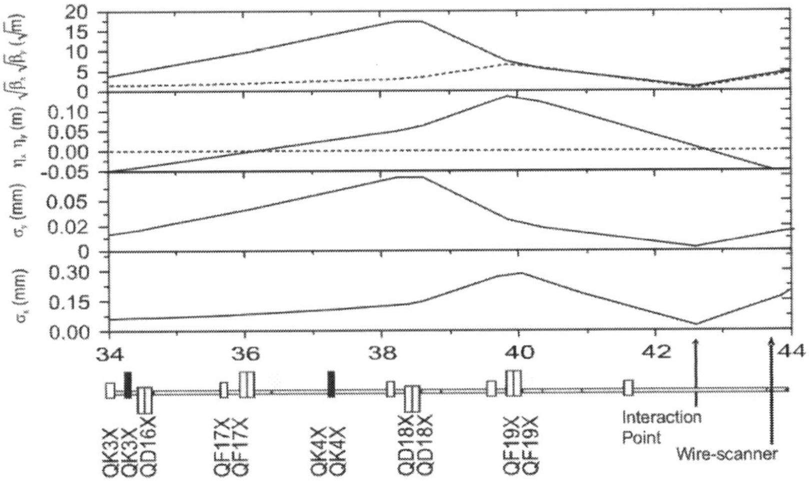

Initial parameters :
α_x =2.047
β_x=14.795 m
α_y=-2.49
β_y=3.656 m
$\Delta E / E$ =6.4E-4
$\varepsilon_x = 1.86E-9$ m rad
$\varepsilon_y = 2.3E-11$ m rad

Table 2.

Fig. 4 SAD simulation of the vertical and horizontal betatron (β_x, β_y), dispersion (η_x, η_y) functions and predicted electron beam sizes (σ_x, σ_y) for the ATF-II extraction line. The arrows shows interaction chamber and wire-scanner locations along the beam line.

All measurements were performed with a single-bunch 1.56 Hz repetition rate operation mode with 0.9 nC bunch charge. The Alta E4000 (100bT Ethernet interface) camera based on the large format interline Kodak KAI-4021M (monochrome) sensor with low dark current, low noise, and high resolution, 2048×2048 pixel array with 7.4 microns pixel size and 16 bit resolution along with the camera control and image acquisition software (MaxIm DL) [7] was applied. A LabView [8] based application for off-line data analysis has been developed. In order to increase the signal-to-noise ratio of the acquired images the 4×4 pixel binding was applied. A single-shot operation mode of the CCD camera allowed us to exclude the shot-by-shot electron beam instability effect such as the beam position jitter.

3. Experimental results and discussions

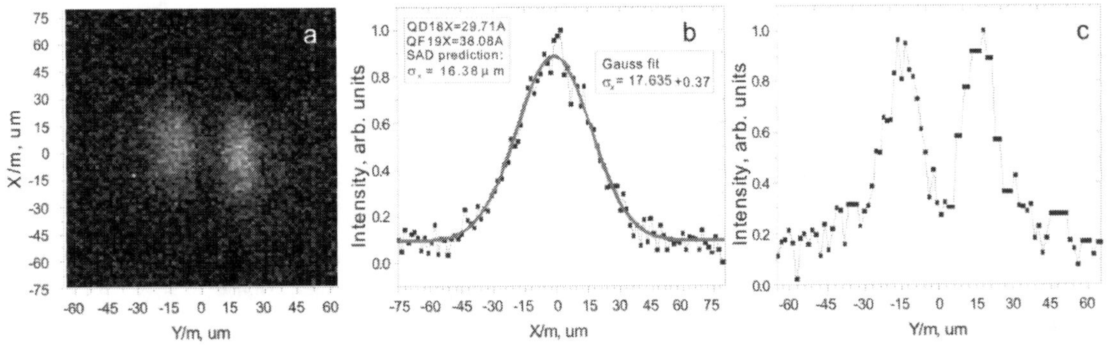

Fig. 5 CCD image of the OTR taken with linear polarizer and 550 nm optical filter (a) and two image projections: horizontal (b) and vertical (c).

Figure 5a shows the CCD image of the OTR taken with linear polarizer set to transmit the vertical polarization component and 550 nm optical filter. Figures 5b and 5c represent two image projections: horizontal (Fig. 5b) and vertical (Fig. 5c) derived from summation of the pixels intensities within the horizontal and vertical broad corridors.

We can clearly see the two lobe distribution in the vertical projection of the OTR spot (Fig. 5c) well described in [2]. The visibility of the distribution significantly depended on the electron beam size. When the vertical beam size was changed from 1.75 μm to 10.16 μm (according to the SAD simulations) the minimum in the centre of the distribution disappeared (see Fig. 6a). Further increase in the beam size transferred our system into a conventional OTR monitor with the resolution of about 22 μm. Such a strong dependence of the visibility on the transverse electron beam size can be applied for the beam size measurements at micrometer level.

Since the horizontal beam size is much larger than the vertical one the horizontal projection shown in Fig. 5b still proportional to the horizontal electron beam size. A simple Gaussian fit was applied to the image projections in order to extract the horizontal RMS beam sizes σ_x. To analyze the vertical projection, especially OTR spot minimum behavior, a special empirically found fit function had been introduced:

$$f(x) = a + \frac{b}{1 + [c(x - \Delta x)]^4}\left[1 - e^{-2c^2\sigma^2}\cos[c(x - \Delta x)]\right] \tag{1}$$

where a, b, c, σ, and Δx are free parameters of the fit function, namely: a is the vertical offset of the distribution with respect to zero which included a constant background; b is the amplitude of the distribution; c is the distribution width; σ is the smoothing parameter dominantly defined by the beam size; and Δx is the horizontal offset of the distribution with respect to zero. An example of the OTR image vertical projection and the fit is represented in Fig. 6b.

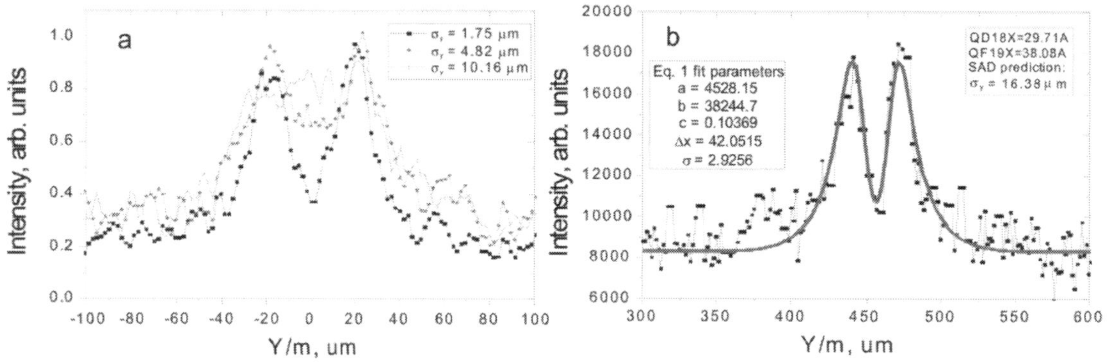

Fig. 6 (a) Three OTR distributions measured with 550nm optical filter for three current settings of QD18X quadruple magnet 29.71A, 29.21A, 28.96A corresponding to SAD calculation of the vertical beam size to be σ_y = 1.75μm, σ_y = 4.82μm and σ_y = 10.16μm respectively. (b) Vertical projection of the OTR image taken for QD18X = 29.71A (1.75 μm calculated vertical beam size) and fit by the function Eq. 1.

From Figure 6b, one can see that the fit function can approximate the experimental data fairly well. To demonstrate the new method application to the beam size measurements, two independent quadrupole scans were performed. At first the strength of the QF19X quadrupole magnet was changed. That corresponded to the horizontal beam size change. The horizontal image projections were fit with Gaussian distribution, and the resulting horizontal RMS beam size as a function of the QF19X strength is represented in Fig. 7a. The data show a good agreement with SAD expectations. A small discrepancy in the right part of the dependence could be explained by the fact that the residual dispersion was not well corrected during the measurements. This effect need to be further investigated.

Nevertheless the data confirm that the monitor can be used for the horizontal beam size measurements as a conventional OTR monitor.

In a similar manner a second quadrupole (QD18X) scan was done. Figure 7b shows the dependence of the smoothing parameter σ (Eq. 1) dominantly defined by the vertical beam size versus QD18X quadrupole magnet strength (red dots) along with SAD predictions of the vertical beam size for the same magnet strengths (black dots). Large error bars at both ends of the scan could be explained by a poor signal-to-noise ratio of the acquired images what degraded its quality and increased the fit errors.

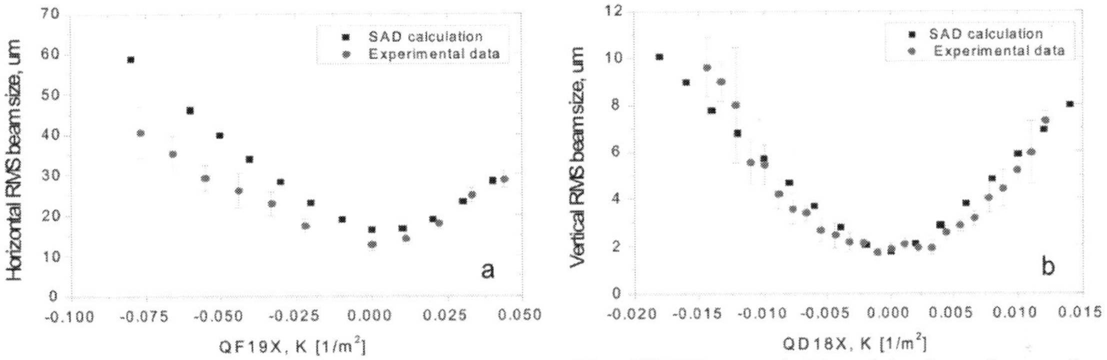

Fig. 7 Horizontal RMS beam size as a dependence of the QF19X strength (a) and the dependence of the smoothing parameter σ (Eq. 1) versus QD18X quadrupole magnet strength. Red dots – measured data, black dots – SAD predictions of the vertical beam size for the same magnet strengths.

4. Conclusion

The present result clearly demonstrates that the method based on the analysis of the PSF structure visibility gives an opportunity to measure the beam size with a micrometer resolution. In [9] a similar method based on measurements of vertical polarization component of synchrotron radiation (SR) PSF was applied at Swiss Light Source. The difficulty related to the SR measurements is that it is difficult to put the optical system close enough to the emission point. As a result the diffraction effect significantly increases the PSF and, therefore, degrades the resolution for beam size measurement. In case of OTR submicrometer resolution might be achieved.

As it was demonstrated in [2] the chromatic aberrations result in significant broadening of the OTR PSF leading to degradation of the potential resolution. In order to improve the beam size measurement technique additional efforts toward the optimization of the optical system, a better understanding of the beam size effect and an increase of the signal-to-noise ratio of the detecting system is required. To be able to achieve our goals and demonstrate sub-micrometer resolution we intend to employ an achromat to minimize the chromatic aberrations in the optical system. We shall also reduce the magnification factor to increase the light density at the CCD camera and improve the signal-to-noise ratio. We shall implement a few more optical filters covering the wavelength range from 350 to 800nm with 50nm step to be able to investigate the spectral characteristics of the OTR PSF in more detail.

5. Acknowledgements

We would like to thank the ATF extraction line Laser-Wire collaboration for giving us an opportunity to conduct the experiment. This research was partially supported by DAIWA Anglo-Japanese Foundation, partial support from the "Grant-in-Aid for Creative Scientific Research of JSPS (KAKENHI 17GS0210)" project and Quantum Beam Technology Program of the Ministry of Education, Science, Sports, Culture and Technology of Japan (MEXT).

References

[1] M. Castellano and V. Verzilov, Physical Review ST-AB 1 (1998) 062801
[2] P.Karataev, et al., The First Observation of the Point Spread Function of Optical Transition Radiation, submitted to Physical Review Letters, October 2009.
[3] A. Seryi, et al., ATF2 Commissioning, in Proceedings of 2009 Particle Accelerator Conference, May 4-8, Vancouver, Canada, 2009.
[4] A. Aryshev, et al., EPAC'08, June 2008, Genoa, Italy, p.1065.
[5] A. Aryshev, et al., PAC'09, May 4-8, Vancouver, Canada, 2009.
[6] SAD, Strategic Accelerator Design, http://acc-physics.kek.jp/SAD/sad.html.
[7] MaximDL: http://www.cyanogen.com/
[8] LabView: http://www.ni.com/labview/
[9] A. Andersson, et al., Nuclear Instruments and Methods A 591 (2008) 437.

Development of microwave and soft X-ray sources based on coherent radiation and Thomson scattering

A. Aryshev*[1], A. Araki[1], M. Fukuda[1], P. Karataev[2], G. Naumenko[3], A. Potylitsyn[4], K. Sakaue[5], L. Sukhikh[4], J. Urakawa[1], D. Verigin[4].

[1] KEK, 1-1 Oho, Tsukuba, Ibaraki 305-0801, Japan

[2] John Adams Institute at Royal Holloway, Egham, Surrey, TW20 0EX, UK

[3] Nuclear Physics Institute at Tomsk Polytechnic University, Tomsk, 634050, Russia

[4] Tomsk Polytechnic University, Tomsk, 634050, Russia

[5] Waseda University, 3-4-1, Okubo, Shinjuku, Tokyo, 169-8555, Japan

E-mail: alar@post.kek.jp

Abstract. Compact, high-brightness and reliable sources in the VUV and the soft X-ray region may be used for numerous applications in medicine and biochemistry. We propose a new approach to produce the intense beams of X-rays in the range of ≤ 500 eV based on compact electron accelerator. We present our first experimental results obtained in the Laser Undulator Compact X-ray facility (LUCX) at KEK: High Energy Accelerator Research Organization devoted to the development of a compact microwave and soft X-ray source based on coherent diffraction radiation and Thompson scattering processes.

1. Introduction

Diffraction radiation (DR) appears when a relativistic charged particle moves in the vicinity of a medium (or target) with impact parameter comparable to or smaller than the electron field radius. The electron field polarizes the target atoms thereby inducing currents changing in time, which in turn give rise to DR. The radiation is emitted coherently (CDR) if the bunch length σ_l is comparable to or shorter than the radiation wavelength ($\lambda \geq \pi\sigma_l$) [2]. In this case the CDR intensity is proportional to a square of the bunch population, N_e^2. When an electron beam passes through an opened resonator made of two mirrors with beam holes, the CDR generated in the system can be stored in it. If the beam is pre-bunched and the length of the resonator is equal to the bunch spacing, the CDR produced by a preceding bunch will stimulate the radiation generated by a subsequent bunch. The similar scheme was investigated in the experiment [3] where stimulated process of coherent transition radiation (CTR) from a resonator consisting of a two flat CTR targets was detected with more than 1 order of magnitude enhancement relative to spontaneous CTR.

In the suggested scheme, the CDR flux produced by the preceding bunch will scatter by the next bunch generating soft X-ray radiation ($\hbar\omega \sim 10 \div 100\,\mathrm{eV}$) via Thomson scattering process [4]. The yield of soft X-ray photons is proportional to a cubic power of the number of electrons in the bunch, N_e^3 [5].

An ultimate goal of our project is to create a compact soft X-ray source based on Thomson scattering of CDR using a small accelerator machine. In contrast to the conventional method based on the inverse Compton scattering of laser light [6, 7], the new method does not require neither any special timing control system, because the timing structures of the photon beam and the electron beam are strongly related to each other, nor the additional cumbersome laser system, which makes the X-ray generation program less challenging and relatively inexpensive.

The experimental results presented in this paper aim to demonstrate both the feasibility of our approach and the performance of our detection system, and also to investigate the coherent diffraction radiation produced by the electron beam generated by the Laser Undulator Compact X-ray facility (LUCX) at KEK: High Energy Accelerator Research Organization.

2. Experimental setup

The LUCX consists of an RF gun and a 3 meter long S-band accelerating tube (Fig. 1) [8]. It produces a multi-bunch electron beam with variable number of bunches (up to 100) with 0.5nC per bunch and 43MeV beam energy [9]. The main beam parameters are summarized in Table1.

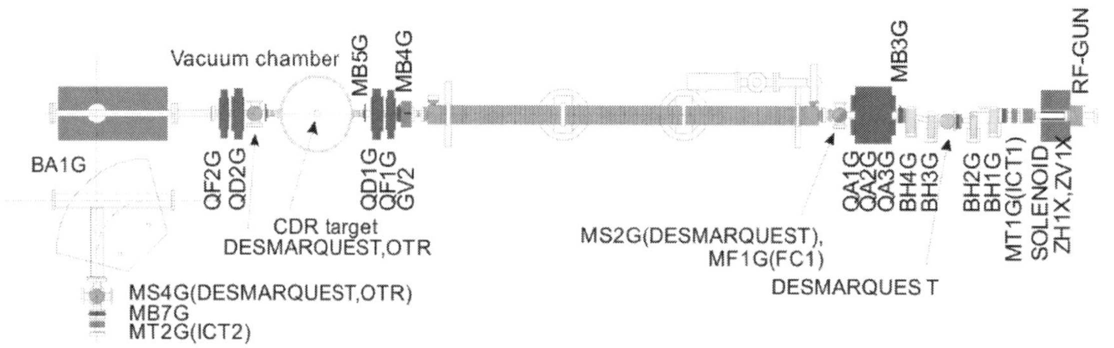

Fig. 1. LUCX schematic layout.

In the first stage experiment we measured the coherent diffraction radiation from a flat conducting target. The purpose was to evaluate the coherency of the radiation in mm-wavelength range, check any background contribution as a function of various electron beam parameters, and check our hardware performance.

Energy	43 MeV
Intensity	0.5nC/bunch
Num. of Bunches	100 bunches
Bunch spacing	2.8ns
Bunch length	10ps
Repetition Rate	12.5 train/sec
Emittance	5 πmm · mrad
(σ_x, σ_y) in vacuum chamber	200μm, 60μm

Table 1: Electron beam parameters of the LUCX facility.

The W30mmxH30mmxT300um aluminum-coated flat silicon target mounted on an air-actuator was used to investigate the radiation properties. The minimal distance between CDR target and the electron beam was 10 millimeters which is greater than the transverse electron beam size and much smaller than the relativistic electron field radius. As a detector we used an ultrafast Schottky Barrier Diode (Fig. 2b) detector along with 30x22mm conical horn antenna. The detector parameters are:

response time ~ 250ps, detection frequency range 60-90 GHz, typical sensitivity 1250 mV/mW (Fig. 2c), typical flatness ±2dB. The basic principles of these detectors are described in [10, 11]. This detector provides an output voltage which is directly proportional to the input power level (Fig. 2c).

This is a room temperature zero-biased detector. That makes the detector very attractive for accelerator physics and many other applications. The SBD detector has a flat frequency response. To collect the radiation energy and to feed it to the metal–semiconductor contact, an impedance matched pyramidal horn antenna was used [12]. This antenna responds to a single mode of the radiation field only. Therefore the detector is polarization sensitive. The horn antenna has a beamwidth of 25° and a typical midband gain of 24 dBi.

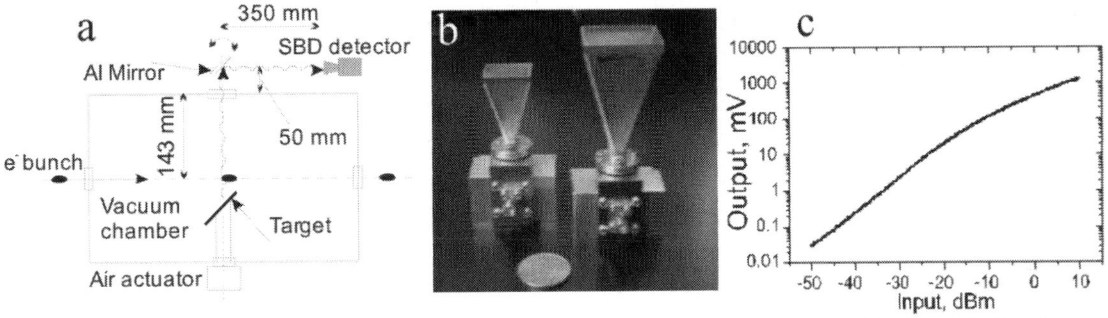

Fig. 2. (a) Schematic layout of the experiment side view with dimensions, (b) photograph of two SBD detectors, (c) SBD detector power calibration curve.

The radiation pulses were extracted out of the vacuum chamber through a 30 millimeter diameter fused silica window located 143 millimeter from the beam line center and reflected into the SBD detector by remotely controlled 50 millimeter diameter rotatable aluminum mirror. The mirror was used to measure CDR angular distributions. Fig. 2a shows a general geometry of the experiment. The Inductive Current Transformer (ICT2, Fig. 1) installed after a bending magnet in front of the beam dump was used to measure the electron beam charge. In order to acquire both signals simultaneously, the output of SBD detector as well as the ICT output was connected to the 1 GHz, 5GS/s Tektronix 684C Oscilloscope.

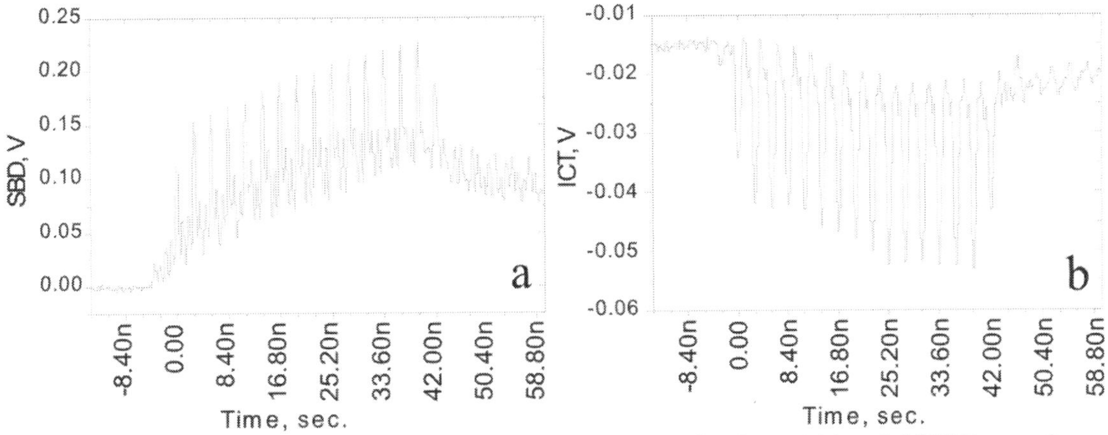

Fig. 3. Typical oscilloscope traces of (a) SBD and (b) ICT for the multi-bunch LUCX operation.

As a demonstration of the ability to monitor multi-bunch DR spikes Fig. 3 shows SBD and ICT oscilloscope traces taken simultaneously for the 16-bunch operation mode of the accelerator. It is clearly seen that we can easily resolve each CDR pulse produced by each of 16 bunches. The time separation between bunches was 2.8 ns. The wide background substratum could be explained by a multiple reflections in the generation system and limited bandwidth of the Oscilloscope.

3. Results

At first we scan the RF Gun laser phase to get the maximum CDR signal (i.e. by changing the RF Gun laser phase we changed the laser arrival time on the photocathode with respect to the accelerating RF field, and, therefore, the longitudinal electron distribution in the bunch). The maximal signal on the oscilloscope at a constant charge actually corresponds to the shortest bunch length. Fig. 4a shows the dependence of the SBD signal normalized by the ICT signal versus RF Gun laser phase. As one can see the maximum of the curve corresponds to 25 degree, which is the nominal operation point of the LUCX gun and, thus, corresponds to a minimal energy spread in the bunch and a shortest bunch length.

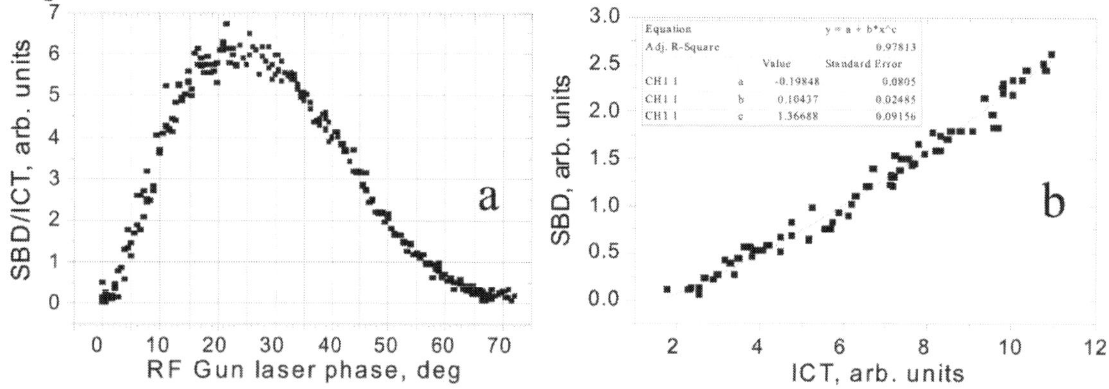

Fig. 4. (a) SBD output normalized by the electron beam current versus RF Gun laser phase, data taken for mirror angle 46.5 degree. (b) Dependence of the CDR intensity on the bunch charge. Data taken for 25 degree of RF Gun laser phase, 46.5 degree mirror angle.

In order to show the coherent nature of registered radiation the dependence of the SBD output signal as a function of the bunch charge was measured at an optimal RF gun phase. The experimental data and the polynomial fit are shown in Fig. 4b. One may see a deviation from a quadratic dependence. One of the potential reasons is that the radiation is generated by a small fraction of the bunch (bunch core) which has some non-linear charge distribution.

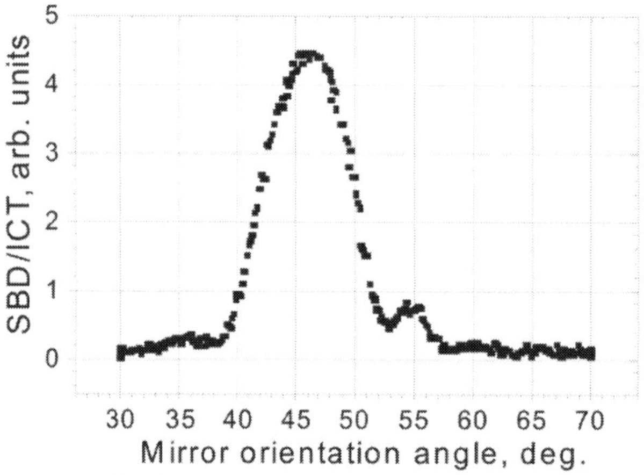

Fig. 5. SBD output normalized on electron beam current versus aluminum mirror angle, data taken for 25 degree of RF Gun laser phase.

Nevertheless, taking into account detection wavelengths range and the polynomial order (1.36 in our case) the dependence presented in Fig. 4b allows us to conclude that the detected radiation is coherent. In the future we shall develop a strategy for increasing the coherency and investigate longitudinal electron distribution in a bunch.

It is worthwhile to mention that all above scans were made while the mirror angle was set to provide the maximum CDR reflection into SBD detector. To be able to set the angle correctly we measured the mirror orientation dependence (the radiation angular distribution) presented in Fig.5. One may see that the angular distribution has a single peak. It agrees with the existing theory [3-4]. The mirror working angle was set in the maximum of the angular distribution.

4. Conclusion and future plans

We have demonstrated the status of the project on development of a novel soft X-ray source based on inverse Thomson scattering of Coherent Diffraction Radiation. We have represented the first CDR measurements at the multibunch beam of LUCX facility and demonstrated the performance of the fast millimeter wavelength range detection system based on the SBD detector. The experimental data demonstrates the feasibility of the LUCX facility for building an intense THz radiation source which can later be used for generating soft X-rays via inverse Thomson scattering process.

In the near future LUCX facility will experience a major upgrade. A new klystron producing a long RF pulse of 24 μs will be installed to create a long multi-bunch beam with 8000 bunches. The new RF gun with high mode separation and high Q value [13] is already installed to produce a high quality multi-bunch electron beam.

Nowadays two LUCX operation modes are planned. One is the low energy (5 MeV) with maximum of 8000 bunches generation. The other one is the high-energy mode (43 MeV) with a maximum of 100 bunches [14]. Both modes are very attractive for our study since the large number of produced bunches in the first mode or higher energy electrons in the second mode will result in increase of the stored radiation power in the microwave open resonator.

The resonator should have a good quality factor and one should be able to change its axis orientation in order to align the system and tune the Thomson scattering process. We are considering a step-by-step approach to achieving our goal. The first step is to have two CDR targets (concave and flat) with slits for the electron beam, forming a microwave cavity of a half length of the bunch-by-bunch spacing. In this case the radiation from preceding bunch will interact with radiation generated by the subsequent bunch because the travel distance of the radiation will be equal to the bunch spacing. However, the interaction point will be close to the upstream target in this case. The lower energy of the electron beam (γ= 84; $1/\gamma$ = 12mrad) is, the lower energy photons (UV) are in the tail of the Thomson photon spatial distribution, which will be cleared by the downstream target. Nevertheless, it is much easier to tune the cavity and demonstrate the microwave power storage.

The second step is to add a third mirror. Two mirrors will be used to generate CDR photons (flat and concave) and the third one will be a movable concave mirror attached to the side of the vacuum chamber. The total length of this cavity will be equal to the bunch spacing and CDR photons produced by the odd bunches will be scattered by the even ones. Since the collision point is very close to the 2nd flat CDR target one can expect low losses in UV and soft X-ray due to target aperture cut and lower collision angle.

To tune the resonator the mirrors on a beam line will be mounted on multi-axes vacuum manipulation systems. The microwave radiation will be detected by Schottky Barrier Diodes capable to resolve the CDR photons produced by each bunch in a train in order to monitor power build-up.

5. Acknowledgements

The work was carried out with the partial support from the "Grant-in-Aid for Creative Scientific Research of JSPS (KAKENHI 17GS0210)" project and Quantum Beam Technology Program of the Ministry of Education, Science, Sports, Culture and Technology of Japan (MEXT) and grant No. 1.81.2006 of the Ministry of Education and Science of Russian Federation.

References

[1] W. Kuulst, O.J. Luisten, van der Wiel et. al. Applied Physics Letters, **79** , (2001), 2999.

[2] Y. Shibata, Sh. Hasebe, *et al.*, Phys. Rev. E **52**, 6787 (1995).

[3] Y. Shibata et.al. NIMA, 528, (2004), 162.

[4] A.P. Potylitsyn, et al., "Generation of soft X-ray radiation using thomson scattering of coherent diffraction radiation by a short electron bunch", Proceedings of RuPAC 2006, Novosibirsk, Russia, p. 147-149.

[5] A.P. Potylitsyn, Physical Review E, vol. 60, 1999, p. 2272.

[6] K. Sakaue et al., Rev. Sci. Instr. to be published "Observation of pulsed X-ray trains produced by laser-electron Compton scatterings".

[7] H. Shimizu et al., J. Phys. Soc. Jpn., Vol. 78, No. 7.

[8] K. Hirano, et al., NIMA, 560 (2006) 233-239.

[9] S. Liu, et al., Nucl. Instr. and Meth. A584, pp1-8 (2008).

[10] K.J. Button (Ed.), Infrared and Millimeter Waves, vol. 6, Academic Press, New York; Tokyo, 1982.

[11] H.A. Watson (Ed.), Microwave Semiconductor Devices and their Circuit Applications, McGraw-Hill, New York, 1969.

[12] S. Silver (Ed.), Microwave Antenna Theory and Design, Dover, New York, 1965.

[13] A. Deshpande, et al., NIMA, 600 (2009) 361-366.

[14] M. Fukuda et al., NIMA. to be published "Upgrade of the accelerator for the Laser Undulator Compact X-ray source (LUCX)".

Transition radiation on semi-infinite plate and Smith-Purcell effect

N F Shul'ga[1] **and V V Syshchenko**[2]

[1] Akhiezer Institute for Theoretical Physics of the NSC "KIPT", Akademicheskaya Street, 1, Kharkov 61108, Ukraine

[2] Belgorod State University, Pobedy Street, 85, Belgorod 308015, Russian Federation

E-mail: `shulga@kipt.kharkov.ua`, `syshch@bsu.edu.ru`, `syshch@yandex.ru`

Abstract. The Smith-Purcell radiation is usually measured when an electron passes over the grating of metallic stripes. However, for high frequencies (exceeding the plasma frequency of the grating material) none material could be treated as a conductor, but ought to be considered as a dielectric with plasma-like permittivity. So for describing Smith-Purcell radiation in the range of high frequencies new theoretical approaches are needed. In the present paper we apply the simple variant of eikonal approximation developed earlier to the case of radiation on the set of parallel semi-infinite dielectric plates. The formulae obtained describe the radiation generated by the particles both passing through the plates (traditionally referred as "transition radiation") and moving in vacuum over the grating formed by the edges of the plates (traditionally referred as "diffraction radiation", and, taking into account the periodicity of the plates arrangement, as Smith-Purcell radiation).

1. Introduction

An interest to Smith-Purcell effect [1] both as a prospective way for generation of electromagnetic radiation from TeraHertz frequencies [2] to soft X-ray range [3] and as a novel method of beam monitoring (see, e.g., [4]) continuously grows during last years. Usually the Smith-Purcell radiation is registered when an electron passes over the grating of metallic stripes. However, for high frequencies (exceeding the plasma frequency ω_p of the grating material) none material could be treated as a perfect conductor, but ought to be considered as a dielectric with permittivity $\varepsilon_\omega = 1 - \omega_p^2/\omega^2$. So for describing Smith-Purcell radiation in the range of high frequencies new theoretical approaches are needed.

A simple variant of eikonal approximation suitable for description of transition radiation on the targets with complex geometry was proposed in [5, 6]. In the present paper we apply that method to the case of radiation on the set of parallel semi-infinite dielectric plates. The formulae obtained describe the radiation generated by the particles both passing through the plates (traditionally referred as "transition radiation") and moving in vacuum over the grating formed by the edges of the plates (traditionally referred as "diffraction radiation", and, taking into account the periodicity of the plates arrangement, as Smith-Purcell radiation).

In our article we use the system of units where the speed of light $c = 1$.

2. Radiation on the semi-infinite plate

The spectral-angular density of the transition radiation (TR) could be expressed in the form [5, 6, 7]

$$\frac{d\mathcal{E}}{d\omega d\Omega} = \frac{\omega^2}{(8\pi^2)^2} \left|\mathbf{k} \times \mathbf{I}\right|^2, \tag{1}$$

where $\mathbf{k}$ is the wave vector of the radiated wave,

$$\mathbf{I} = \int (1 - \varepsilon_\omega(\mathbf{r}))\, \mathbf{E}_\omega(\mathbf{r}) e^{-i\mathbf{kr}} d^3r, \tag{2}$$

$\mathbf{E}_\omega$ is the Fourier component by time,

$$\mathbf{E}_\omega(\mathbf{r}) = \int_{-\infty}^{\infty} \mathbf{E}(\mathbf{r}, t) e^{i\omega t} dt,$$

of the electric field produced by the moving particle in the substance of the target with the dielectric permittivity $\varepsilon_\omega(\mathbf{r})$.

If $|1 - \varepsilon_\omega(\mathbf{r})| \ll 1$, the precise value of the field in the target in (2) could be replaced in the first approximation by non-disturbed Coulomb field of the uniformly moving particle in vacuum. Although this approximation permits to calculate the TR characteristics for the targets with complex geometry [7], its applicability is restricted by the range of extremely high frequencies,

$$\omega \gg \gamma\omega_p, \tag{3}$$

where $\gamma = \sqrt{1 - v^2}$ is the particle's Lorentz factor. For investigation of TR in the range of more soft photons, some different approximate method is needed.

A simple variant of eikonal approximation in TR theory was developed in [5, 6]. In that approach the component of $\mathbf{I}$ perpendicular to the particle's velocity $\mathbf{v}$ could be written in the form

$$\mathbf{I}_\perp^{(eik)} = i\frac{4e}{v^2\gamma} \int d^2\rho\, e^{-i\mathbf{k}_\perp \boldsymbol{\rho}} \frac{\boldsymbol{\rho}}{\rho} K_1\left(\frac{\omega\rho}{v\gamma}\right) \left\{ \exp\left[-i\frac{\omega}{2} \int_{-\infty}^{\infty} (1 - \varepsilon_\omega(\mathbf{r}))\, dz \right] - 1 \right\}, \tag{4}$$

where $K_n(x)$ is the modified Bessel function of the third kind, $\boldsymbol{\rho}$ is the component of $\mathbf{r}$ perpendicular to $\mathbf{v}$. The eikonal approximation in TR theory is valid for

$$\omega \gg \omega_p \tag{5}$$

(compare with the condition(3)), but only for small radiation angles,

$$\theta \ll 1/\sqrt{a\omega}, \tag{6}$$

where a is the thickness of the target (the last constraint leads to the possibility of taking into account only the transverse component of $\mathbf{I}$).

Consider the radiation arising under the interaction of the moving particle with the semi-infinite dielectric plate of the thickness a (figure 1). In this case the application of (4) and (1) leads to the following result for the spectral-angular density of the radiation:

$$\frac{d\mathcal{E}}{d\omega d\Omega} = \frac{e^2\gamma^2}{2\pi^2} \left[1 - \cos\left(\frac{a\omega}{2} \frac{\omega_p^2}{\omega^2} \right) \right] F(\theta_x, \theta_y), \tag{7}$$

where θ_x and θ_y are the components of the two-dimensional radiation angle $\boldsymbol{\theta}$,

$$F(\theta_x, \theta_y) = \frac{1 + 2\gamma^2\theta_y^2}{(1 + \gamma^2\theta_y^2)(1 + \gamma^2\theta^2)} \exp\left(-2\left|x_0\right| (\omega/\gamma) \sqrt{1 + \gamma^2\theta_y^2} \right), \tag{8a}$$

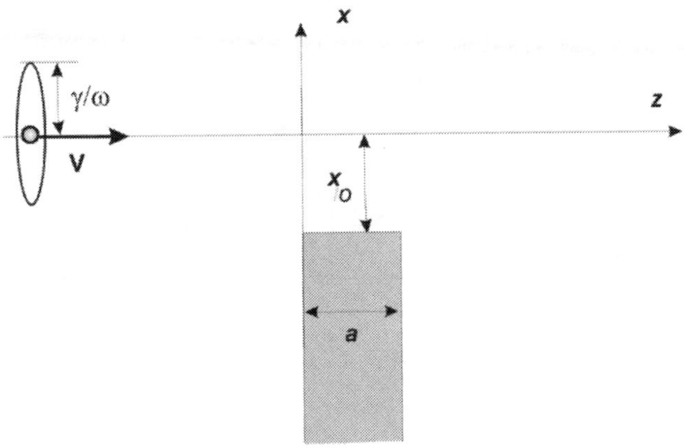

Figure 1. Particle's motion near the edge of the plate. The value γ/ω is the characteristic transverse dimension of the Fourier component of the particle's Coulomb field corresponding to the frequency ω. Positive values of the impact parameter x_0 correspond to the case of penetration of the particle through the plate.

under $x_0 \leq 0$, and

$$F(\theta_x, \theta_y) = \frac{4\gamma^2\theta^2}{(1+\gamma^2\theta^2)^2} + \tag{8b}$$

$$+\frac{1+2\gamma^2\theta_y^2}{(1+\gamma^2\theta_y^2)(1+\gamma^2\theta^2)}\exp\left(-2|x_0|(\omega/\gamma)\sqrt{1+\gamma^2\theta_y^2}\right) -$$

$$-\frac{4\exp\left(-|x_0|(\omega/\gamma)\sqrt{1+\gamma^2\theta_y^2}\right)}{(+\gamma^2\theta^2)^2}\left(\frac{\gamma\theta_x\sin(x_0\omega\theta_x)}{\sqrt{1+\gamma^2\theta_y^2}} + \gamma^2\theta^2\cos(x_0\omega\theta_x)\right),$$

under $x_0 > 0$. This formula is in agreement with the result of the paper [8] obtained using the method [9].

Surface plots of the function $F(\theta_x, \theta_y)$ for some values of the impact parameter x_0 are presented on the figure 2.

3. Radiation on the grating

Consider now the radiation arising under incidence on the periodic set of such plates (figure 3).

Direct account of the periodic structure of the target is impossible because of the integration along the z axis in the exponent in (4). Nevertheless, if the formation length of the radiation on the single plate do not exceed the distance between two plates,

$$l_{coh} = \frac{2\gamma^2/\omega}{1+\gamma^2\theta^2+\gamma^2\omega_p^2/\omega^2} < b - a, \tag{9}$$

the problem is easy for computation: the effect of the target's complexity will consist in the interference of the electromagnetic waves emitted under the interaction of the moving particle with separated plates.

The radiation in this case would be described by equation (7) multiplied by the factor

$$2\pi N \sum_{m=-\infty}^{\infty} \delta\left\{\omega b\left(\frac{1}{v} - \cos\theta\right) - 2\pi m\right\}, \tag{10}$$

where N is the total number of plates, $N \gg 1$ (like in the most part of the papers devoted to TR, we neglect the refraction of the emitted radiation in the substance of the target). These

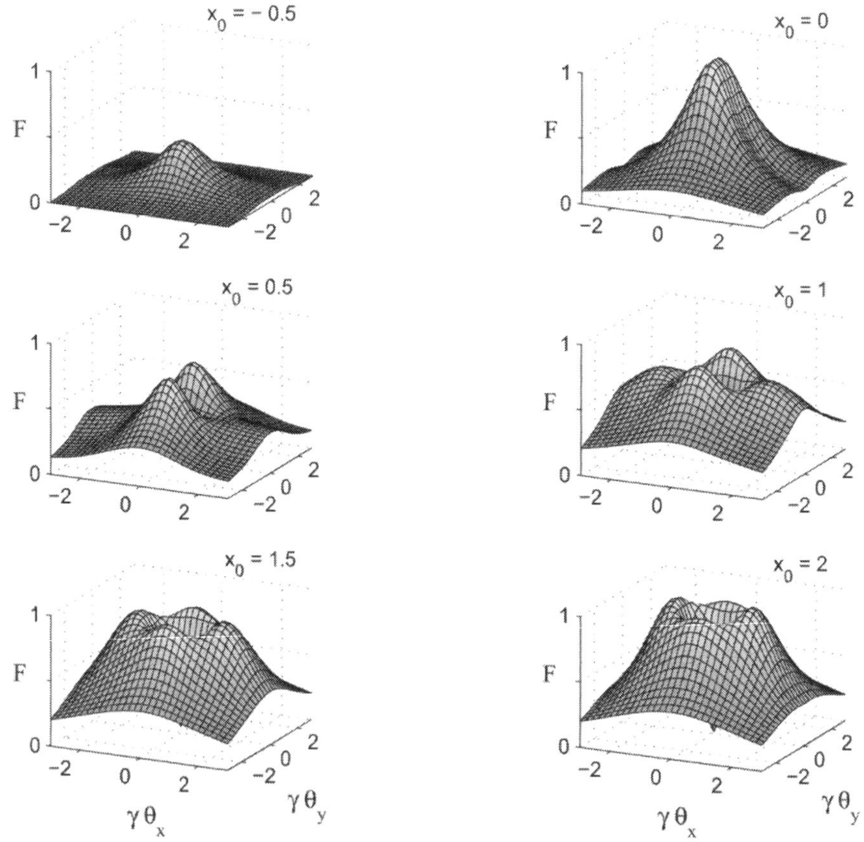

Figure 2. Angular distribution of radiation on a single semi-infinite plate according to (8). Impact parameter x_0 is given in the units of γ/ω.

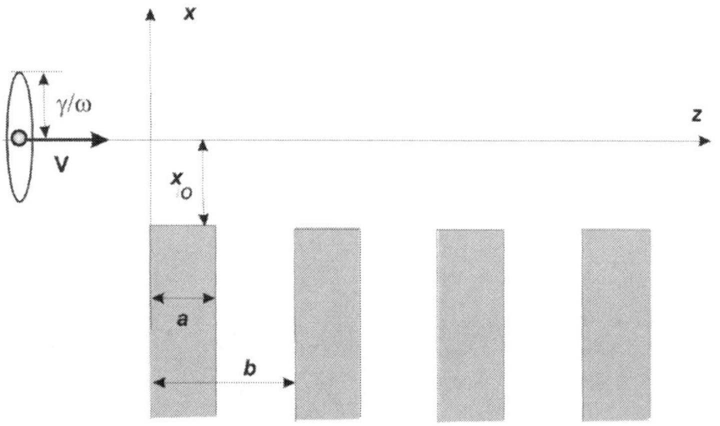

Figure 3. Grazing incidence of the particle onto the grating

δ-functions mean that the radiation under angle θ would take the place only for the frequencies satisfying the condition

$$\omega = \frac{2\pi m}{b\left(\dfrac{1}{v} - \cos\theta\right)} \tag{11a}$$

(m is the positive integer) or, for small angles,

$$\omega_m = \frac{2\gamma^2}{1+\gamma^2\theta_m^2}\frac{2\pi}{b}m. \tag{11b}$$

This is well known Smith-Purcell condition [1].

4. Conclusion

Equations (7), (8a), (10) describe the Smith-Purcell radiation in the high-frequency limit,

$$\omega \gg \omega_p.$$

The exponent in (8a) describes the radiation intensity dependence on the impact parameter. The characteristic values of the last one, on which the radiation is substantial, are

$$|x_0|_{eff} = \gamma/2\omega.$$

Note that such the exponent is present in the formulae describing Smith-Purcell effect independently on the range of frequencies and the computation method (see, e.g., [10]; the results of different models are distinct from each other only by pre-exponential factors).

In the region $x_0 > 0$ when the particle's trajectory penetrates the plates, the radiation arising is described by equations (7), (8b), (10). The first term in (8b) describes the transition radiation on the infinite plate in the eikonal approximation [5, 6], the second term accounts the edge of the plate (its contribution evidently coincides with Smith-Purcell radiation), the third term describes the interference of that two mechanisms.

In the frames of our approach, for both negative and positive x_0 the angular distribution of the radiation is symmetric in relation to the (y, z) plane.

Acknowledgments

This work is supported in part by the internal grant of Belgorod State University.

References

[1] Smith S J and Purcell E M 1953 *Phys. Rev.* **92** 1069

[2] Korbly S E, Kesar A E, Sirigiri J R and Temkin R J 2005 *Phys. Rev. Lett.* **94** 054803

[3] Moran M J 1992 *Phys. Rev. Lett.* **69** 2523

[4] Blackmore V, Doucas G, Perry C, Ottewell B, Kimmitt M F, Woods M, Molloy S and Arnold R 2009 *Phys. Rev. ST Accel. Beams* **12** 032803

[5] Syshchenko V V and Shul'ga N F 2007 *Journal of Surface Investigation* **1** 217

[6] Shul'ga N F and Syshchenko V V 2006 *Advanced Radiation Sources and Applications* ed. H Wiedemann (Springer Netherlands) pp 129-148

[7] Shulga N F and Syshchenko V V 2003 *Phys. Lett.* A **313** 307

[8] Tishchenko A A, Strikhanov M N and Potylitsyn A P 2005 *Nucl. Instr. and Meth.* B **227** 63

[9] Durand L 1975 *Phys. Rev.* D **11** 89

[10] Karlovets D V and Potylitsyn A P 2006 *Phys. Rev. ST Accel. Beams* **9** 080701

Backward Transition Radiation in EUV-region as a possible tool for beam diagnostics

L.G. Sukhikh, S.Yu. Gogolev, A.P. Potylitsyn
Tomsk Polytechnic University, Lenin avenue, 30, Tomsk, 634050, Russia
E-mail: Sukhikh@interact.phtd.tpu.ru

Abstract. In this paper we discuss the possibility of proof-of-principle experiment to observe backward transition radiation in EUV region as a first step for beam diagnostics. We have carried out simulations of backward transition radiation angular distributions in the EUV region for molybdenum target for different geometries. The simulation results are promising for beam diagnostics purposes.

Transition radiation (TR) appearing when a charged particle crosses a vacuum-metal interface have a very broad spectrum of applications, for example, for electron beam diagnostics. Nowadays, transverse beam profile monitors based on backward TR (BTR) in optical region are widely used in modern accelerators. Simple and reliable construction and a possibility of single-shot measurements should be mentioned as the great advantages of this technique. However, the use of BTR in optical region has some principle physical limitations which restrict applications at modern accelerators.

The first limitation is the spatial resolution of an optical system that is proportional to BTR wavelength λ. The best BTR monitor resolution obtained in the experiment is equal to $\sigma = 2\mu$m for optical wavelengths [1]. Such resolution do not allow to measure profiles of submicron beams of such accelerator as KEK-ATF2 [2] with a proper accuracy.

The second limitation is connected with so-called "pre-wave" zone effect [3]. This effect causes the broadening of the BTR radiation cone from the well known γ^{-1} value if the detector is situated closer than the $\gamma^2\lambda$ distance (γ is the Lorentz-factor). If the detector is set up closer the broadening of the radiation cone may cause the waste of useful information in the optical path.

The third limitation is connected with the coherent radiation. One knows, that the radiation is coherent when a bunch length is comparable with the radiation wavelength. It seems, that it is not a problem because it is almost impossible to obtain the bunch with longitudinal length comparable with optical wavelength but indeed some accelerator instabilities may create some bunch microstructure of such size. It is a big problem for modern X-ray Free Electron Lasers as it was shown at LCLS in recent paper [4]. These microstructures that may vary from shot to shot making it impossible to measure the bunch transverse profile. In the European X-FEL project the optical BTR was planned to be used as a main tool for bunch transverse profile measurement. But the recent LCLS results shows that there may be real problems in optical region. The vacuum chambers for European X-FEL were designed for 22.5 deg geometry that makes one to try to use them to use money sparingly.

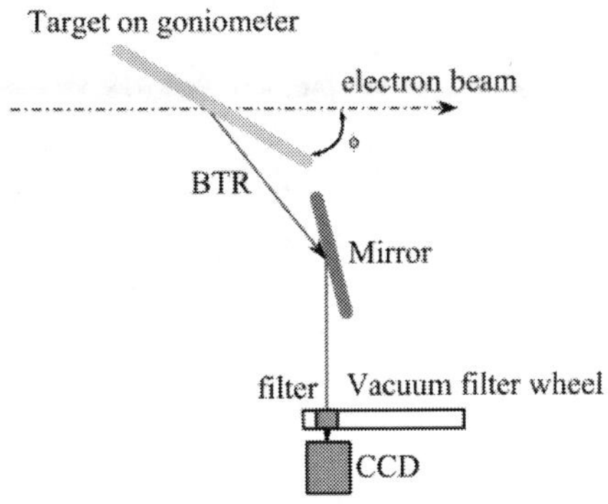

Figure 1. The scheme of the planned setup.

Taking into account all mentioned above problems with use of optical BTR for the bunch profile measurement one needs some simple method to solve all problems. We propose to use BTR in EUV region ($\lambda \simeq 13.5\,\text{nm}$) as a tool for beam profiles measurement. Even a simple estimation shows that the change of the radiation wavelength from 550 nm to 13.5 nm allows to improve the resolution at approximately 40 times. The pre-wave zone for such wavelength is equal to 84 mm ($\gamma = 2500, \lambda = 13.5\,\text{nm}$) that is not significant. The size of microstructure should also be 40 times smaller to radiate coherently.

Some preliminary simulations were presented in TIPP09 Conference and will be published in [5]. In this paper we present the plan of proof-of-principle experiment that is planned to be carry out in the electron beam of Mainz microtron MAMI.

The MAMI electron beam has energy of 855 MeV and the beam size (rms) may be varied in a wide range from tens of microns to millimeters. We plan to use the setup that was previously used for PXR measurements and have several outputs at different angles and vacuum goniometer. The scheme of the planned setup is shown in Fig. 1.

Two targets will be installed at the target holder in goniometer. The first target is planned to be an "optical" one and the second — the "EUV" one. The target material is the main problem. The target material should have high BTR production efficiency and a long lifetime under intensive electron beam. The latter may be achieved if one uses the silicon metal-coated target. For example, a target that was 300 μm Si coated by 2 μm gold was successfully used in KEK-ATF OTR experiments with no damage under intensive electron beam with micron dimensions [6]. As the "optical" target the aluminum one may be chosen because the reflectivity is of about 95%, as the "EUV" target — the molybdenum one. The latter has the best reflectivity in the $80 \div 90\,\text{eV}$ region [7]. The molybdenum reflectivity in the spectral region under consideration is shown in Fig. 2. The radiation from a target will be reflected by the mirror M that is planned to be used to better the background conditions. The filter wheel will allow to choose the spectral region that will be detected by the CCD-camera that is planned to be Andor DO 434 BN. The camera is sensitive in a wide spectral range and have the efficiency in $80 \div 90\,\text{eV}$ region of about 40%. The angular acceptance of the vacuum tract should be larger then $4\gamma^{-1} = 2.4\,\text{mrad}$ in order to loose no information. The size of the sensitive area of the CCD-camera ($13.3 \times 13.3\,\text{mm}^2$) allows to install it on a distance of 2 m from the target. The whole tract from the target to the detector should be the vacuum one.

We plan to compare the angular distributions of BTR and the integral photon yield of both targets and theoretical simulation results. We also plan to investigate the background conditions for EUV region.

For our simulations we used the well-known model [8]. It is possible to use the BTR model for

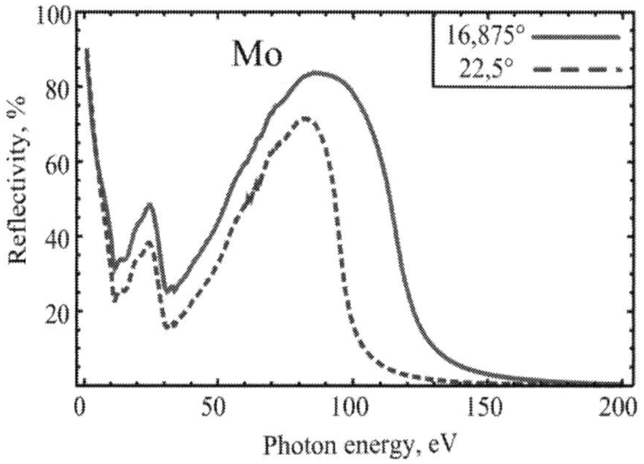

Figure 2. The reflectivity of molybdenum in the region of interest [7].

the infinite target because the planned target will have the dimensions of $10 \times 10\,\text{mm}^2$ (projection on the beam direction) while the effective beam field size is of about $\gamma\lambda$ that for EUV region is equal to $30\,\mu$m. In our simulations we are interested in the radiation in photon energy region from 30 to 150 eV. All simulations have been carried out for an electron with $\gamma = 1675$ Lorentz-factor and infinite flat target. The reflection coefficients for most materials rapidly decrease in the EUV region with the photon energy increase as one may see in Fig. 2. That is why one should install the target at small grazing angle. We have carried out our simulations for target orientation angles $\phi = 10$ deg, $\phi = 16.75$ deg and $\phi = 22.5$ deg. The first angle is as an example, second one is planned for our proof-of-principle experiment and the third one is very convenient for practical use in European X-FEL project.

Figure 3 shows the angular distribution for the mentioned above target orientation angles for spectral region $30 \div 150$ eV. The upper energy limit was chosen because the molybdenum reflectivity is almost zero after this value. The lower integration limit was chosen because we plan to use thin filter to block optical and UV parts of BTR. From Fig. 3 one may see that the number of photons decreases rapidly with increase of orientation angle. The central figure 3 shows the angular density of radiation photons that are expected in our experiment.

For planned electron beam diagnostics one needs to focus BTR to obtain the source image. For EUV region the best variant is to use focusing multilayer mirror that also operates as a bandpass filter with central line defined by the layers period and with spectral width of about

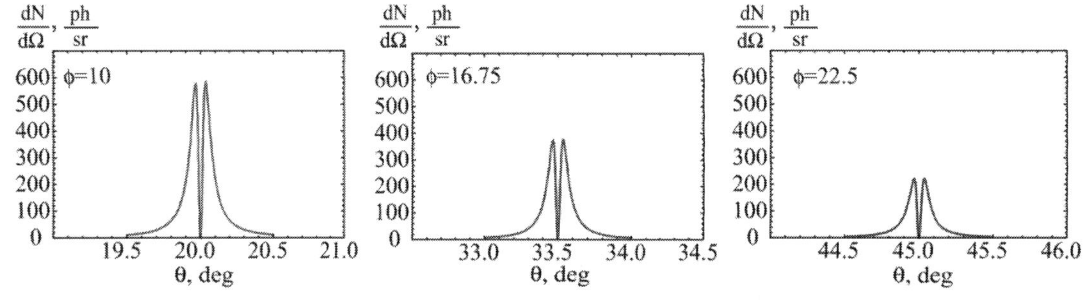

Figure 3. The angular distribution of BTR from Mo target for different orientation angles integrated over spectral region $30 \div 150$ eV

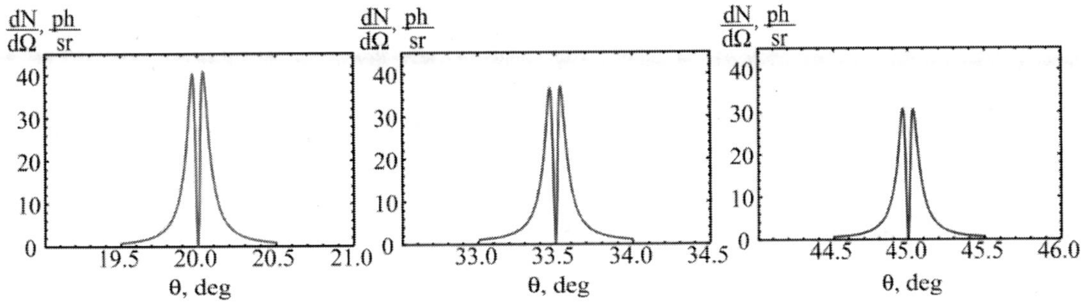

Figure 4. The angular distribution of BTR from Mo target for different orientation angles integrated over spectral region $80 \pm 3\,\mathrm{eV}$

$5 \div 8\,\mathrm{eV}$. That is why we also simulated the angular distributions of BTR in spectral region of $80 \pm 3\,\mathrm{eV}$. The latter are shown in Fig. 4. One may see from this figure, that the angular photon density is much smaller that for previous spectral region. However, the decrease of the angular photon density is not so fast and if we will be able to obtain the image of the beam in the second stage of our proof-of-principle experiment for 16.75 deg geometry we may hope that the technique will also be efficient for European X-FEL project.

The BTR angular distribution shown in Figs. 3 and 4 allows confident identifying of the radiation process in the experiment. The form of the measured distributions will allow to estimate the background conditions for future diagnostics. The photon yield for 16.75 deg geometry and spectral region $30 \div 150\,\mathrm{eV}$ is equal to $4.73 \cdot 10^{-3}$ per electron that gives for MAMI current ($100\,\mu$A) $3 \cdot 10^{12}$ ph/sec. For spectral region of $80 \pm 3\,\mathrm{eV}$ one may obtain in 22.5 deg geometry the photon yield $3.6 \cdot 10^{-4}$. For the typical bunch population ($5 \cdot 10^9$ electrons/bunch) this value gives $1.8 \cdot 10^6$ ph/bunch that seems to be enough to carry out single shot measurements of the bunch transverse profiles in European X-FEL project.

Finally, we can say that the EUV BTR seems to be a promising tool for a single shot high resolution diagnostics of electron beam profiles. The estimations carried out using well known models show the possibility of such diagnostics. The simulations carried out for 22.5 deg geometry shows that in principle it may be possible to use European X-FEL vacuum chambers with new filling to use money sparingly. Now we hope that the proof-of-principle experiment will be successful and will prove our simulations.

Reference

[1] M. Ross et al., SLAC-Pub-9280, (2002).
[2] ATF2 Group, ATF2 Proposal, **1**, (2006)
[3] V.A. Verzilov, Phys. Lett. A **273** (2000) 135.
[4] H. Loos et al., SLAC-PUB-13395, (2008).
[5] L.G. Sukhikh, S.Yu. Gogolev, A.P. Potylitsyn, NIMA, to be published.
[6] P. Karataev, PhD thesis, Tokyo Metropolitan University, 2004.
[7] Henke B. L., Gullikson E. M. and Davis J. C., Atomic Data and Nuclear Data Tables **54**(2) (1993) 181
[8] V.E. Pafomov, Trudy Fiz. Inst. Akademii Nauk USSR XLIV (1969) 28 [Proc. P.N. Lebedev Phys. Inst. **44** (1971) 25].

Radiation from a charged particle-in-flight from a laminated medium to vacuum

L Sh Grigoryan[1,3], **A R Mkrtchyan**[1], **H F Khachatryan**[1], **S R Arzumanyan**[1] **and W Wagner**[2]

[1]Institute of Applied Problems in Physics, 25 Hr. Nersessian Str.,0014 Yerevan, Armenia

[2]Forschungszentrum Dresden-Rossendorf, Institute of Radiation Physics, POB 510119, 01328 Dresden, Germany

E-mail: levonshg@mail.ru

Abstract. The radiation from a charged particle-in-flight from a semi-infinite laminated medium to vacuum and back,- from vacuum to the laminated medium, has been investigated. Expressions for the spectral-angular distribution of radiation energy in vacuum (at large distances from the boundary of laminated medium) were obtained for both the cases with no limitations on the amplitude and variation profile of the laminated medium permittivity. The results of appropriate numerical calculations are presented and possible applications of the obtained results are discussed.

Keywords: Periodical structures, relativistic particles, radiation

1. Introduction

The period of existing periodical structures varies from ~ 10^{-8} (crystals) to microns (photonic crystals) and up to ~ 1 *cm* (undulators). It is natural that the radiation from charged particles in the above periodical structures has multiple manifestations that are of interest for practical applications [1, 2].

For instance, it is specific that in the periodic structures there are forbidden bands on dispersion curves for electromagnetic waves (see, e.g. [3]). The presence of such bands shall influence the radiation from a charged particle in periodical structures and for this reason these may serve as additional means for controlling the particle radiation parameters. This mechanism was first investigated in the case of an infinite laminated medium composed of two alternating plates having different values of permittivity ε [4]. In [5, 6] the infinite stack of plates was replaced with a laminated medium, the permittivity of which is an arbitrary one-dimensional periodic function (the problem of a stack of the finite number of plates has been investigated in [7], see also [8]). However, the calculation methods used in [5,6] (quasi-classical approximation, perturbation theory) failed to allow for the influence of forbidden bands on the emitted electromagnetic waves. The present paper deals with the solution of this problem.

The problem is formulated in section 2. Main stages of analytical calculations are given in section 3, the results of numerical calculations are analyzed in section 4 and the main results of the paper are summarized in the last section.

---

[3] To whom any correspondence should be addressed.

2. Formulation of the problem

Consider the motion of charge q along the positive direction of OZ axis (see figure 1*a*) at its travel from a semi-infinite laminated medium ($z<0$, range 1) to vacuum ($z>0$, range 2). The plane $z=0$ is taken as an interface between the laminated medium with vacuum.

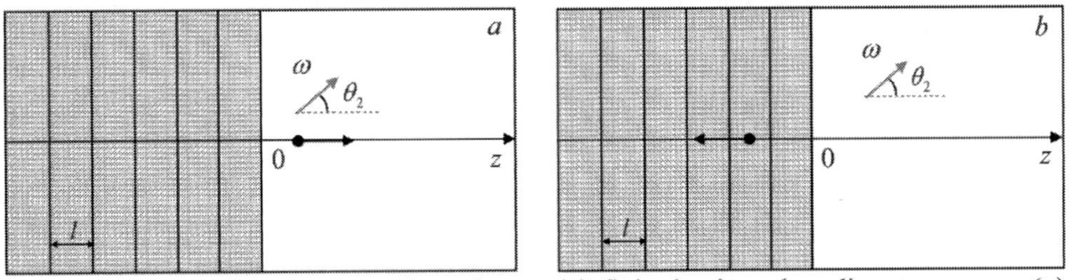

Figure 1. The particle emerging from the semi-infinite laminated medium to vacuum (*a*) and vice versa, – incoming from vacuum to the laminated medium (*b*).

We shall assume that the electrical permittivity ε_1 and magnetic permeability μ_1 of the laminated medium are independent of $(x,y)\equiv\vec{\chi}$ and periodically vary along the direction of particle motion:

$$\varepsilon_1(z-l)=\varepsilon_1(z), \qquad \mu_1(z-l)=\mu_1(z), \tag{1}$$

with arbitrary profiles (l is the period of laminated medium). As a result, the electrical permittivity ε and magnetic permeability μ of the system on the whole may be written as

$$\begin{aligned} &\varepsilon(z)=\varepsilon_1(z), \quad \mu(z)=\mu_1(z) \quad \text{for } z<0 \quad \text{(range 1)} \\ &\varepsilon(z)=1, \quad \mu(z)=1 \quad \text{for } z>0 \quad \text{(range 2).} \end{aligned} \tag{2}$$

If we represent the energy W_2 of radiation transmitted through the surface $z=z_0$ distant from $z=0$ in the range 2 as

$$W_2=\int I_2(\omega,\theta_2)d\omega d\theta_2 \tag{3}$$

during the whole time of particle motion, then $I_2(\omega,\theta_2)$ shall describe the spectral-angular distribution of the energy of radiation that propagates in vacuum along the direction of particle motion (forward radiation). We aim at an analysis of this quantity.

Note that at the reversal of the direction of particle motion $\mathrm{v}\rightarrow\text{-}\mathrm{v}$ (see figure 1*b*), we arrive at the problem of radiation from particle incoming into the laminated medium. In this case $I_{2,\mathrm{v}\rightarrow-\mathrm{v}}(\omega,\theta_2)$ shall describe the radiation in vacuum, propagating in the backward direction to the particle motion (backward radiation).

3. The stages of analytical calculations and the final formula

Taking into account the azimuthal symmetry of problem and using a Fourier transform

$$f_{u\omega}(z)=\frac{1}{(2\pi)^3}\int f(\vec{r},t)\exp[i(\omega t-\vec{u}\vec{\chi})]d\vec{\chi}dt \tag{4}$$

the solution of the set of Maxwell equations is reduced to the solution of a single equation [4]

$$[\varepsilon\frac{\partial}{\partial z}\left(\frac{1}{\varepsilon}\frac{\partial}{\partial z}\right)+\frac{\omega^2}{c^2}\varepsilon\mu-u^2]\varepsilon E_{u\omega}=4\pi[\varepsilon\frac{\partial}{\partial z}\left(\frac{\rho_{u\omega}}{\varepsilon}\right)-i\frac{\omega}{c^2}\varepsilon\mu j_{u\omega}] \tag{5}$$

(the basic equation of the problem under consideration). In this equation $\varepsilon\equiv\varepsilon_\omega$ and $\mu\equiv\mu_\omega$ are Fourier transformations of the electric and magnetic permittivities (taking into account the frequency dispersion, the spatial dispersion being assumed negligibly small). Here we assume that $\varepsilon_1=\varepsilon'_{1\omega}+i\varepsilon''_{1\omega}$ and $\mu_1=\mu'_{1\omega}+i\mu''_{1\omega}$ are complex valued functions (making allowance for the absorption of radiation by the material of laminated medium). Having determined the solution $E_{u\omega}$ of the equation (5), one may calculate the strengths of electric and magnetic fields by means of the following formula:

$$\vec{E}_{u\omega} = E_{u\omega}\vec{n}_z + \frac{i\vec{u}}{\varepsilon u^2}\left(\frac{\partial}{\partial z}\varepsilon E_{u\omega} - 4\pi\rho_{u\omega}\right), \qquad \vec{H}_{u\omega} = \frac{\omega}{cu^2}(\varepsilon E_{u\omega} + 4\pi\frac{i}{\omega}j_{u\omega})\vec{u}\times\vec{n}_z . \quad (6)$$

Here

$$\rho(\vec{r},t) = q\delta(x)\delta(y)\delta(z - \mathrm{v}t) \quad \text{and} \quad \mathrm{v}\rho_{u\omega}(z) = \frac{q}{(2\pi)^3}\exp(i\frac{\omega}{\mathrm{v}}z) = j_{u\omega}(z) \quad (7)$$

is the density of current connected with the charged particle.

In range 2 equation (5) shall have the following solution:

$$E_{u\omega} = E^q_{u\omega} + \frac{iq}{2\pi^2\omega}a_2\exp(ik_2 z), \qquad k_2 = \sqrt{\omega^2/c^2 - u^2}, \quad (8)$$

where the first term is the known field of charge in vacuum [6], and the 2nd one describes the free field (the radiation field). At large distances from the semi-infinite medium the spectral-angular distribution of radiation energy is determined by the well-known expression [6,7]

$$I_2(\omega,\theta_2) = \frac{2q^2}{\pi c}\frac{\cos^2\theta_2}{\sin\theta_2}|a_2(\omega,u)|^2, \qquad \text{where} \quad u = \frac{\omega}{c}\sin\theta_2, \quad (9)$$

and θ_2 is the angle showing the direction of radiation with respect to OZ axis (see figures 1*a,b*). For calculation of a_2 it would be enough to equate (8) to the complete solution of equation (5) that is valid for all $-\infty < z < \infty$. This solution is determined most easier by means of the method of Green functions, because the Green function is the solution of a homogeneous equation, which is obtained from (5) by means of substitution $\rho_{u\omega} = j_{u\omega} = 0$. In this case the general solution of homogeneous equation inside the laminated medium may be represented according to the Bloch theorem (see, e.g. [9]) as a superposition $b_1L_+(z) + c_1L_-(z)$ of running waves $\exp(\pm ik_1 z)$ modulated with the period of laminated medium, i.e.,

$$L_\pm(z) = w_\pm(z)\exp(\pm ik_1 z), \quad \text{where} \quad w_\pm(z\text{-}l) = w_\pm(z). \quad (10)$$

Here

$$\cos k_1 l = \frac{\chi_a(\tau)\dot{\chi}_b(0) + \chi_a(0)\dot{\chi}_b(\tau) - [\dot{\chi}_a(\tau)\chi_b(0) + \dot{\chi}_a(0)\chi_b(\tau)]}{2[\chi_a(0)\dot{\chi}_b(0) - \dot{\chi}_a(0)\chi_b(0)]} \quad (11)$$

and

$$L_\pm(z) = [\chi_b(0) - \exp(\pm ik_1 l)\chi_b(\tau)]\chi_a(z) - [\chi_a(0) - \exp(\pm ik_1 l)\chi_a(\tau)]\chi_b(z), \quad (12)$$

where $\chi_a(z)$ and $\chi_b(z)$ is the pair of any linearly independent solutions of the homogeneous equation, $\tau = -l$, and the point over the function means the differentiation with respect to its argument. The quasi-wave number $k_1(\omega,u)$ is determined from (11) with an accuracy of the sign. To fix the sign we shall assume that

$$k_1 = k_1' + ik_1'', \qquad \text{where} \quad \omega k_1''(\omega,u) \geq 0 \quad (13)$$

($k_1'' \neq 0$ as the absorption of radiation in the material of laminated medium was taken into account). The forbidden bands of the laminated medium mentioned in the Introduction correspond to the values of ω,u, for which $\omega k_1''(\omega,u)$ stays positive even when $\varepsilon_1''/\varepsilon_1' \to 0$. The electromagnetic oscillations with these values of ω,u can not freely propagate in a laminated medium as the corresponding solutions $L_\pm(z)$ of the homogeneous equation exponentially decrease along one of the directions of OZ axis.

Omitting the intermediate calculations we give the final expression

$$a_2 = \frac{\omega}{2k_2 \mathrm{v}} \left\{ \frac{1-\mathrm{v}^2/c^2}{1-k_2\mathrm{v}/\omega} - \frac{1-\mathrm{v}^2/c^2}{1+k_2\mathrm{v}/\omega} \cdot \frac{k-\varepsilon_1 k_2}{k+\varepsilon_1 k_2} - \left[\varepsilon_1 - 1 + \frac{\sigma}{1-\exp[i(k_1-\omega/\mathrm{v})l]} \right] \frac{2k_2}{k+\varepsilon_1 k_2} \right\}_{z=0} . \qquad (14)$$

Entering into this expression are the values of $\varepsilon_1(z)$ and $k(z)$ in point $z=0$,

$$k = i\frac{\dot{L}_-(0)}{L_-(0)} \quad \text{and} \quad \sigma = i\frac{\omega\varepsilon_1(0)}{\mathrm{v}L_-(0)} \int_{-l}^{0} \left(1 - \varepsilon_1\mu_1 \frac{\mathrm{v}^2}{c^2} + \frac{i\mathrm{v}\dot{\varepsilon}_1}{\omega\varepsilon_1} \right) \frac{L_-}{\varepsilon_1} \exp(i\frac{\omega}{\mathrm{v}}z)\, dz . \qquad (15)$$

Three terms in (14) correspond to three types of waves: the 1st term corresponds to the wave emitted in vacuum in the forward direction, the 2nd term – to the wave emitted in vacuum primarily in backward direction and then reflected from the interface between vacuum and the laminated medium, and the 3rd term – to the perturbation emitted in the laminated medium in the forward direction and traversed through the interface between the laminated medium and vacuum. In the partial case of $\varepsilon_1(z)$, $\mu_1(z) = const$ in (14), there follows the well known expression for transition radiation at the passage of particle from a homogeneous medium to vacuum [5-7].

To obtain the expressions for radiation in the backward direction (see figure 1*b*) one has to replace v by - v in above formulae.

4. Numerical results

We confine ourselves to the consideration of a model of laminated medium (see figure 2) with non-harmonic variations of $\varepsilon_1(z)$ and large modulation depth $\Delta\varepsilon_1' \sim \overline{\varepsilon_1'(z)}$ so, that

$$\Delta\varepsilon_1' = 1, \qquad \varepsilon_1'(0) = \overline{\varepsilon_1'(z)} = 1.5, \qquad \varepsilon_1''/\varepsilon_1' = 0.01 \quad \text{and} \quad \mu_1 = 1. \qquad (16)$$

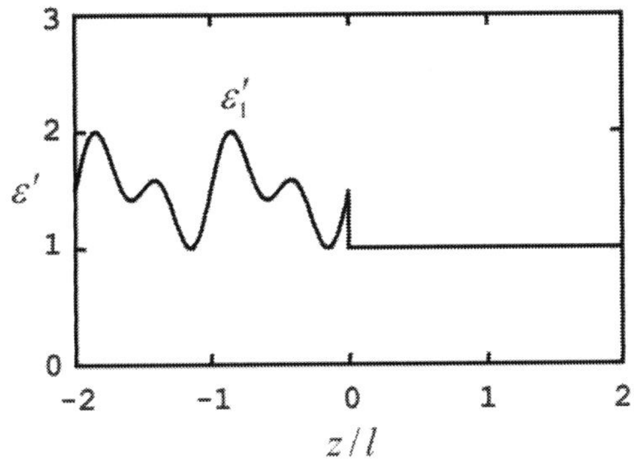

Figure 2. The model of z-dependence of the real part of electric permittivity $\varepsilon(z)$ of the system. The laminated medium with $\varepsilon = \varepsilon_1$ is to the left and vacuum is to the right of $z=0$.

It was assumed that in the range of ω under consideration one can neglect the dispersion of electromagnetic waves in the laminated medium. If we take the energy of particle (electron) equal to $30 MeV$, then the Cherenkov condition in the form

$$\cos\overline{\theta}_1 = c/\mathrm{v}\sqrt{\overline{\varepsilon}_1'} \qquad (17)$$

is satisfied. Here the value $\overline{\theta}_1 \cong 35°$ specifies the characteristic direction of Cherenkov radiation (CR) in the laminated medium. Now introduce a dimensionless parameter

$$\omega_l = \omega l / \pi \mathrm{v} . \quad (18)$$

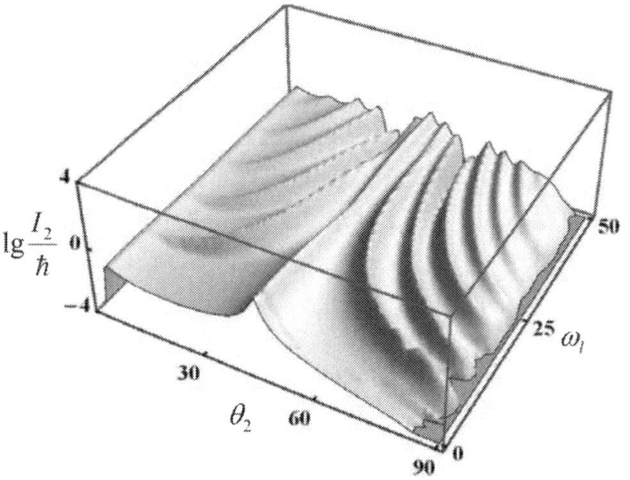

Figure 3. Spectral-angular distribution of radiation from $30 MeV$ electron emerging from the laminated medium with non-harmonic variations of $\varepsilon_1(z)$ (see figure 2, (16)) to vacuum (radiation in the forward direction).

The spectral-angular distribution $I_2(\omega,\theta_2)$ of the energy of particle radiation is shown in figure 3. The central part of this figure with

$$\theta_2 \approx \arcsin(\sqrt{\bar{\varepsilon}_1'} \sin \bar{\theta}_1) \cong 45^\circ \quad (19)$$

corresponds to the main beam of particle CR. On either sides of this central part the lines of diffracted radiation split up (the beams of the 1st, 2nd and higher orders of parametric radiation). The results of our numerical calculations indicate that the distribution $I_2(\omega,\theta_2)$ on a separate branch is determined by the amplitude and profile of ε_1 variations.

It is noteworthy that some part of the branches of spectral-angular distribution reaches the range of $\theta_2 < 10^\circ$ and superimposes on the existing small peak in the angular distribution of radiation in this range. It is clearly seen in the lower left angle of figure 3 and corresponds to the transition radiation (TR) of particle at the interface of laminated medium with vacuum. Thus, the peaks from diffracted beams of parametric radiation superimpose on the spectral distribution of TR.

Now consider the case of particle flying into the laminated medium from vacuum (backward radiation in figure 1*b*).

As was mentioned above, not all the waves may propagate in the periodic medium due to the presence of forbidden bands (see figure 4). Particularly, the part of TR emitted in the forward direction by the particle flying into the laminated medium (main part of TR), that fits the range of forbidden bands of laminated medium (B_1, B_2 in figure 4), also cannot propagate. This part of TR shall be reflected by the laminated medium to vacuum. Such a throwing over of particle TR shall be followed by the appearance of appropriate maxima in the spectral-angular distribution $I_{2,\mathrm{v}\to-\mathrm{v}}(\omega,\theta_2)$ of radiation energy in backward direction (to vacuum).

Here (in figure 5), the spectral-angular distribution $I_{2,\mathrm{v}\to-\mathrm{v}}(\omega,\theta_2)$ of the energy of backward radiation from $30 MeV$ electron in flight from vacuum to laminated medium is for the same values of parameters as those in figure 3 (TR range).

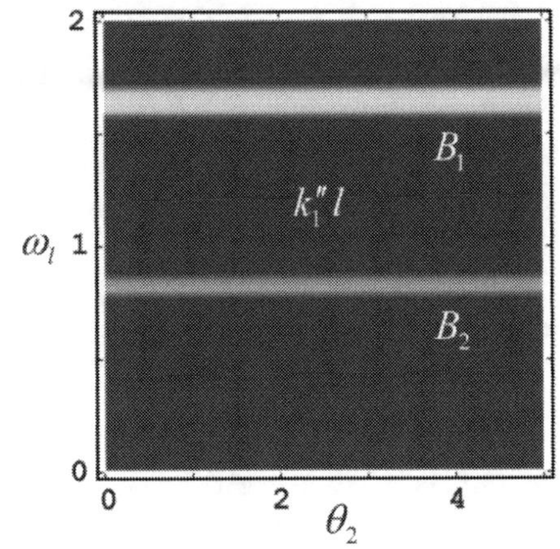

Figure 4. The imaginary part of quasi-wave number $k_1(\omega, u)$ (see (11)) versus ω и θ_2 (see (9)) for a laminated medium with non-harmonic variations of $\varepsilon_1 = \varepsilon_1'$ (refer to figure 2, $\mu = 1$). The light-colored areas of the figure correspond to forbidden bands B_1, B_2 with $k_1'' > 0$ (the lighter the area, the higher is the value of k_1'').

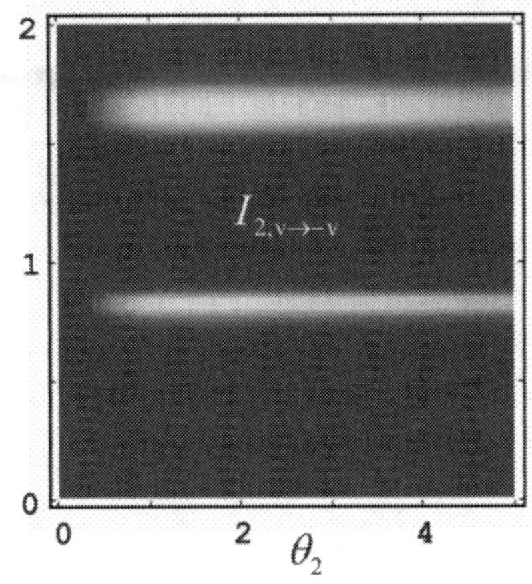

Figure 5. Spectral-angular distribution of the energy of backward radiation from electron in flight from vacuum to the laminated medium. Calculations are based on the model of non-harmonic variations of ε_1. The values of system parameters are the same as those in figure 3. The value of $I_{2,v\to -v}$ is the higher, the brighter is the point representing this value.

As was to be expected, to two forbidden bands B_1, B_2 in figure 4 there almost exactly correspond two ranges of maxima of $I_{2,v\to -v}(\omega, \theta_2)$ in figure 5. For both the maxima

$$I_{2,v\to -v}(\omega, \theta_2) \cong I^{(i)}_{2,v\to -v}(\theta_2), \quad \text{when} \quad (\omega, \theta_2) \in B_i \quad \text{and} \quad i = 1,2. \tag{20}$$

The plots of functions $I^{(i)}_{2,v\to -v}(\theta_2)$ are shown in figure 6. It is noteworthy that as the absorption in the laminated medium decreases

$$I^{(i)}_{2,v\to -v}(\theta_2) \to I^{(0)}_2(\theta_2) \quad \text{at} \quad \varepsilon_1''/\varepsilon_1' \to 0 \quad \text{and} \quad i = 1,2. \tag{21}$$

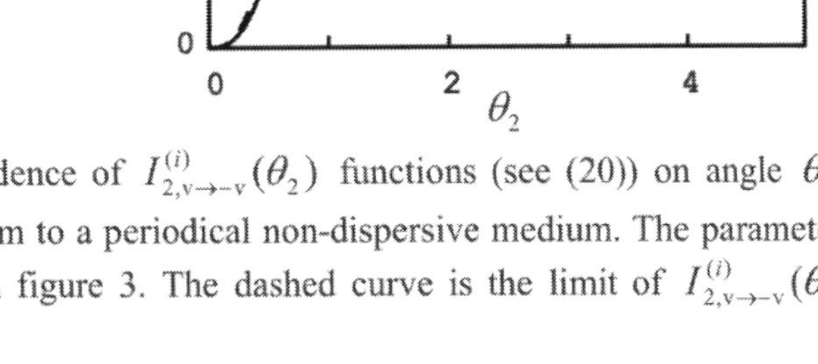

Figure 6. Dependence of $I^{(i)}_{2,v\to -v}(\theta_2)$ functions (see (20)) on angle θ_2 for an electron in flight from vacuum to a periodical non-dispersive medium. The parameters of system are the same, as those in figure 3. The dashed curve is the limit of $I^{(i)}_{2,v\to -v}(\theta_2)$ in the absence of

absorption.

Here $I_2^{(0)}(\theta_2)$ (dashed curve in figure 6) is the well known [5-7] angular distribution of TR in the forward direction for a particle flying to vacuum out of non-absorbing and non-dispersive medium with $\varepsilon' = 1.5$, $\varepsilon'' = 0$ and $\mu = 1$ (confer with (16)). Expressions (20), (21) are confirmed by our numerical calculations. In reality $I_{2,v\to-v}^{(i)} \cong I_2^{(0)}$ already for $\varepsilon_1''/\varepsilon_1' = 0.001$. So, the inequality $I_{2,v\to-v}^{(i)} < I_2^{(0)}$ in figure 6 is due to the absorption of radiation by the laminated medium on the one side, and the presence of formation zone for the radiation reflected by the periodical structure on the other side.

5. Conclusions

In the present work we have investigated the radiation from a charged particle flying out of a semi-infinite laminated medium to vacuum and back, - from vacuum to the laminated medium. In both the cases expressions were obtained for the spectral-angular distribution of radiation energy emitted to vacuum. In these expressions there were no limitations on the amplitude and variation profile of the dielectric permittivity $\varepsilon_1(z)$ of the laminated medium.

The results of appropriate numerical calculations indicate that:

- The spectral-angular distribution of the energy of parametric radiation from a relativistic particle along any separate branch of distribution is determined by the amplitude and profile of the variations of $\varepsilon_1(z)$.
- The spectral distribution of the energy of transition radiation of particle may be superimposed by the peaks from the diffracted beams of parametric radiation.
- Some part of transition radiation that is emitted in the forward direction at the flight of particle from vacuum to the laminated medium and which fall within forbidden bands of the periodical structure (except for the absorbed part of radiation) is completely thrown over by this structure back to vacuum.

The obtained results may be of use for controlling the parameters of charged particle radiation in laminated media, as well as for electron probing of periodical structures, for instance, of photonic crystals.

References

[1] Rullhusen P, Artru X and Dhez P 1998 *Novel Radiation Sources Using Relativistic Electrons* (Singapore: World Scientific)

[2] Potylitsyn A P 2009 *Radiation from Electrons in Periodic Structures* (Tomsk: NTL Press) (in Russian)

[3] Inoue K and Ohtaka K eds 2004 *Photonic Crystals: Physics, Fabrication and Applications* (Berlin: Springer-Verlag)

[4] Fainberg Y B and Khizhniak N A 1957 *Zh. Eksp. Teor. Fiz.* **32** 883 (1957 *Sov. Phys. JETP* **5** 720)

[5] Ter-Mikaelian M L 1972 *High Energy Electromagnetic Processes in Condensed Matter* (New York: Wiley)

[6] Ginzburg V L and Tsytovich V N 1990 *Transition Radiation and Transition Scattering* (Bristol: Adam Hilger)

[7] Garibian G M and Yang C 1983 *Transition Radiation* (Yerevan: AN Arm. SSR Press) (in Russian)

[8] Grigoryan L Sh, Mkrtchyan A H and Saharian A A 1998 *Nucl. Instr. and Meth.* B **145** 197

[9] Flugge S 1971 *Practical Quantum Mechanics 1* (Berlin: Springer-Verlag)

Acknowledgments

L Sh Grigoryan is thankful to X Artru, K A Ispirian, A P Potylitsyn, N F Shulga and A A Tishchenko for valuable comments. This work was supported in part by the Grant No.077 from Ministry of Education and Science of RA.

Relativistic electron PXR and FPXR yield ratio

S Blazhevich, A Noskov
Belgorod State University, Belgorod, Russia

E-mail: blazh@bsu.edu.ru

Abstract. The parametric X-radiation of relativistic electron crossing a single crystal plate is considered on the basis of dynamic diffraction theory in Laue geometry both in Bragg (PXR) and forward (FPXR) directions. Analytical expressions for spectral-angular density of these radiations in indicated directions are derived in general case of electron coulomb field asymmetric reflection from the target surface. The ratio of contributions of these radiation mechanisms into the total yield of the radiation is investigated. It is shown that the ratio of contributions of these radiation mechanisms can change sharply depending on the value of the reflection asymmetry.

1. Introduction

When a fast charged particle crosses a single crystal, its coulomb field scatters on the system of parallel atomic planes generating parametric X-radiation (PXR) [1-3]. The theory of PXR of relativistic particle in a crystal forecasts the radiation not only in Bragg scattering direction, but also in the direction of radiating particle velocity (FPXR) [4-6], which represents an appearance of dynamical diffraction in PXR. It is known that some attempts of the FPXR experimental research were made [7-11], but the first report about its experimental discovery has appeared recently in ref. [10], where the method of TR background suppression was used in vicinity of Bragg frequency based on destructive interference of TR generated on the entrance and on the escape surfaces of the crystal plate (as target) accordingly. But a spectrally narrow peak arisen near Bragg frequency due to dynamical diffraction [12], which could take place in the FPXR experiment was not analyzed in the work [10]. In the experiment [11] the relativistic electron X-radiation was registered from a thick single non-absorptive crystal target in the condition of FPXR generation, nevertheless the sought-for peak became weakly apparent on the background of the radiation generated by scattered electrons in the structural material of the experimental setup. Thus the theoretical investigations of FPXR properties and the optimization of viewing conditions for experimental observation of this dynamical effect remain topical up to date.

The detailed theoretical description of dynamical effect of FPXR and its accompanying transition radiation (TR), as a background, was done for the case of symmetrical reflection in the works [13-15]. Under condition of asymmetrical reflection of the relativistic electron coulomb field the PXR and TR were theoretically investigated in ref.-s [16-18] and FPXR in Bragg geometry in ref. [19]. The above-mentioned works showed that the spectral-angular density of these radiation mechanisms depends essentially on the reflection asymmetry and revealed the effects associated with it.

In the present work the analytical expressions for spectral-angular density of FPXR is derived in the framework of two-wave approximation of dynamic diffraction theory [20] in general case of asymmetrical field reflection. An expression for spectra-angular density of PXR was derived recently in the work [17] describing the dependence of the radiation parameters on reflection asymmetry. In the present work on the basis of the expressions obtained the ratio of FPXR and PXR yield was investigated depending on the angle between the reflecting atomic planes and crystal plate surface, i.e. on the reflection asymmetry.

2. FPXR amplitude

Let us consider the radiation of a fast charged particle crossing a single crystal plate with a constant velocity **V** (look in figure 1.).We will use the same table of symbols as in the work [17].

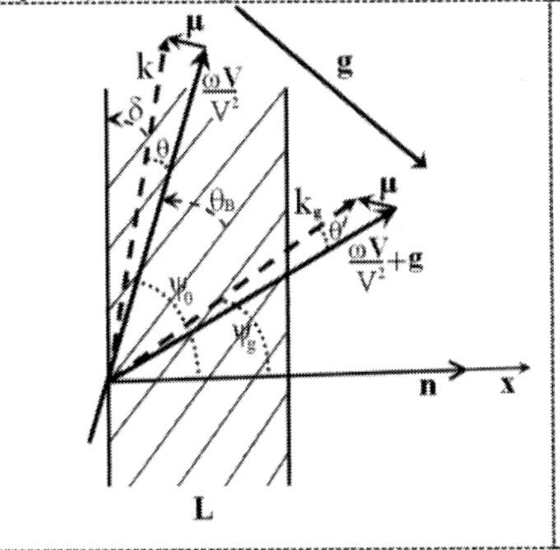

Figure 1. Radiation process geometry. θ' is the angle of radiation of a photon, θ_B is Bragg angle (angle between velocity **V** of the electron and diffracting atomic planes), δ is the angle between the plate surface and diffracting atomic planes of the crystal, **k** and $\mathbf{k}_g = \mathbf{k} + \mathbf{g}$ are wave vectors of incident and diffracted photons.

While solving the problem, let us consider an equation for a Fourier image of an electromagnetic field

$$\mathbf{E}(\mathbf{k},\omega) = \int dt\, d^3r\, \mathbf{E}(\mathbf{r},t)\exp(i\omega t - i\mathbf{kr}) \qquad (1)$$

Since the field of a relativistic particle could, to a good accuracy, be taken as being transverse, the incident $\mathbf{E}_0(\mathbf{k},\omega)$ and diffracted $\mathbf{E}_\mathbf{g}(\mathbf{k},\omega)$ electromagnetic waves are determined by two amplitudes with different values of transverse polarization

$$\begin{aligned}\mathbf{E}_0(\mathbf{k},\omega) &= E_0^{(1)}(\mathbf{k},\omega)\mathbf{e}_0^{(1)} + E_0^{(2)}(\mathbf{k},\omega)\mathbf{e}_0^{(2)},\\ \mathbf{E}_\mathbf{g}(\mathbf{k},\omega) &= E_\mathbf{g}^{(1)}(\mathbf{k},\omega)\mathbf{e}_1^{(1)} + E_\mathbf{g}^{(2)}(\mathbf{k},\omega)\mathbf{e}_1^{(2)}.\end{aligned} \qquad (2)$$

The unit vectors of polarization $\mathbf{e}_0^{(1)}$, $\mathbf{e}_0^{(2)}$, $\mathbf{e}_1^{(1)}$ and $\mathbf{e}_1^{(2)}$ are chosen in the following way. Vectors $\mathbf{e}_0^{(1)}$ and $\mathbf{e}_0^{(2)}$ are perpendicular to vector $\mathbf{k}$, and vectors $\mathbf{e}_1^{(1)}$ and $\mathbf{e}_1^{(2)}$ are perpendicular to vector $\mathbf{k}_g = \mathbf{k} + \mathbf{g}$ .Vectors $\mathbf{e}_0^{(2)}$, $\mathbf{e}_1^{(2)}$ are situated on the plane of vectors $\mathbf{k}$ и $\mathbf{k}_g$ (π -polarization) and $\mathbf{e}_0^{(1)}$, $\mathbf{e}_1^{(1)}$ are perpendicular to this plane (σ -polarization); g is vector of the reciprocal lattice, defining a set of reflecting atomic planes. The system of equation for the Fourier transform images of electromagnetic field in two-wave approximation of dynamic theory of diffraction has the following view [21]:

$$\begin{cases}(\omega^2(1+\chi_0) - k^2)E_0^{(s)} + \omega^2\chi_{-\mathbf{g}}C^{(s,\tau)}E_\mathbf{g}^{(s)} = 8\pi^2 ie\omega\theta VP^{(s)}\delta(\omega - \mathbf{kV}),\\ \omega^2\chi_\mathbf{g}C^{(s,\tau)}E_0^{(s)} + (\omega^2(1+\chi_0) - k_\mathbf{g}^2)E_\mathbf{g}^{(s)} = 0,\end{cases} \qquad (3)$$

where $\chi_0 = \chi_0' + i\chi_0''$ is the average dielectric susceptibility, $\chi_\mathbf{g}$, $\chi_{-\mathbf{g}}$ are the coefficients of the Fourier expansion of the dielectric susceptibility of a crystal over the reciprocal lattice vectors $\mathbf{g}$.

The values $C^{(s)}$ and $P^{(s)}$ are defined in the system (3) as

$$C^{(s)} = \mathbf{e}_0^{(s)}\mathbf{e}_1^{(s)},\ C^{(1)} = 1,\ C^{(2)} = \cos 2\theta_B,\quad P^{(s)} = \mathbf{e}_0^{(s)}(\boldsymbol{\mu}/\mu),\ P^{(1)} = \sin\varphi,\ P^{(2)} = \cos\varphi. \qquad (4)$$

$\mathbf{\mu} = \mathbf{k} - \omega\mathbf{V}/V^2$ is the virtual photon momentum vector component perpendicular to particle velocity vector V ($\mu = \omega\theta/V$, where $\theta \ll 1$ is the angle between vectors k and V), θ_B is the angle between electron velocity and a set of atomic planes in the crystal (Bragg angle), φ is the azimuth angle, counted off from the plane formed by vectors V and g, the value of the reciprocal lattice vector is shown by expression $g = 2\omega_B \sin\theta_B / V$, ω_B is Bragg's frequency. The angle between vector $\frac{\omega\mathbf{V}}{V^2} + \mathbf{g}$ and diffracted wave vector k_g is defined as θ'. The equation system (3) under $s = 1$ describes the fields of σ - polarization, and under $s = 2$ the fields of π -polarization.

Let us solve the dispersion equation for x-waves in crystal following from the system (3) :

$$(\omega^2(1+\chi_0) - k^2)(\omega^2(1+\chi_0) - k_g^2) - \omega^4 \chi_{-g}\chi_g C^{(s)^2} = 0, \tag{5}$$

using standard methods of dynamic theory [20].

Let us search for the wave vectors projection k and k_g to the axis X, aligned with the vector n (see figure 1.) as:

$$k_x = \omega\cos\psi_0 + \frac{\omega\chi_0}{2\cos\psi_0} + \frac{\lambda_0}{\cos\psi_0}, \quad k_{gx} = \omega\cos\psi_g + \frac{\omega\chi_0}{2\cos\psi_g} + \frac{\lambda_g}{\cos\psi_g}. \tag{6}$$

For this purpose we will use the well-known relation, connecting dynamic addition agents λ_0 and λ_g for X-waves [20]:

$$\lambda_g = \frac{\omega\beta}{2} + \lambda_0 \frac{\gamma_g}{\gamma_0}, \tag{7}$$

where $\beta = \alpha - \chi_0\left(1 - \frac{\gamma_g}{\gamma_0}\right)$, $\alpha = \frac{1}{\omega^2}(k_g^2 - k^2)$, $\gamma_0 = \cos\psi_0$, $\gamma_g = \cos\psi_g$, ψ_0 is the angle between incident wave vector k and vector normal to the plate surface n, ψ_g is the angle between wave vector k_g and the vector n (see figure 1). The modules of vectors k and k_g are:

$$k = \omega\sqrt{1+\chi_0} + \lambda_0, \quad k_g = \omega\sqrt{1+\chi_0} + \lambda_g. \tag{8}$$

Assumed that $k_{||} \approx \omega\sin\psi_0$ and $k_{g||} \approx \omega\sin\psi_g$, we will obtain

$$\lambda_0^{(1,2)} = \omega\frac{\gamma_0}{4\gamma_g}\left(-\beta \pm \sqrt{\beta^2 + 4\chi_g\chi_{-g}C^{(s)^2}\frac{\gamma_g}{\gamma_0}}\right), \tag{9}$$

$$\lambda_g^{(1,2)} = \frac{\omega}{4}\left(\beta \pm \sqrt{\beta^2 + 4\chi_g\chi_{-g}C^{(s)^2}\frac{\gamma_g}{\gamma_0}}\right). \tag{10}$$

As $|\lambda_0| \ll \omega$ and $|\lambda_g| \ll \omega$, we can show that $\theta \approx \theta'$ (see in figure 1), and hereinafter will use θ in the all occasions.

The solution of the combined equations (3) gives us the relativistic particle field in the vacuum in front of the crystal as

$$E_0^{(s)vacI}=\frac{8\pi^2 ieV\theta P^{(s)}}{\omega}\frac{1}{(\gamma_0/\gamma_g)\left(-\chi_0-2\lambda_g\gamma_0/\omega\gamma_g+\beta\gamma_0/\gamma_g\right)}\delta\left(\lambda_g-\lambda_g^*\right), \quad (11)$$

the same field inside the crystal as

$$E_0^{(s)cr}=\frac{8\pi^2 ieV\theta P^{(s)}}{\omega}\frac{-\omega^2\beta-2\omega\lambda_0\gamma_g/\gamma_0}{4(\gamma_g/\gamma_0)\left(\lambda_0-\lambda_0^{(1)}\right)\left(\lambda_0-\lambda_0^{(2)}\right)}\delta(\lambda_0-\lambda_0^*)+$$
$$+E_0^{(s)^{(1)}}\delta(\lambda_0-\lambda_0^{(1)})+E_0^{(s)^{(2)}}\delta(\lambda_0-\lambda_0^{(2)}) \quad (12)$$

and the field in the vacuum behind the crystal as

$$E_0^{(s)vacII}=\frac{8\pi^2 ieV\theta P^{(s)}}{\omega}\frac{1}{-\chi_0-2\lambda_0/\omega}\delta(\lambda_0-\lambda_0^*)+E_0^{(s)Rad}\delta\left(\lambda_0+\omega\chi_0/2\right), \quad (13)$$

where $E_0^{(s)Rad}$ is the coherent radiation amplitude in the electron velocity direction.

The following expression bounding the diffracted and incident fields inside of the crystal results from second equation of system (3):

$$E_0^{(s)cr}=\left(2\omega\lambda_g/\omega^2\chi_g C^{(s)}\right)E_g^{(s)cr}. \quad (14)$$

For definition of amplitude $E_0^{(s)Rad}$ we will use the ordinary boundary conditions on the inlet and outlet surfaces of the crystal plate:

$$\int E_0^{(s)vacI}d\lambda_0=\int E_0^{(s)cr}d\lambda_0, \quad \int E_g^{(s)cr}d\lambda_0=0,$$
$$\int E_0^{(s)cr}\exp\left(i(\lambda_0/\gamma_0)L\right)d\lambda_0=\int E_0^{(s)vacII}\exp\left(i(\lambda_0/\gamma_0)L\right)d\lambda_0. \quad (15)$$

Hence we will obtain the expression for the field of the radiation:

$$E_0^{(s)Rad}=\frac{8\pi^2 ieV\theta P^{(s)}}{\omega}\frac{\exp\left[i\left(\omega\chi_0/2+\lambda_g^*\right)L/\gamma_g\right]}{2\left(\gamma_g/\gamma_0\right)\left(\lambda_0^{(1)}-\lambda_0^{(2)}\right)}\times$$

$$\times\Big[\left(-\omega^2\beta-2\omega\lambda_0^{(2)}\gamma_g/\gamma_0\right)\left(\left(-\omega\chi_0-2\lambda_0^*\right)^{-1}+\left(2(\lambda_0^*-\lambda_0^{(2)})\right)^{-1}\right)\left(\exp\left(iL(\lambda_0^{(2)}-\lambda_0^*)/\gamma_0\right)-1\right)-$$

$$-\left(-\omega^2\beta-2\omega\lambda_0^{(1)}\gamma_g/\gamma_0\right)\left(\left(-\omega\chi_0-2\lambda_0^*\right)^{-1}+\left(2(\lambda_0^*-\lambda_0^{(1)})\right)^{-1}\right)\left(\exp\left(iL(\lambda_0^{(1)}-\lambda_0^*)/\gamma_0\right)-1\right)\Big]. \quad (16)$$

The terms in square brackets in (16) correspond to two branches of solution for X-ray waves excited in the crystal.

To make the FPXR reflex appear, at least one of the following inequalities must be hold:

$$\mathrm{Re}\left(\lambda_0^*-\lambda_0^{(1)}\right)=0, \quad \mathrm{Re}\left(\lambda_0^*-\lambda_0^{(2)}\right)=0, \quad (17)$$

i.e., in (16) the real part of denominator of at least one of the terms in square bracket must be vanishing. For the further analyses we will represent the dynamic addition (9) in the following view:

$$\lambda_0^{(1,2)}=\left(\omega\left|\chi_g'\right|C^{(s)}/2\varepsilon\right)\left(-\xi^{(s)}+i\rho^{(s)}(1-\varepsilon)/2\pm\right.$$

$$\left.\pm\sqrt{\xi^{(s)^2}+\varepsilon-2i\rho^{(s)}\left((1-\varepsilon)\xi^{(s)}/2+\kappa^{(s)}\varepsilon\right)-\rho^{(s)^2}\left((1-\varepsilon)^2/4+\kappa^{(s)^2}\varepsilon\right)}\right), \quad (18)$$

where

$\rho^{(s)} = \chi_0'' / |\chi_{\mathbf{g}}'| C^{(s)}$, $\quad \varepsilon = \cos\psi_{\mathbf{g}} / \cos\psi_0$, $\quad \kappa^{(s)} = \chi_{\mathbf{g}}'' C^{(s)} / \chi_0''$, $\quad \xi^{(s)} = \eta^{(s)}(\omega) + (1+\varepsilon)/2\nu^{(s)}$,

$\eta^{(s)}(\omega) = (2\sin^2\theta_B / |\chi_{\mathbf{g}}'| C^{(s)})(\omega_B(1+\theta\cos\varphi\cot\theta_B)/\omega - 1)$.

The asymmetry parameter ε can be represented as $\varepsilon = \sin(\delta+\theta_B)/\sin(\delta-\theta_B)$, where θ_B is the angle between electron velocity and diffracting atomic plane. Hence we can see that parameter ε increases when the angle of electron incidence $\delta-\theta_B$ decreases and vice versa (see figure 2).

Since in the region of X-ray frequencies the inequality $2\sin^2\theta_B / V^2 |\chi_{\mathbf{g}}'| C^{(s)} >> 1$ is satisfied, then $\eta^{(s)}(\omega)$ is a fast function of the frequency ω, so for further analysis of the properties of the PXR spectra it is convenient to treat $\eta^{(s)}(\omega)$ as a spectral variable characterizing the frequency ω.

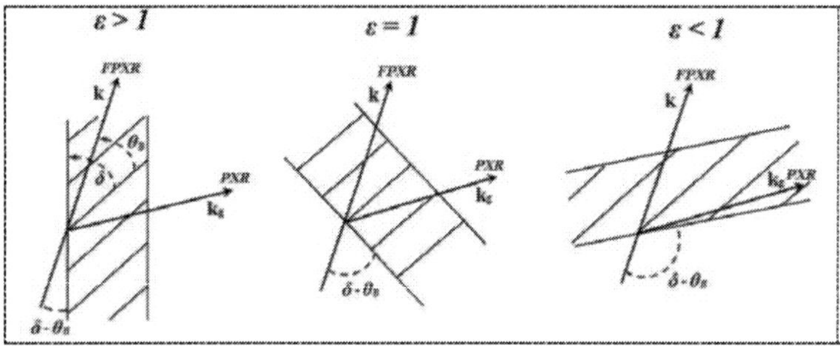

Figure 2. Circuitry of asymmetric ($\varepsilon > 1$, $\varepsilon < 1$) reflection of the radiation from single crystal plate. The case of $\varepsilon = 1$ correspond with symmetric reflection

Using (18) let us represent the expression for field amplitude of radiation along the electron velocity vector as two summands:

$$E_0^{(s)Rad} = E_{FPXR}^{(s)} + E_{TR}^{(s)}, \tag{19a}$$

$$E_{FPXR}^{(s)} = \frac{4\pi^2 ieV}{\omega} \frac{\theta P^{(s)}}{\gamma^{-2}+\theta^2-\chi_0} \frac{1}{\sqrt{\xi^{(s)2}+\varepsilon}} \times$$

$$\times \left(\frac{1-\exp\left[-ib^{(s)}\Sigma^{(2)} - b^{(s)}\rho^{(s)}\Delta^{(2)}\right]}{\Sigma^{(2)} - i\rho^{(s)}\Delta^{(2)}} - \frac{1-\exp\left[-ib^{(s)}\Sigma^{(1)} - b^{(s)}\rho^{(s)}\Delta^{(1)}\right]}{\Sigma^{(1)} - i\rho^{(s)}\Delta^{(1)}} \right) \exp\left[i\omega L(\gamma^{-2}+\theta^2)/2\gamma_0\right]$$

(19b)

$$E_{TR}^{(s)} = \frac{4\pi^2 ieV\theta P^{(s)}}{\omega} \left(\frac{1}{\gamma^{-2}+\theta^2} - \frac{1}{\gamma^{-2}+\theta^2-\chi_0} \right) \times$$

$$\left(\left(1 - \xi^{(s)} / \sqrt{\xi^{(s)^2}+\varepsilon}\right)\left(1-\exp\left[-ib^{(s)}\Sigma^{(2)} - b^{(s)}\rho^{(s)}\Delta^{(2)}\right]\right) + \right.$$

$$\left. + \left(1 + \xi^{(s)} / \sqrt{\xi^{(s)^2}+\varepsilon}\right)\left(1-\exp\left[-ib^{(s)}\Sigma^{(1)} - b^{(s)}\rho^{(s)}\Delta^{(1)}\right]\right) \right) \exp\left[i\omega L(\gamma^{-2}+\theta^2)/2\gamma_0\right] \tag{19c}$$

where

$$\Delta^{(1,2)} = (\varepsilon+1)/2\varepsilon \mp (1-\varepsilon)\xi^{(s)} / 2\varepsilon\sqrt{\xi^{(s)^2}+\varepsilon} \mp \kappa^{(s)} / \sqrt{\xi^{(s)^2}+\varepsilon},$$

$$\Sigma^{(1)} = \sigma^{(s)} + \left(\xi^{(s)} - \sqrt{\xi^{(s)2}+\varepsilon}\right)/\varepsilon, \quad \Sigma^{(2)} = \sigma^{(s)} + \left(\xi^{(s)} + \sqrt{\xi^{(s)2}+\varepsilon}\right)/\varepsilon.$$

$$\sigma^{(s)} = \left(\theta^2 + \gamma^{-2} - \chi_0'\right) / |\chi_{\mathbf{g}}'| C^{(s)} \equiv \left(\theta^2 / |\chi_0'| + \gamma^{-2} / |\chi_0'| + 1\right) / \nu^{(s)}, \nu^{(s)} = |\chi_{\mathbf{g}}'| C^{(s)} / |\chi_0'|$$

$$b^{(s)} = \omega |\chi_{\mathbf{g}}'| C^{(s)} L / 2\gamma_0. \tag{20}$$

Parameter $b^{(s)}$ can be presented in view

$$b^{(s)} = L / 2\sin(\delta - \theta_B) L_{ext}^{(s)} .$$

Hence it follows that parameter $b^{(s)}$ is equal to half of the relativistic electron path length in the plate expressed in extinction lengths.

The expression (19b) represents the field amplitude for FPXR, and (19c) represents the TR generated on inlet and outlet surfaces of the plate. The FPXR yield is formed basically at the expense of one of the branches in the solution for X-ray waves generated in the crystal that corresponds to the second summand in (19b) because only there the real part of the denominator must be vanishing:

$$\Sigma^{(1)} \equiv \sigma^{(s)} + \left(\xi^{(s)} - \sqrt{\xi^{(s)2} + \varepsilon} \right) / \varepsilon = 0 . \qquad (21)$$

The equation (21) solution defines a frequency ω_* in whose vicinity the spectrum of FPXR photons radiated at a fixed observation angle is concentrated.

We will consider the case of a thin target $\left(b^{(s)} \rho^{(s)} << 1 \right)$ when the absorption coefficient $\rho^{(s)}$ can be neglected. For the graphic extraction of dynamical effects in coherent radiation processes we will use a crystal plate of such a thickness L that the electron path in the plate $L / \sin(\delta - \theta_B)$ will much exceed the extinction length $L_{ext}^{(s)} = 1 / \omega |\chi_{\mathbf{g}}'| C^{(s)}$ of x-ray waves in the crystal, i.e. $b^{(s)} >> 1$.

3. Spectral-angular density of FPXR

Substituting (19b) into the well known [21] expression for spectral-angular X-radiation density

$$\omega \frac{d^2 N}{d\omega d\Omega} = \omega^2 (2\pi)^{-6} \sum_{s=1}^{2} \left| E_0^{(s)Rad} \right|^2 \qquad (22)$$

we will obtain for thin crystal the expressions, accordingly describing the contributions of FPXR:

$$\omega \frac{d^2 N_{FPXR}^{(s)}}{d\omega d\Omega} = \frac{e^2}{4\pi^2} P^{(s)2} \frac{\theta^2}{(\theta^2 + \gamma^{-2} - \chi_0')^2} R_{FPXR}^{(s)} , \qquad (23a)$$

$$R_{FPXR}^{(s)} = \frac{4}{\xi^{(s)2} + \varepsilon} \sin^2 \left(\frac{b^{(s)}}{2} \left(\sigma^{(s)} + \left(\xi^{(s)} - \sqrt{\xi^{(s)2} + \varepsilon} \right) / \varepsilon \right) \right) \left(\sigma^{(s)} + \left(\xi^{(s)} - \sqrt{\xi^{(s)2} + \varepsilon} \right) / \varepsilon \right)^{-2} , \qquad (23b)$$

The expressions obtained (23) are the main result of the present work, as they let us investigate all the spectral-angular characteristics of FPXR.

4. Spectral-angular density of the radiation in Bragg direction

The expression describing the contributions of FPXR was obtained in the work [17]:

$$\omega \frac{d^2 N_{PXR}^{(s)}}{d\omega d\Omega} = \frac{e^2}{4\pi^2} P^{(s)2} \frac{\theta^2}{(\theta^2 + \gamma^{-2} - \chi_0')^2} R_{PXR}^{(s)} , \qquad (24a)$$

$$R_{PXR}^{(s)} = 4 \left(1 - \frac{\xi^{(s)}}{\sqrt{\xi^{(s)2} + \varepsilon}} \right)^2 \sin^2 \left(\frac{b^{(s)}}{2} \left(\sigma^{(s)} + \left(\xi^{(s)} - \sqrt{\xi^{(s)2} + \varepsilon} \right) / \varepsilon \right) \right) \left(\sigma^{(s)} + \left(\xi^{(s)} - \sqrt{\xi^{(s)2} + \varepsilon} \right) / \varepsilon \right)^{-2} . \qquad (24b)$$

5. Ratio of the PXR and FPXR yields

The PXR and FPXR spectra described accordingly by $R_{PXR}^{(s)}$ and $R_{FPXR}^{(s)}$ depend on parameter ε, therefore the effect of asymmetry on the ratio of these radiation yields derived from (23b) and (24b) is of interest:

$$R^{(s)}_{PXR} / R^{(s)}_{FPXR} = \left(\sqrt{\xi^{(s)2} + \varepsilon} - \xi^{(s)} \right)^2 \qquad (25)$$

When the asymmetry parameter ε decreases, i.e. the angle $\delta - \theta_B$ of the relativistic electron incident on the crystal target surface increases (see in figure 2), the PXR in forward direction (FPXR) intensity increases and can considerably exceed the PXR in Bragg direction.

$$R^{(s)}_{PXR} << R^{(s)}_{FPXR} \, . \qquad (26)$$

Let us consider how the intensity of each of the radiations changes separately. The curves calculated by formulas (23b) and (24b) and represented in figure 3 show that the FPXR peak amplitude increases and that of PXR decreases when parameter ε decreases. The curves in figure 3 are plotted for fixed values of the observation angle, the length of path and the energy of the radiating particle.

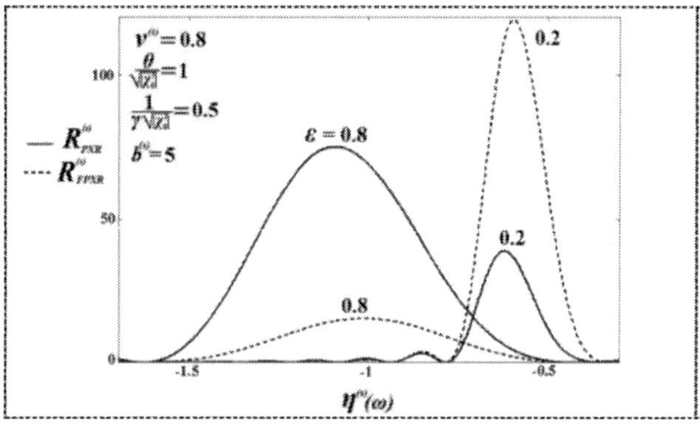

Figure 3. Ratio of PXR and FPXR spectra under different reflection asymmetry parameter ε.

The ratio (25) also depends on observation angle θ, radiating particle energy defined by Lorentz factor γ and parameter $\nu^{(s)}$. Using (21) we will find the expression for frequency ω_* near which spectrum of FPXR is concentrated and write the ratio (25) for that frequency:

$$R^{(s)}_{PXR} / R^{(s)}_{FPXR} = \varepsilon^2 \left(\theta^2 / |\chi_0'| + \gamma^{-2} / |\chi_0'| + 1 \right)^2 / \mathbf{v}^{(s)2} \, . \qquad (27)$$

In maximum of the parametric radiation angular density, when $\theta = \sqrt{\gamma^{-2} + |\chi_0'|}$ this ratio will have the following view:

$$R^{(s)}_{PXR} / R^{(s)}_{FPXR} = 2\varepsilon^2 \left(\gamma^{-2} / |\chi_0'| + 1 \right)^2 / \mathbf{v}^{(s)2} \, . \qquad (28)$$

As the value of parameter $\nu^{(s)}$ is always less then unity, it results from (28) that when $\gamma^2 |\chi_0'| << 1$ than PXR yield will exceed the yield of FPXR. In the case when $\gamma^2 |\chi_0'| \geq 1$ (or even some less then unity) the ratio (28) will strongly depend on asymmetry parameter ε.

To consider the influence of the reflection asymmetry on the ratio of angular densities of these radiations we will integrate the expression (23) and (24) over frequency function $\eta^{(s)}(\omega)$:

$$\frac{dN_i^{(s)}}{d\Omega} = \frac{e^2 P^{(s)^2}}{8\pi^2 \sin^2 \theta_B} F_i^{(s)}, \quad F_i^{(s)} = \nu^{(s)} \left[\frac{\theta^2}{|\chi_0'|} \Big/ \left(\frac{\theta^2}{|\chi_0'|} + \frac{1}{\gamma^2 |\chi_0'|} + 1 \right)^2 \right] \int_{-\infty}^{+\infty} R_i^{(s)} d\eta^{(s)}(\omega) \qquad (29)$$

$i = PXR, FPXR.$

The curves describing the angular density of PXR and FPXR plotted by formulas (29) are presented in figure 4 and figure 5. It is evident from these figures, that under condition $\varepsilon << 1$ the angular density of FPXR considerably exceeds the density of PXR.

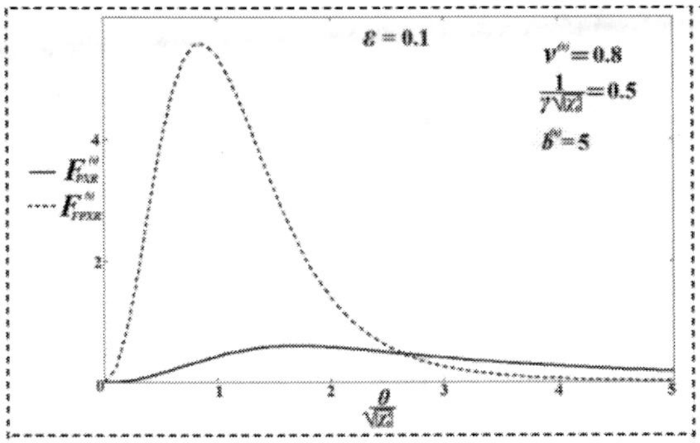

Figure 4. PXR and FPXR angular density under condition of reflection asymmetry (parameter $\varepsilon = 0.1$).

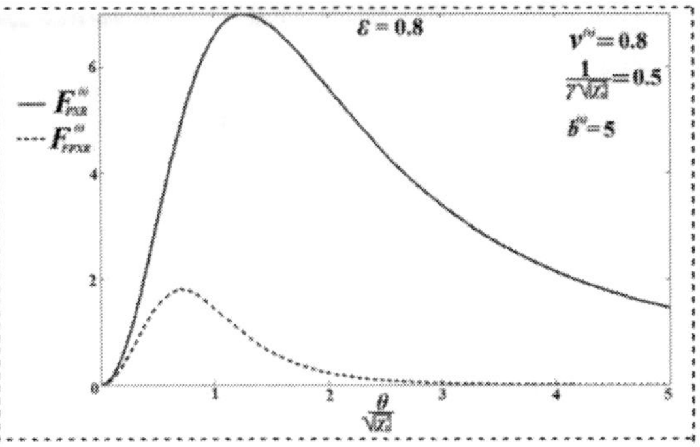

Figure 5. The same as in Fig.4 for another reflection asymmetry ($\varepsilon = 0.8$).

Conclusion

Based on two-wave approximation of the dynamic scattering theory, forward parametric X-ray radiation (FPXR) of a relativistic electron is investigated in a single crystal plate in a Laue scattering geometry. Analytical expressions for the spectral-angular distribution of FPXR and transition radiation (TR) are derived in general case of asymmetric reflection. It is shown that under fixed Bragg angle θ_B and path length $2b^{(s)}$ of relativistic electron in the crystal plate the ratio of FPXR and PXR yields considerably depends on the angle between reflected atomic planes and inlet surface of the plate δ, i.e. on reflection asymmetry parameter $\varepsilon = \sin(\delta+\theta_B)/\sin(\delta-\theta_B)$, at that in the case where the angle of the electron incidence on the surface of the crystal plate $\delta-\theta_B$ increases (i.e. ε decreases) the spectral-angular density of PXR decreases and that of FPXR increases and can exceed the density of PXR. In the present work the ratio of FPXR and PXR yields was investigated and it was shown that this ratio significantly depends on the reflection asymmetry, at that with decreasing asymmetry parameter ε the FPXR relative contribution increases for any observation angle. The results obtained can be used when preparing the experimental investigations of FPXR.

References

[1] Ter-Mikaelian M 1972 *High-Energy Electromagnetic Process in Condensed Media* (New York : Wiley)
[2] Garibian G, Yang C 1971 J. Exp. Theor. Phys. **61** 930
[3] Baryshevsky V, Feranchuk I 1971 J. Exp. Theor. Phys. **61** 944
[4] Garibian G, Yang C 1972 J. Exp. Theor. Phys. **63** 1198
[5] Baryshevsky V, Feranchuk I 1976 Phys. Lett. A **57** 183
[6] Baryshevsky V, Feranchuk I 1983 J. Physique (Paris) **44** 913
[7] Yuan Luke C, Alley P, Bamberger A et al. 1985 Nucl. Instr. and Meth. In Phys. Res. A **234** 426
[8] Kalinin B, Naumenko G, Padalko D et al. 2001 Nucl. Instr. and Meth. In Phys. Res. B **173** 253
[9] Kube G, Ay C, Backe H, Clawiter N at al. 2001 in Abstracts V International Symposium "Radiation from Relativistic Electrons in Periodic Structures", Lake Aya, Altai Mountains, Russia, 10-14 September
[10] Backe H, Clawiter N 2003 at al, in Proc. Of the Intern. Symp. on Channeling –Bent Crystals – Radiation Processes, , Frankfurt am Main, Germany, EP Systema Bt., Debrecen, p.41.
[11] Aleinik A, Baldin A, Bogomasova E, Vnukov I at al. 2004 J. Exp. Theor. Phys Letters **80** 447
[12] Imanishi N, Nasonov N, and Yajima K 2001 Nucl. Instr. And Meth. In Phys. Res. B **173** 227
[13] Kubankin A, Nasonov N, Sergienko V, Vnukov I 2003 Nucl. Instr. and Meth. In Phys. Res. B **201** 97

[14] Nasonov N, Noskov A 2003 Nucl. Instr. Meth. In Phys. Res. B **201** 67
[15] Kubankin A, Nasonov N, Noskov A, 2002 in Proc. 7 Int. Russian-Japanese Symposium "Interaction of fast charged particles with solids", Nov. 24-30, Kyoto, Japan, p.217-225
[16] Blazhevich S, Noskov A 2006 Nucl. Instr. Meth. In Phys. Res. B **252** 69
[17] Blazhevich S, Noskov A 2008 Nucl. Instr. Meth. In Phys. Res. B **266** 3770
[18] Blazhevich S, Noskov A 2008 Nucl. Instr. Meth. In Phys. Res. B **266** 3777
[19] Blazhevich S, Noskov A 2006 Russian Physics Journal **50** 574
[20] Pinsker Z 1984 *Dynamic Scattering of X-rays in Crystals* (Berlin: Springer)
[21] Bazylev V, Zhevago N 1987 *Emission From Fast Particles Moving in a Medium and External Fields* Moscow Nauka, 1987

Appropriateness of kinematical approach in description of parametric X-radiation of relativistic electron in a single crystal

S Blazhevich, A Noskov
Belgorod State University, Belgorod, Russia

E-mail: blazh@bsu.edu.ru

Abstract. Contrastive analysis of the formulas for parametric x-ray radiation (PXR) in the dynamic and kinematic approximation is made in the present work. The criteria for the usage of the kinematic formulas are determined on the basis of the dynamic theory. Specifically it was shown that the dynamic effects of PXR must be taken into account even in case of a thin non-absorptive crystal. The published results of the experiment on relativistic electron PXR on Mainz microtron MAMI are discussed. The manifestation of the effect of abnormal photo absorption (the Borrmann effect) in PXR is demonstrated in the experiment.

1. Introduction

When a fast charged particle crosses a single crystal, its Coulomb field scatters on a system of parallel atomic planes of the crystal, generating parametric X-radiation (PXR) [1-3].

Nowadays there are two approaches of PXR description: kinematic [4,5] and dynamic [2,3,6]. It should be noted that kinematic approach takes into account the interaction of each atom only with a primary or refracted wave in the crystal. In contrast to the dynamic one this approach does not take into account multiwave scattering. A considerable progress has been made recently in the description of coherent radiation of relativistic electrons in crystals using dynamic approach [7-11]. The parametric radiation reflection in the direction of electron velocity [12], observed in the experiments and unpredicted by the kinematic theory, clearly proves the validity of the dynamic theory of the coherent x-radiation. A considerable influence of the asymmetry of the radiation field reflection relative to the surface of the crystal plate on the spectral-angular density of the radiation should be noted. According to the kinematic theory the asymmetry influences only the relation of the path of the charged particle and the radiated photon in the plate. The PXR dynamic theory predicts a considerable influence of the reflection asymmetry on the process of radiation, which leads to a change in PXR spectrum [10]. Thus, the problem of the limits of applicability of the kinematic theory of relativistic electron PXR the paper deals with is rather topical.

Previously an expression was derived in two-wave approximation describing PXR spectral-angular density for the case of asymmetric reflection on the crystal plate [10] and the peculiarities of the process of coherent radiation connected with it were revealed.

In the present work a contrastive analysis of the formulas for dynamic and kinematic approach is made both for a thin and thick crystal in order to determine the criteria limiting the applicability of the kinematic theory formulas.

2. PXR spectral – angular distribution

Let us consider the radiation of a fast charged particle crossing a single crystal plate with a constant velocity V (the radiation process geometry look in figure 1. in [13]).

Let us write the expression for PXR spectral-angular density based on the dynamic diffraction theory [10] in the following convenient form:

$$\omega\frac{d^2N^{(s)}_{ПРИ}}{d\omega d\Omega}=\frac{e^2}{4\pi^2}\frac{P^{(s)^2}\theta^2}{(\theta^2+\gamma^{-2}-\chi'_0)^2}R^{(s)}_{ПРИ}, \tag{1a}$$

$$R^{(s)}_{ПРИ}=\left(1-\xi^{(s)}/\sqrt{\xi^{(s)2}+\varepsilon}\right)^2\frac{1+e^{-2b^{(s)}\rho^{(s)}\Delta^{(s)}}-2e^{-b^{(s)}\rho^{(s)}\Delta^{(s)}}\cos\left(b^{(s)}\left(\sigma^{(s)}+\left(\xi^{(s)}-\sqrt{\xi^{(s)2}+\varepsilon}\right)/\varepsilon\right)\right)}{\left(\sigma^{(s)}+\left(\xi^{(s)}-\sqrt{\xi^{(s)2}+\varepsilon}\right)/\varepsilon\right)^2+\rho^{(s)2}\Delta^{(s)2}}, \tag{1b}$$

where

$$\Delta^{(s)}=(\varepsilon+1)/2\varepsilon-(1-\varepsilon)\xi^{(s)}/2\varepsilon\sqrt{\xi^{(s)^2}+\varepsilon}-\kappa^{(s)}/\sqrt{\xi^{(s)^2}+\varepsilon}$$

$$\sigma^{(s)}=\left(\theta^2/|\chi'_0|+\gamma^{-2}/|\chi'_0|+1\right)/\nu^{(s)},\nu^{(s)}=|\chi'_{\rm g}|C^{(s)}/|\chi'_0|,\ b^{(s)}=\omega|\chi'_{\rm g}|C^{(s)}L/2\gamma_0. \tag{2}$$

Having integrated (1) on frequency function $\eta^{(s)}(\omega)$, let us write the following expression, describing PXR angular density:

$$\frac{dN^{(s)}_{PXR}}{d\Omega}=\frac{e^2|\chi'_{\rm g}|C^{(s)}}{8\pi^2\sin^2\theta_B}\frac{P^{(s)^2}\theta^2}{(\theta^2+\gamma^{-2}-\chi'_0)^2}\int_{-\infty}^{+\infty}R^{(s)}_{PXR}d\eta^{(s)}(\omega). \tag{3}$$

The expression, describing PXR spectrum (1b) in case of a thin target $\left(b^{(s)}\rho^{(s)}<<1\right)$, when the absorption coefficient $\rho^{(s)}$ can be neglected, takes on the following form:

$$R^{(s)}_{PXR}=4\left(1-\frac{\xi^{(s)}}{\sqrt{\xi^{(s)2}+\varepsilon}}\right)^2\sin^2\left(\frac{b^{(s)}}{2}\left(\sigma^{(s)}+\left(\xi^{(s)}-\sqrt{\xi^{(s)2}+\varepsilon}\right)/\varepsilon\right)\right)\left(\sigma^{(s)}+\left(\xi^{(s)}-\sqrt{\xi^{(s)2}+\varepsilon}\right)/\varepsilon\right)^{-2}. \tag{4}$$

For brighter manifestation of the dynamic effects, we shall consider a crystal plate of such a thickness when the electron path length in the plate $L/\sin(\delta-\theta_B)$ exceeds manifold the ray wave extinction length in the crystal $L^{(s)}_{ext}=1/\omega|\chi'_{\rm g}|C^{(s)}$ i.e. $b^{(s)}>>1$. As under the condition $b^{(s)}>>1$ the PXR spectrum peak is very narrow, to integrate (8) a well-known approximation $\sin^2(Tx)/x^2\to\pi T\delta(x)$ can be used. The formula for PXR angular distribution, resulting from (3) takes on the following form:

$$\frac{dN^{(s)din}_{PXR}}{d\Omega}=\frac{e^2\omega_B\chi'^2_{\rm g}C^{(s)2}P^{(s)2}}{4\pi\sin^2\theta_B}\cdot\frac{\varepsilon^2\theta^2}{\chi'^2_{\rm g}C^{(s)2}+\varepsilon(\theta^2+\gamma^{-2}-\chi'_0)^2}\cdot\frac{L}{\sin(\delta-\theta_B)}. \tag{5}$$

Let us write the following well-known expression for PXR angular density for kinematic approximation (for instance, see [5,14])

$$\frac{dN^{(s)kin}_{PXR}}{d\Omega}=\frac{e^2\omega_B\chi'^2_{\rm g}C^{(s)2}P^{(s)2}}{4\pi\sin^2\theta_B}\cdot\frac{\theta^2}{(\theta^2+\gamma^{-2}-\chi'_0)^2}\Gamma(\varepsilon),\ \Gamma(\varepsilon)=\frac{\varepsilon}{\omega_B\chi''_0}\left(1-e^{-\frac{\omega_B\chi''_0L}{\varepsilon\sin(\delta-\theta_B)}}\right). \tag{6}$$

One can see that the asymmetry coefficient ε which is contained in the kinematic expression, is only present in the geometrical factor $\Gamma(\varepsilon)$. In case of a thin non-absorptive crystal the geometrical factor is the electron path in the crystal plate

$$\Gamma(\varepsilon) = \frac{L}{\sin(\delta - \theta_B)}. \tag{7}$$

Hence the kinematic expression for PXR angular density takes on the following form:

$$\frac{dN_{PXR}^{(s)kin}}{d\Omega} = \frac{e^2 \omega_B \chi_g'^2 C^{(s)^2} P^{(s)^2}}{4\pi \sin^2 \theta_B} \cdot \frac{\theta^2}{(\theta^2 + \gamma^{-2} - \chi_0')^2} \frac{L}{\sin(\delta - \theta_B)}. \tag{8}$$

It is important to underline that in the dynamic formula in contrast to the kinematic one, the asymmetry (angle δ) is present not only in the expression for electron path in the plate, but also in the PXR angular distribution expression, which can be the basis for the comparison of kinematic and dynamic formulas for the case of a thin crystal.

3. Contrastive analysis of kinematic and dynamic formulas for a thin crystal

For further analysis it is convenient to present the expressions (5) and (8) in the following way:

$$\frac{dN_{PXR}^{(s)din}}{d\Omega} = \frac{e^2 \omega_B |\chi_g'| C^{(s)} P^{(s)^2}}{4\pi \sin^2 \theta_B} \cdot \frac{L}{\sin(\delta - \theta_B)} F^{din}, \tag{9a}$$

$$F^{din} = \nu^{(s)} \varepsilon^2 \frac{\theta^2}{|\chi_0'|} \left(\nu^{(s)^2} + \varepsilon \left(\frac{\theta^2}{|\chi_0'|} + \frac{1}{\gamma^2 |\chi_0'|} + 1 \right)^2 \right)^{-1}, \tag{9b}$$

$$\frac{dN_{PXR}^{(s)kin}}{d\Omega} = \frac{e^2 \omega_B |\chi_g'| C^{(s)} P^{(s)^2}}{4\pi \sin^2 \theta_B} \cdot \frac{L}{\sin(\delta - \theta_B)} F^{kin}, \tag{10a}$$

$$F^{kin} = \nu^{(s)} \frac{\theta^2}{|\chi_0'|} \left(\frac{\theta^2}{|\chi_0'|} + \frac{1}{\gamma^2 |\chi_0'|} + 1 \right)^{-2}. \tag{10b}$$

Functions (9b) and (10b) describe PXR angular density. It should be noted that in contrast to the kinematic approach, the dynamic one takes into account the influence of asymmetry on PXR angular part.

3.1 Symmetric reflection

At first let us consider a case of symmetric reflection ($\varepsilon = 1$). In case of low energy of the radiating particle $\gamma \leq 1/\sqrt{|\chi_0'|} \approx \omega/\omega_p$ (ω_p -plasma frequency) we can use the approximate equality $F_{din} \approx F_{kin}$ for all the observation angles θ. For relatively high energy $\gamma >> \omega/\omega_p$ for weak reflections (e.g. $\nu^{(s)} \approx 0{,}3$) $F_{din} \approx F_{kin}$. In case of strong reflections (e.g. $\nu^{(s)} \approx 0{,}9$) angular dependences, plotted on the basis of kinematic and dynamic formulas differ, which can be judged by the curves in figure 1. In this case the kinematic formulas will result in the error in photon absolute yield, therefore the results of rather accurate experiments on PXR absolute yield should be compared with the dynamic formula (5).

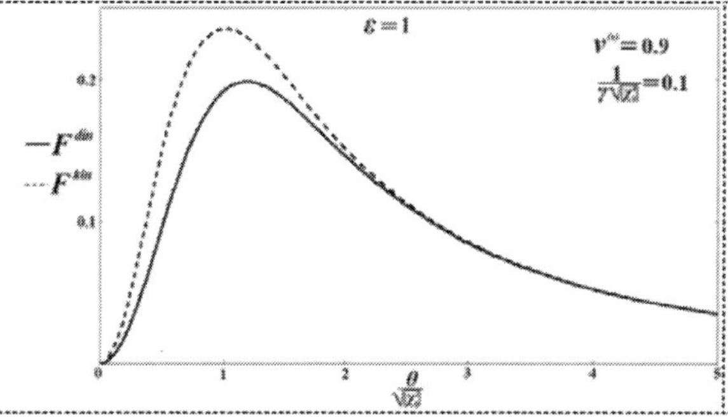

Figure 1. The comparison of PXR angular densities calculated on the basis of dynamic and kinematic formulas. Under the given circumstances the kinematic formula results in an error.

3.2 Asymmetric reflection

When the asymmetry growth which is caused by the decrease (when $\varepsilon > 1$) or increase (when $\varepsilon > 1$) of the electron incidence angle $\delta - \theta_B$ (see in figure 2 in [13]) the kinematic formula error increases, which is demonstrated by the curves represented in figure 2.

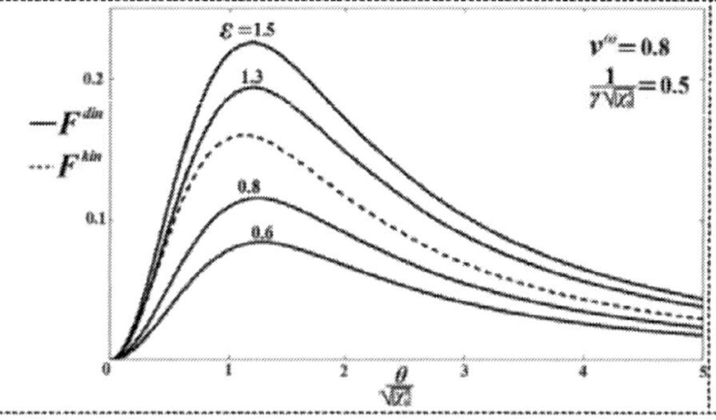

Figure 2. PXR angular density for different asymmetry. It demonstrates an increase of kinematic formula error under the growth of asymmetry.

In case of a strong asymmetry ($\varepsilon >> 1$), when the electron incidence angle to the plate surface ($\delta - \theta_B$) is relatively small, the expression $F^{din} / F^{kin} = \varepsilon$ follows from (9b) и (10b) both when relativistic electron energy is less or more than ω / ω_p.

Thus, in case of a strong asymmetry the PXR angular density is ε times more than it was predicted by the kinematic formula (see in figure 3).

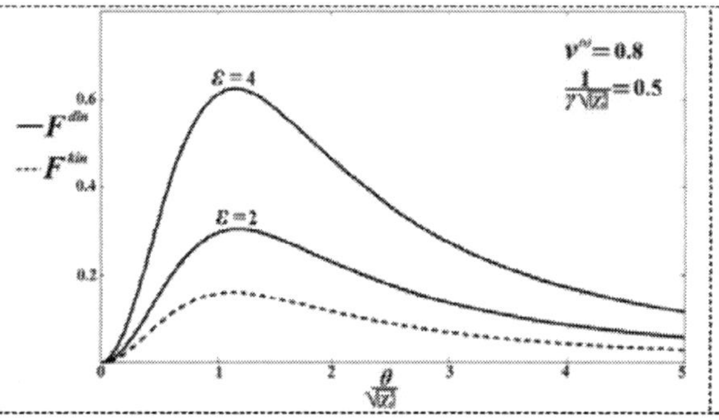

Figure 3. Under the strong asymmetry PXR angular density is ε times more than it was predicted by the kinematic formula PXR angular density for different asymmetry.

4. Contrastive analysis of kinematic and dynamic formulas for a thick crystal

For further analysis it is convenient to present the expressions (1) and (6) in the following way:

$$\frac{dN^{(s)din}_{PXR}}{d\Omega}=\frac{e^2\omega_B|\chi'_{\mathbf{g}}|C^{(s)}P^{(s)2}}{4\pi\sin^2\theta_B}F^{din}_{abs} \tag{11a}$$

$$F^{din}_{abs}=\nu^{(s)}\frac{\theta^2}{|\chi'_0|}\left(\frac{\theta^2}{|\chi'_0|}+\frac{1}{\gamma^2|\chi'_0|}+1\right)^{-2}\frac{L_{ext}}{2\pi}\int_{-\infty}^{+\infty}R^{(s)}_{PXR}\,d\eta^{(s)}(\omega) \tag{11b}$$

$$R^{(s)}_{PXR}=\left(1-\frac{\xi^{(s)}}{\sqrt{\xi^{(s)2}+\varepsilon}}\right)^2\frac{1+e^{-L_f\mu^{(s)}}-2e^{-\frac{L_f\mu^{(s)}}{2}}\cos\left(\frac{L_{e^-}}{2L_{ext}}\left(\sigma^{(s)}+\frac{\xi^{(s)}-\sqrt{\xi^{(s)2}+\varepsilon}}{\varepsilon}\right)\right)}{\left(\sigma^{(s)}+\frac{\xi^{(s)}-\sqrt{\xi^{(s)2}+\varepsilon}}{\varepsilon}\right)^2+\frac{L^2_{ext}\mu^{(s)2}}{\varepsilon^2}} \tag{11c}$$

$$\frac{dN^{(s)kin}_{PXR}}{d\Omega}=\frac{e^2\omega_B|\chi'_{\mathbf{g}}|C^{(s)}P^{(s)2}}{4\pi\sin^2\theta_B}F^{kin}_{abs}, \tag{12a}$$

$$F^{kin}_{abs}=\nu^{(s)}\frac{\theta^2}{|\chi'_0|}\left(\frac{\theta^2}{|\chi'_0|}+\frac{1}{\gamma^2|\chi'_0|}+1\right)^{-2}\frac{\varepsilon}{\mu_0}\left(1-e^{-L_f\mu_0}\right), \tag{12b}$$

where

$$\mu^{(s)}=\mu_0\cdot\left(\frac{\varepsilon+1}{2}-\frac{1-\varepsilon}{2}\frac{\xi^{(s)}}{\sqrt{\xi^{(s)^2}+\varepsilon}}-\frac{\varepsilon\kappa^{(s)}}{\sqrt{\xi^{(s)^2}+\varepsilon}}\right),\quad \mu_0=\omega_B\chi''_0 .$$

$\mu^{(s)}$ is the effective coefficient of photon absorption in crystals, dependent on the asymmetry parameter ε. The parameter $\kappa^{(s)}$ determines the degree of manifestation of the abnormal photo absorption low (the Borrmann effect) [15] in x-ray waves crossing the crystal. This effect lies in the formation of a standing wave by incident and diffracted waves whose antinodes are situated in the middle of the space between neighboring atomic planes, where the electron density and consequently the photo absorption are minimal. Herewith two waves are formed in the crystal, one of which absorbed abnormal strong and other absorbed abnormal weak. The expression (11) describes the PXR branch which is absorbed anomalously weak.

As well in the case of free x-ray waves, the imaginary part value proximity in the appropriate decomposition coefficients of dielectric susceptibility of the crystal in the Fourier series on the reciprocal lattice vectors ($\kappa^{(s)}\approx 1$) is a necessary condition for the manifestation of this effect in PXR. It necessary to remind that this parameter depends on the selection of the system of diffracted atomic planes in the crystal, radiation frequency and its polarization.

As the solution of the equation

$$\sigma^{(s)}+\left(\xi^{(s)}-\sqrt{\xi^{(s)2}+\varepsilon}\right)/\varepsilon=0, \tag{13}$$

determines frequency ω_*, in whose vicinity the spectrum of PXR photons radiated under fixed observation angle is concentrated,

$$\xi^{(s)}(\omega_*)=(1-\varepsilon\sigma^{(s)2})/2\sigma^{(s)}, \tag{14}$$

than the effective absorption coefficient also depends on the observation angle θ. Thus, the account of absorption in the dynamic theory may cause the deformation of PXR angular density (see in figure 4).

Once again it should be noted that in the kinematic formula only the geometry of process depends on the reflection asymmetry, while the absorption coefficient μ_0 corresponds to a normal one in amorphous medium. In the dynamic formula (11) the effective absorption coefficient $\mu^{(s)}$ depends not only on the selection of the system of diffracted crystal atomic planes (parameter $\kappa^{(s)}$), but also on the reflection asymmetry and the observation angle.

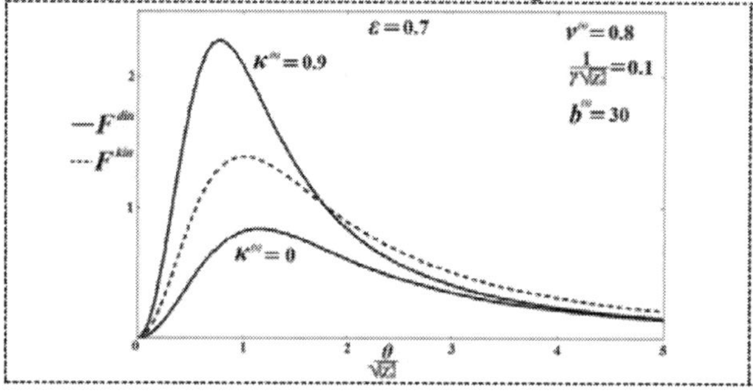

Figure 4. PXR angular density at the expense of different effective absorption coefficient for different observation angles θ.

5. Interpretation of the results of the experiment

In the work [14] the results of the experimental research of spectral angular distribution of parametric x-ray radiation (PXR) by electrons with the energy of 855 MeV are published. We would underline the high quality of the accelerator beam (small lateral dimensions and divergence) and that of the detecting equipment which allowed the authors [14] to obtain x-ray photon angular distribution in absolute values of high accuracy and high angular resolution. We compared the published experimental results with the angular distribution calculated by us based on the well-known kinematic formula (see e.g. [5, 14]) and then with the dynamic formula which we obtained [10]. As one can see in figure 7 the curve plotted on the basis of the kinematic formula (dot line 1), does not correspond to the results of the experiment. The curve plotted on the basis of our dynamic formula, which takes into account photon reflection from the system of atomic planes (111) asymmetrically situated with respect to the target surface (solid curve 2), completely coincides with the data of the experiment. It is important to note that the experiment conditions also correspond to conditions of a significant manifestation of anomalously low photo absorption (Borrmann effect) in PXR in the case of asymmetric refection [11]. To illustrate the presence of this effect the curve calculated by our dynamic formula without taking into account the Borrmann effect is plotted in figure 5 (solid curve3).

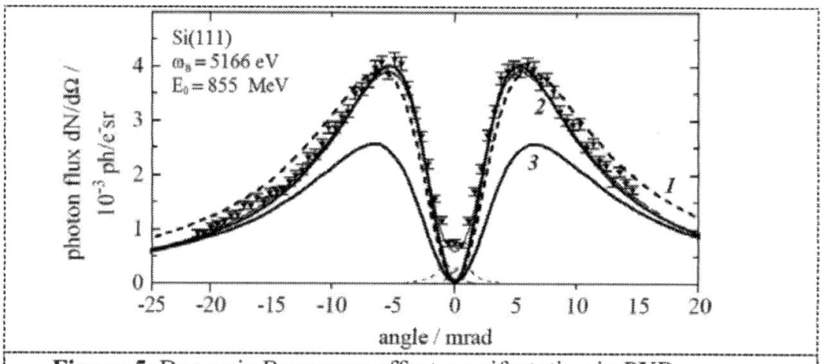

Figure 5. Dynamic Borrmann effect manifestation in PXR.

Conclusion

In the present work the contrastive analysis the kinematic and dynamic formulas for PXR was made for the cases of symmetric and asymmetric reflection of coulomb field of a relativistic electron. It was shown for a thin crystal that in the case of symmetric reflection in the region of radiating particle energy $\gamma >> \omega / \omega_p$ kinematic formula results in error even for thin non-absorptive crystal, and for the case $\gamma < \omega / \omega_p$ both the kinematic and dynamic formulas give the same result. It was shown that for the increase in reflection asymmetry the error resulting by kinematic formula for PXR increases and for strong asymmetry ($\varepsilon >> 1$) the angular density calculated by kinematic formula turn out in ε times less than the density calculated by means of dynamic formula.

It is shown that in the case of thick absorbing crystal the absorption coefficient in dynamic theory, in contrast to kinematic formula, depends on reflection asymmetry in the direction of photon radiation, which can cause the deformation of angular distribution of PXR. In the present work in the network of the developed theory it is done the interpretation of the results obtained in the experiment on registration of PXR on Mainz electron microtron MAMI and it was shown that the effect of abnormal photo absorption (Borrmann effect) was considerably manifested in that experiment.

References

[1] Ter-Mikaelian M 1972 *High-Energy Electromagnetic Process in Condensed Media* (New York : Wiley)
[2] Garibian G, Yang C 1971 J. Exp. Theor. Phys. **61** 930
[3] Baryshevsky V, Feranchuk I 1971 J. Exp. Theor. Phys. **61** 944
[4] Nitta H 1991 Phys.Lett.A **158** 270
[5] Feranchuk I, Ivashin A 1985 J. Physique **46** 1981
[6] Baryshevsky V, Feranchuk I 1983 J. Physique (Paris) **44** 913
[7] Kubankin A, Nasonov N, Sergienko V, Vnukov I 2003 Nucl. Instr. and Meth. In Phys. Res. B **201** 97
[8] Nasonov N, Noskov A 2003 Nucl. Instr. Meth. In Phys. Res. B **201** 67
[9] Blazhevich S, Noskov A 2006 Nucl. Instr. Meth. In Phys. Res. B **252** 69
[10] Blazhevich S, Noskov A 2008 Nucl. Instr. Meth. In Phys. Res. B **266** 3770
[11] Blazhevich S, Noskov A 2008 Nucl. Instr. Meth. In Phys. Res. B **266** 3777
[12] Aleinik A, Baldin A, Bogomasova E, Vnukov I at al. 2004 J. Exp. Theor. Phys Letters **80** 447
[13] Blazhevich S, Noskov A "Relativistic electron PXR and FPXR yield ratio" (see in this issue)
[14] K.-H. Brenzinger, C. Herberg, B.Limburg, H. Backe and al, Phys. A **358**, 107 (1997).
[15] G. Borrmann, Zh. Phys. 42, **157** (1941).

Laser induced x-ray radiation under the grazing incidence geometry

A I Benediktovitch and I D Feranchuk

Department of Physics, Belarusian University, Nezavisimosti Ave., 4, 220030, Minsk, Belarus

E-mail: aib@tut.by

Abstract. In conventional schemes of the parametric x-ray radiation (PXR) the beam of charged particles moves through the crystal and destruction of target and beam takes place. It is possible to avoid the destruction and to increase the interaction length if the scheme of grazing incidence diffraction (GID) is used: the beam of charged particles propagates in vacuum near the crystal surface while the photons are emitted under conditions of GID. In this case a highly collimated and precisely positioned beam is required. he restrictions on beam parameters can be avoided if intense laser radiation is backscattered by the electron beam. In the paper a theory of laser induced x-ray radiation under conditions of grazing incidence diffraction is presented and distributions of emitted quanta are calculated. Properties of the radiation and possible applications are discussed.

1. Introduction

One of the ways to get higher output of the parametric x-ray radiation (PXR) is to increase the interaction length and beam current. These parameters could be much higher if electrons move along the surface of the crystal without penetrating it. To obtain PXR output from such electrons the structure of eigen electromagnetic field near crystal should have a wave propagating at grazing angle to the surface. This kind of the electromagnetic field belongs to special cases of electromagnetic fields considered in the x-ray diffractometry (XRD) for which traditional two-wave simplification of the dispersion equation is not valid and additional roots and waves should be taken into consideration [1].

There are two main special cases of XRD: extremely asymmetric diffraction (EAD) and grazing incidence diffraction (GID). In case of EAD diffracted (or incident) wave makes a grazing angle to the surface and it is necessary to consider the reflected wave also. It should be mentioned that the pioneering PXR experiments in Tomsk were done in this geometry and were completely described theoretically [2], [3].

The second case is GID. It is widely used nowadays in XRD (due to its surface sensitivity). Here the reciprocal space vector lies in the plane of crystal surface and all waves propagate at grazing angles. In this conditions reflected waves play essential role and complete fourth-order dispersion equation should be used. The spectral-angular distribution of PXR in GID geometry was considered before in the series of the papers [4] - [6]. The integral characteristics of PXR under GID geometry (GID-PXR) in dependence of the electron beam parameters and the influence of the external radiation source (laser field) on the GID-PXR properties are considered in the present paper.

2. Calculation of GID-PXR intensity

Consider the next experimental geometry: high-energy electrons move almost parallel to the crystal surface, the distance between the beam and the surface is z_0, the velocity vector $\vec{v}$ is inclined on angle α against the surface and makes angle θ_B with considered set of crystallographic planes (Figure 1). Under this conditions PXR is concentrated along the direction of vector $-\vec{k}$, which is parameterized via out-of-plane angle ϕ and in-plane deviation θ. The wave vector $\vec{k}+\vec{h}$ of the diffracted wave is almost antiparallel to the electron velocity $\vec{v}$. Under such conditions no destruction of the crystal and the beam are occurred. It enables to increase the interaction length from the absorption length, which is typical for the conventional schemes of PXR, to the crystal size. Also, it becomes possible to generate PXR directly inside the accelerator ring and to take advantage of high current and high energy of particles there [4], [7].

The spectral density of PXR can be calculated on the basis of classical electrodynamics according to the next formula [8], system of units $\hbar = c = 1$ is used:

$$W_{\vec{n}\omega s} = \frac{\omega^2}{4\pi^2} \left| \int d^3\vec{r}\, \vec{j}_\omega(\vec{r})\, \vec{E}^{(-)}_{\vec{k}\omega s}(\vec{r}) \right|^2, \tag{1}$$

where $\vec{n} = -\vec{k}/k$ is the unit vector directed towards the detector of radiation, $\vec{j}_\omega(\vec{r})$ is the temporal Fourier-transform of the beam current, index $s = \sigma, \pi$ defines the radiation polarization. $\vec{E}^{(-)}_{\vec{k}\omega s}(\vec{r})$ is the electric field of the time-reversed eigen wave which satisfies the homogeneous Maxwell equations taking into account the presence of crystal and the boundary conditions $\vec{e}_s e^{i\vec{k}\vec{r}}$ at the detector side [8]. It replaces the plane wave in vacuum electrodynamics and accounts for interaction with the media.

In order to find out the field $\vec{E}^{(-)}_{\vec{k}\omega s}(\vec{r})$ the XRD problem should be solved. When the incoming wave vector $\vec{k}$ is close to the Bragg condition for the reflex defined by the reciprocal lattice vector $\vec{h}$ the two-wave approximation can be used, which leads to the next structure of the electric field above the crystal:

$$\vec{E}^{(-)}_{\vec{k}\omega s}(\vec{r}) = \vec{e}_{\vec{k}s} T_0\, e^{i\vec{k}\vec{r}} + \vec{e}_{\vec{k}_R s} R e^{i\vec{k}_R\vec{r}} + \vec{e}_{\vec{k}_h s} H e^{i\vec{k}_h\vec{r}}, \tag{2}$$

where $T_0 = 1$, R, H, are the amplitudes of the incident, reflected, diffracted waves; $\vec{k}, \vec{e}_{\vec{k}s}$; $\vec{k}_R, \vec{e}_{\vec{k}_R s}$; $\vec{k}_h, \vec{e}_{\vec{k}_h s}$ are the wave vectors and unit polarization vectors correspondingly. In the chosen reference frame, wave vectors components are following:

$$\begin{aligned} &\vec{h} = \{-h_x, 0, 0\}; \quad \vec{k} = \{\vec{k}_\parallel, \omega\gamma_0\}; \quad \vec{k}_R = \{\vec{k}_\parallel, -\omega\gamma_0\}; \quad \vec{k}_h = \{\vec{k}_\parallel + \vec{h}, -\omega\gamma_h\}; \\ &\vec{k}_\parallel = \omega\{\cos\varphi \sin(\theta_B + \theta), \cos\varphi\cos(\theta_B + \theta), 0\}; \quad \gamma_0 = \sin\varphi. \end{aligned} \tag{3}$$

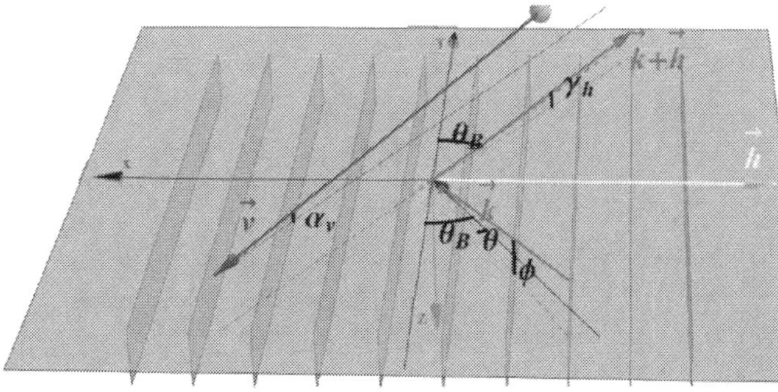

Figure 1. GID-PXR geometry, see text for the details.

Here $\vec{h}$ is assumed to belong the interface plane (Figure 1), i.e. no miscut is considered. From the condition $k = k_R = k_h = \omega$ the useful relation between γ_0 and γ_h follows

$$\gamma_h = \sqrt{\gamma_0^2 - \alpha}, \quad \alpha = \frac{2\vec{k}_{\parallel}\vec{h} + h^2}{\omega^2}, \tag{4}$$

where α is deviation from Bragg condition.

To find the wave field inside the crystal, consider the wave field in the form of

$$\vec{D}_s(\vec{r}) = \sum_i D_i(\vec{e}_{\vec{k},s} e^{i(\vec{k}_{\parallel}\vec{r}_{\parallel} + \omega u_i z)} + v_i \vec{e}_{\vec{k}_h,s} e^{i([\vec{k}_{\parallel} + \vec{h}]\vec{r}_{\parallel} + \omega u_i z)}), \tag{5}$$

where the longitudinal wave vector component $\vec{k}_{\parallel}$ is the same as above, the interface due to boundary conditions and the value u_i describes the wave refraction. The values v_i and u_i can be found from a system of algebraic equations which follows from the Maxwell equations taking into account crystal susceptibilities $\chi_0, \chi_{\vec{h}}, \chi_{-\vec{h}}$ in the two-wave approximation:

$$(u^2 - \gamma_0^2 - \chi_0) - vC\chi_{-\vec{h}} = 0; \quad (u^2 - \gamma_h^2 - \chi_0)v - C\chi_{\vec{h}} = 0, \tag{6}$$

here $C = 1$ for σ polarization and $C = \cos 2\theta_B$ for π polarization. The condition of nontrivial solution of (6) provides the dispersion equation of the fourth-order for the value u, which can be solved here analytically; after u_i is known, v_i are found trivially:

$$u_{1,2} = \sqrt{\gamma_0^2 + \chi_0 - \frac{\alpha \pm \sqrt{\alpha^2 + 4C^2\chi_{\vec{h}}\chi_{-\vec{h}}}}{2}}, \quad v_{1,2} = \frac{u_{1,2}^2 - \gamma_0^2 - \chi_0}{C\chi_{-\vec{h}}}, \tag{7}$$

the solutions $u_{3,4} = -u_{1,2}$ are not considered as they do not agree with boundary conditions.

To determine the wave filed completely, the amplitudes R, H, D_1, D_2 should be found from the boundary conditions. In the case of x-rays standard conditions for magnetic and electric field are reduced to the continuity conditions for amplitudes and their derivatives for each longitudinal wave vector. With definitions (2), (5) we obtain:

$$\begin{aligned} &1 + R = D_1 + D_2; \quad \gamma_0(1 - R) = u_1 D_1 + u_2 D_2; \\ &H = v_1 D_1 + v_2 D_2; \quad -\gamma_h H = u_1 v_1 D_1 + u_2 v_2 D_2. \end{aligned} \tag{8}$$

For the calculation of the PXR intensity only the diffracted wave $\vec{e}_{\vec{k}_h s} H e^{i\vec{k}_h\vec{r}}$ is of interest, its amplitude is found from (8):

$$H = \frac{2(u_1 - u_2)v_1 v_2 \gamma_0}{((v_1 - v_2)(u_1 u_2 + \gamma_0\gamma_h) + (u_1 v_1 - u_2 v_2)\gamma_0 + (u_2 v_1 - u_1 v_2)\gamma_h)}, \tag{9}$$

and unit polarization vectors can be chosen as

$$\vec{e}_\sigma = \vec{k} \times \vec{k}_h / \left\|\vec{k} \times \vec{k}_h\right\|, \; \vec{e}_\pi = \vec{e}_\sigma \times \vec{k}_h / \left\|\vec{e}_\sigma \times \vec{k}_h\right\|. \tag{10}$$

The second important term in the equation for PXR spectral density is the spectral density of beam current. Consider the next model parameters describing the beam: small angle α between electron velocity and crystal surface, mean distance between the beam and the surface z_0, beam width (dispersion of beam position) σ_z. Suppose also that inclination angle α and beam position z_0 are interconnected through a parameter $\delta_{z,\alpha}$ called emittance. Then the beam current spectral density normalized on one electron is:

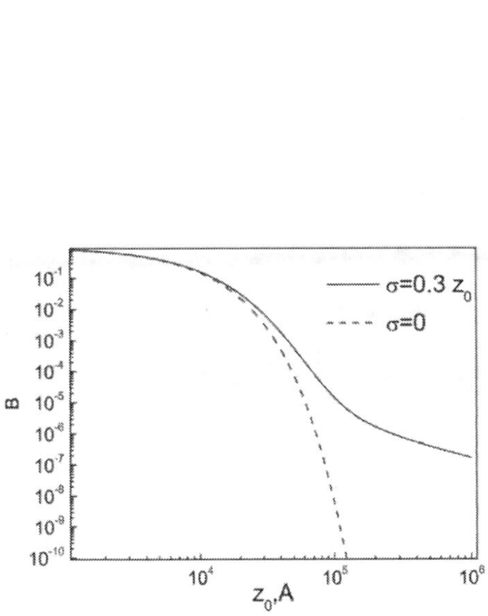

Figure 2. Attenuation of PXR intensity with distance between the beam and the crystal surface for narrow beam (exponential dependence, dashed line) and beam distributed in z direction (described via factor B, solid line).

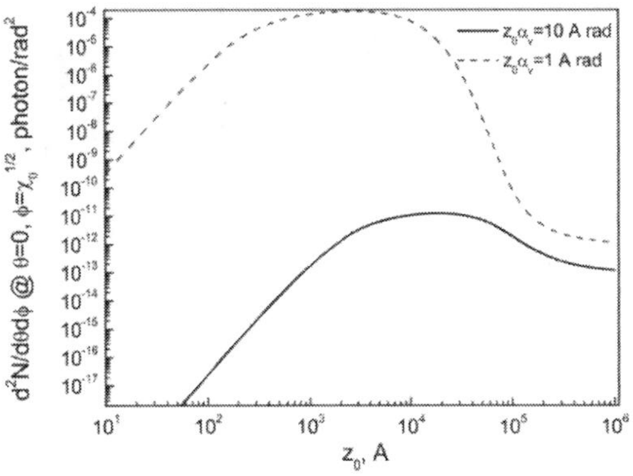

Figure 3. PXR photon density at maximum for different values of emittance.

$$\vec{j}_\omega(\vec{r}) = \frac{e}{2\pi}\int dt dz \frac{1}{\sqrt{\pi\sigma_z}} \exp\left(-\frac{(z-z_0)^2}{\sigma_z}\right) e^{i\omega t}\vec{v}\delta(\vec{r}-\vec{v}t-\vec{r}_0). \tag{11}$$

After inserting field and current in the spectral density expression and integrating over the frequency we obtain the next expression for angular distribution of PXR photons:

$$\begin{aligned} \frac{\partial^2 N_s}{\partial\theta\partial\varphi} &= \frac{\alpha}{2\pi} L_c \omega^2 \left|H(\theta,\varphi)\,\vec{e}_s\vec{v}\,B\right|^2 \\ B &= \frac{1}{2} e^{i\omega\gamma_h z_0 - \frac{1}{4}\sigma_z\omega^2\gamma_h^2} \mathrm{Erfc}\left(\frac{z_0}{\sqrt{\sigma_z}} + \frac{1}{2} i\omega\gamma_h\sqrt{\sigma_z}\right), \end{aligned} \tag{12}$$

where L_c is the coherent length that is equal here to the crystal length, B is the factor originating from distribution in z direction, Erfc is the error function, $\alpha = 1/137$ is the fine structure constant.

The PXR frequency ω is found from a Cherenkov-like condition $\delta(\omega - \vec{k}_h\vec{v})$ and has the following form:

$$\omega = \frac{h_x\beta\cos\alpha_v\sin\theta_B}{1-\beta\cos\alpha_v\cos(2\theta_B+\theta)\cos\varphi}, \tag{13}$$

From the dispersion equation for electromagnetic field in vacuum the next expression for the diffracted wave vector component normal to the surface follows

$$\omega\gamma_h = \sqrt{\omega^2\sin^2\varphi - h_x^2 + 2h_x\omega\cos\varphi\sin(\theta_B+\theta)}. \tag{14}$$

This component proved to be always imaginary. It means that intensity from a single electron decreases exponentially when the distance from the crystal surface increases, the typical distance of attenuation can be estimated as

$$z_0 \sim \frac{\lambda_{\text{x - ray}}}{\sqrt{1/\gamma^2+\alpha_v^2}}. \tag{15}$$

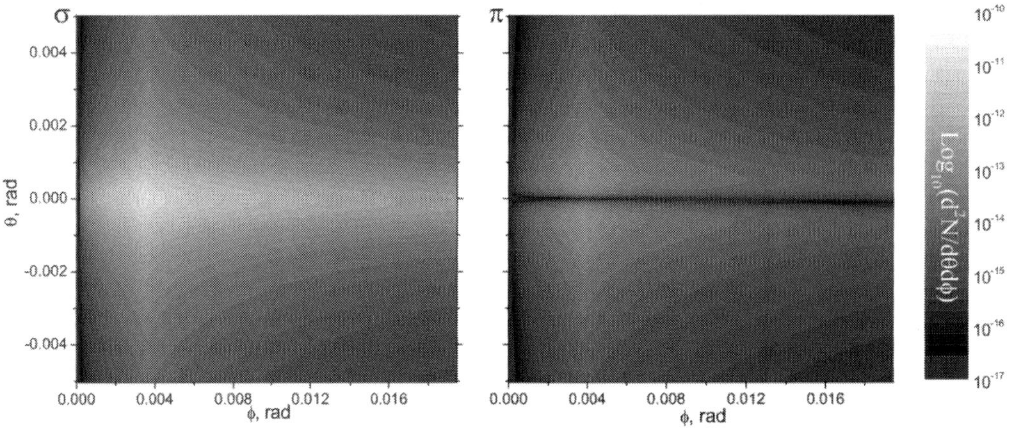

Figure 4. Angular distribution of PXR photon density for σ and π polarizations for $L_c = 0.1cm, \alpha_v = 10^{-3}rad, z_0 = 1\mu m$, Si crystal, reflex (220).

However, for a beam of particles with Gaussian distribution this dependence is described by factor B in (12) and is not so strong. Figure 2 shows the dependence of the factor B on the distance of beam center from the surface for electron energy $E = 5GeV$ and $\sigma_z = 0.3z_0, \theta_B = \pi/3$, reflex (220) of Si crystal (this parameters are used below as well) at angles $\theta = 0, \varphi = \sqrt{|\chi_0|}, \alpha_v = 0$. For comparison the curve calculated for infinitely narrow beam is also shown. One can see that for these realistic parameters PXR intensity decreases rapidly with distance z_0 and in case of beam with distribution in z mainly the "tail" of the distribution close to the interface contributes to PXR formation. Hence, the alignment of the beam as close as possible to the crystal surface is of primary importance.

The dependence on beam emittance comes out to be very strong. Actually, at given beam position z_0 the emittance determines the coherent length and normal wave vector component that enters the factor B. Figure 3 shows dependence of PXR intensity at maximum on beam distance from crystal for two values of emittance. One can see that increase of one order in emittance causes many orders of decrease in intensity.

One of distinctive PXR features is narrow angle and spectral distribution. Figure 4 shows the angular distribution of PXR photons density for σ and π polarizations calculated for (220) reflex of Si crystal. The width of peak in out-of-plane direction has the order of $\sqrt{|\chi_0|}$ that is typical for GID, the width in in-plane direction depends on factor $\gamma = E/m$ and angle α. The absolute value of intensity depends greatly on beam emittance.

3. Laser induced GID-XR radiation

From mathematical point of view, to overcome the exponential dependence on the distance from crystal one can add a small additional item in the Cherenkov-condition. It would make the normal wave vector component real in some angle area and remove the dependence of intensity on beam distance. Physically, it corresponds to some periodical influence. For example, consider an external linearly-polarized laser field $\vec{E}(\vec{r},t) = \vec{E}_0 e^{i(\vec{k}_L\vec{r}-\Omega t)}$ of optical frequency Ω acting on the electron. This field makes the electron to move with the oscillating velocity $\vec{u}$:

$$\vec{v}(t) = \vec{v}_0 + \vec{u}\,e^{-i\Omega(1+\beta)t}; \quad \vec{u} = \frac{e\vec{E}_0}{\gamma m \Omega}, \tag{16}$$

the resulting law of motion being

$$\vec{r}(t) = \vec{r}_0 + \vec{v}_0 t - \frac{\vec{u}\,e^{-i\Omega(1+\beta)t}}{i\Omega(1+\beta)}. \tag{17}$$

Inserting the law of motion obtained in the expression for the beam current and performing the integration over frequency in general expression for spectral-angular distribution we get the next angular distribution of emitted XR quanta:

$$\frac{\partial^2 N_s}{\partial\theta\partial\varphi} = \frac{\alpha}{2\pi} L_c \omega \left| H(\theta,\varphi)\,\vec{e}_{\vec{k}_h,s} \left(\vec{u} - \vec{v}_0 \frac{\vec{k}_h \vec{u}}{\Omega(1+\beta)}\right) e^{i\omega\gamma_h z_0} \right|^2 \frac{1}{1+\vec{k}\vec{v}_0} \tag{18}$$

The frequency of electron oscillations $\Omega(1+\beta)$ enters the Cherenkov-condition which has the form $\delta(\omega + \vec{k}_h\vec{v} - \Omega(1+\beta))$. The frequency of emitted XR photons is

$$\omega = \frac{h_x \beta \cos\alpha_v \sin\theta_B + \Omega(1+\beta)}{1 - \beta\cos\alpha_v \cos(2\theta_B + \theta)\cos\varphi}, \tag{19}$$

and is slightly changed compared to (13). This change makes the value of normal wave vector component (14) real when the in-plane angle θ is less than θ_{cr} equal to

$$\theta_{cr} = \sqrt{2\left(\frac{\Omega(1+\beta)}{\omega_B} - \frac{1}{2\gamma^2}\right)}. \tag{20}$$

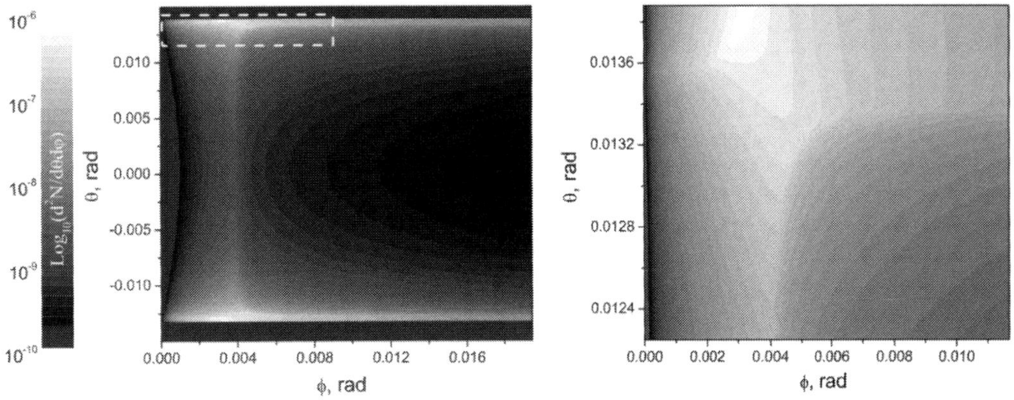

Figure 5. Angular distribution of PXR photons density at $L_c = 0.1cm, \alpha_v = 10^{-3}rad, z_0 = 1\mu m$, Si crystal, reflex (220), $\gamma = 30, \lambda_{\text{x - ray}} = 1.9\,A, \lambda_{\text{laser}} = 1.026\,\mu m$, laser power $W_{\text{laser}} = 1\,MWt$, focal diameter $D_{\text{focal laser}} = 50\,\mu m$. The vicinity of the peak bordered by dashed line in the left panel is shown on the right panel.

For optical range of laser frequency the necessary electron energy has order of tens MeV, and the dependence on beam collimation becomes quite weak.

This behavior can be understood if we consider the process of XR quanta emission as Compton backscattering of laser photons and their diffraction in crystal. The photons are generated in vacuum and the crystal acts like a monochromator.

The angular distribution is presented on Figure 5. It has two almost symmetric maxima placed at in-plane angle θ_{cr} (Figure 5; left panel). The width of the peaks in both directions has the order of $\sqrt{|\chi_0|}$. The shape of the peak is typical for GID diffraction (Figure 5; right panel). The absolute value of XR intensity is proportional to square of induced velocity and hence to the laser power density. For the parameters used the quantum yield of x-ray quanta is approximately $10^{-11} photons/electron$. For more powerful laser it will be higher, correspondingly.

4. Conclusions

A new scheme for generation of parametric x-ray radiation is proposed. The beam of ultrarelativistic charged particles propagates in vacuum near the crystal surface while the photons are emitted under conditions of GID. Under such conditions no destruction of crystal and beam takes place. Hence, the interaction length can be increased to the lateral size of the crystal and no restrictions on beam current are imposed. However, the analysis shows that ultra relativistic electrons are required (tens GeV) and GID-PXR is very sensitive to beam position and emittance. This sensitivity can be utilized to perform the transverse beam profile diagnostics, such application will be discussed elsewhere.

Application of the laser field enables to avoid strong dependence on the beam parameters and to obtain XR quanta at electron energies of tens MeV. The intensity of XR then becomes proportional to the laser power density.

A promising application of GID-PXR is construction of a diffraction resonator without the absorption of the emitted photons. It can enhance parametric beam instability [9] and lead to the coherent radiation in the x-ray range.

References

[1] Pietsch U, Holy V and Baumbach T 2004 *X-ray Scattering: from Thin Films to Lateral Nanostructures* (Heidelberg: Springer Verlag)

[2] Adishchev Y, Baryshevsky V, Vorobiev S, Danilov V, Pak S, Potylizyn A, Safronov P and Feranchuk I 1985 *Sov. Phys. JETP Lett.* **41** 361

[3] Didenko A, Kalinin B, Pak S, Potylizyn A, Vorobiev S, Baryshevsky V, Danilov V and Feranchuk I 1985 *Phys. Lett. A* **110** 177

[4] Baryshevsky V 1988 *Doklady Akademy of Science USSR* **299** 1363

[5] Andriyanchik A, Baryshevsky V and Kaminsky A 1993 *NIM B* **83** 482

[6] Andriyanchik A, Baryshevsky V and Kaminsky A 1994 *Phys.Stat.Sol. B* **184** 543

[7] Feranchuk I and Feranchuk S 2007 *Eur. Phys. J. Appl. Phys.* **38** 135

[8] Baryshevsky V, Feranchuk I and Ulyanenkov A 2006 *Parametric X-ray Radiation in Crystalls* (Berlin: Springer Verlag)

[9] Baryshevsky V and Feranchuk I 1984 *Phys. Lett. A* **102** 141

Angular Distributions of DCR from Axially Channeled Electrons in ‹110› LiF Crystal

K B Korotchenko, E I Fiks, Yu L Pivovarov and T A Tukhfatullin

Tomsk Polytechnic University, Tomsk, 634050 Russia

korotchenko@tpu.ru

Abstract. The formulae and new computation method for calculation of angular distributions of diffracted channeling radiation (DCR) from relativistic electrons axially channeled in a crystal are presented. Numerical results are obtained for $\gamma = 20$ electrons channeled in $\langle 110 \rangle$ LiF crystal.

1. Introduction

Parametric X–radiation (PXR) from relativistic electrons is explained in detail in the framework of kinematic and dynamics theories [1]. More detailed theory was proposed in Ref.[2] which takes into account both the channeling of electrons and diffraction of emitted photons. The new combined effect in X-radiation was called in [2] Diffracted Channeling Radiation (DCR). Later, the theory was extended to take into account the band structure of transverse energy levels and initial population of levels (states) for planar channeled electrons in Refs.[3–4]. In Ref.[3] calculations of DCR angular distributions were performed for electrons with relativistic factor γ = 18,20, 25, 35, 50, 75, and 100 channeled along (110) and (111) Si planes , while in Ref.[4] for γ = 107 and (111) LiF planes.

Here, we present new method of calculations and obtained results for DCR angular distributions from axially channeled electrons. This problem is more complicated compared to the case of DCR from planar channeled electrons [2–4], since one needs:

- to calculate 2D periodic potential for axial channeling;
- to solve 2D eigenvalue problem instead of 1D in order to determine transverse energy band structure and eigenfunctions $\Psi(\boldsymbol{r}_\perp, \boldsymbol{k}_\perp)$ for the Schrödinger equation with relativistic mass of electron γm_e;
- to define initial populations of states and dipole matrix elements using determined 2D eigenfunctions;
- to modify a formula for DCR angular distribution derived in [2–4] for planar channeling to the case of axial channeling.

2. Electron axial channeling in a crystal: a theoretical background

As is well known [5], the wave function $\Psi(\boldsymbol{r},t)$ of electron with an energy E > 1 MeV, channeled along the crystal axis may be written as

$$\Psi(\boldsymbol{r},t) = \exp(-\mathrm{i}(p_z z - Et)/\hbar)\phi(\boldsymbol{r}_\perp) \ . \qquad (1)$$

Here, $\boldsymbol{r}_\perp$ is the radius vector perpendicular to the channeling axes, p_z is the longitudinal momentum. For the stationary states of the transverse motion along XY–plane the wave function of electron is the solution of the Klein–Gordon equation, which transforms into Schrödinger equation with relativistic mass γm_e [5] when $2E_\parallel >> (E_\perp - U(\boldsymbol{r}_\perp))$, with $E_\parallel = \sqrt{(cp_\parallel)^2 + m_e^2 c^4}$ and $E_\perp$ is the energy of electron transverse motion. In a periodic potential $U(\boldsymbol{r}_\perp)$ which is the case of axial channeling, the wave function $\phi(\boldsymbol{r}_\perp)$ and energy of transverse motion of electron $E_\perp$ are continuous functions of electron wave vector $\boldsymbol{k}_\perp$, which lies in the first Brillouin zone (e.g., see[6]). That is:

$$\phi^i(\boldsymbol{r}_\perp, \boldsymbol{k}_\perp) = \sum_m C^i(\boldsymbol{g}_{\perp m}, \boldsymbol{k}_\perp) \exp\{\mathrm{i}(\boldsymbol{k}_\perp + \boldsymbol{g}_{\perp m})\boldsymbol{r}_\perp\}, \tag{2}$$

Using the Fourier expansion of periodic function $U(\boldsymbol{r}_\perp) = \sum_m U(\boldsymbol{g}_{\perp m}) \exp\{\mathrm{i}(\boldsymbol{k}_\perp + \boldsymbol{g}_{\perp m})\boldsymbol{r}_\perp\}$, we obtain the eigensystem problem [5], [7–8]:

$$\sum_m A_{mn} C^i(\boldsymbol{g}_{\perp m}, \boldsymbol{k}_\perp) = E^i(\boldsymbol{k}_\perp) C^i(\boldsymbol{g}_{\perp n}, \boldsymbol{k}_\perp), \;\; A_{mn} = U_{mn} + \delta(\boldsymbol{g}_{\perp m}, \boldsymbol{g}_{\perp n})(\hbar^2 |\boldsymbol{g}_{\perp m} + \boldsymbol{k}_\perp|^2 / 2m_e\gamma). \tag{3}$$

Here, $\boldsymbol{g}_{\perp m}$ are the transverse reciprocal lattice vectors, which lie in a plane perpendicular to channeling axes, $C^i(\boldsymbol{g}_{\perp m}, \boldsymbol{k}_\perp)$ are the Fourier components of electron wave function $\phi^i(\boldsymbol{r}_\perp, \boldsymbol{k}_\perp)$, $U_{mn} = U(\boldsymbol{g}_{\perp m} - \boldsymbol{g}_{\perp n})$, $\delta(a,b)$ is the Kronecker delta.

3. The solution of eigensystem problem for the case of axial channeling

The numerical methods of solving an eigenvalue problem (3) are well known [7–8]. Unfortunally, the result of the solution of the Eq. (3) is very sensitive to even small variation of potential function $U(\boldsymbol{r}_\perp)$. That's why at first we build up the potential $U(\boldsymbol{r}_\perp)$, and then numerically find its Fourier components.

The experience of using this method for solving the problem (3) with the help of the Mathematica© [3–4] let us develop sufficiently simple and high–performance algorithm. For simulating of potential $U(\boldsymbol{r}_\perp)$ we use the formula obtained by Doyle and Turner [9] for the potential due to a single atom

$$U_{At}(\vec{r}) = -(2\hbar^2 / \sqrt{\pi} m_e) \sum_j a_j c_j^{-3/2} \exp(-r^2 / c_j), \tag{4}$$

where a_j and $b_j = 4\pi^2 c_j$ are the coefficients, obtained by fitting of the electron scattering form factor, $r^2 = z^2 + \boldsymbol{r}_\perp^2$.

Further, we choose Z–axis as channeling axis and integrate formula (4) along this axis. Next we may calculate the Fourier components of the potential using embedded Mathematica© package, but for required accuracy this is a time–consuming operation. It is easy to obtain the following equation for the Fourier components $U_{mn} = U(\boldsymbol{g}_{\perp m} - \boldsymbol{g}_{\perp n})$ in the Cartesian coordinates (for $\langle 110 \rangle$ LiF crystal):

$$U_{mn} = [1 + \exp(-\mathrm{i}m\pi)\exp(-\mathrm{i}n\pi)] \sum_j F_{xmj} F_{ynj} + [\exp(\mathrm{i}m\pi) + \exp(-\mathrm{i}n\pi)] \sum_j L_{xmj} L_{ynj}, \tag{5}$$

where F_{xmj} and F_{ymj} (the symbols F and L denote the values characterizing fluorine and lithium atoms respectively) are:

$$F_{xmj} = \frac{\sqrt{A_d \pi a_{Fj}}}{2a} \exp\left(-\frac{m^2\pi^2}{a^2 b_{Foj}}\right) \left[-Erfi\left(\frac{2m\pi - \mathrm{i}a^2 b_{Foj}}{2a\sqrt{b_{Foj}}}\right)\right] + \left[Erfi\left(\frac{2m\pi + \mathrm{i}a^2 b_{Foj}}{2a\sqrt{b_{Foj}}}\right)\right],$$

$$F_{ynj} = \frac{\sqrt{A_d \pi a_{Fj}}}{\sqrt{2}a} \exp\left(-\frac{2n^2\pi^2}{a^2 b_{Foj}}\right)\left[-Erfi\left(\frac{4n\pi - ia^2 b_{Foj}}{2\sqrt{2}a\sqrt{b_{Foj}}}\right)\right] + \left[Erfi\left(\frac{4n\pi + ia^2 b_{Foj}}{2\sqrt{2}a\sqrt{b_{Foj}}}\right)\right], \quad (6)$$

here, $Erfi(x)$ is imaginary error function; m and n are integers, that determine the numbers of the components U_{mn}, $A_d = 2\times 27.2116 a_o^2 d^{-1}$, d is the distance between atoms the axis, a is a lattice constant, a_o is the Bohr radius, $4\pi^2 b_{Foj} = b_j + 8\pi^2 \rho_{1F}^2$, where ρ_{1F} is the amplitude of thermal vibrations of fluorine atoms. The formulas for lithium, i.e. L_{xmj} and L_{ynj}, are similar to (6).

It should be noted that formulas (5–6) are the exact expressions for Fourier components of 2D–periodic potential $U(\boldsymbol{r}_\perp)$. The evaluation of formulas (5–6) using Mathematica© takes only several minutes. The solving of the problem (3) using the formulae (5–6) is performed further using a special procedure in Mathematica© for solution of the linear algebraic eigenvalue problem. Though the very large size of the matrix A_{mn}, the calculation of its values takes about one minute.

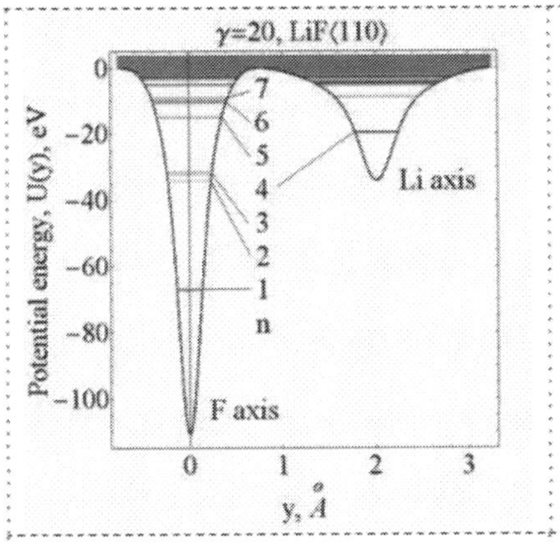

Figure 1a. Schematics of subbarrier energy levels (with number n) of electron transverse motion.

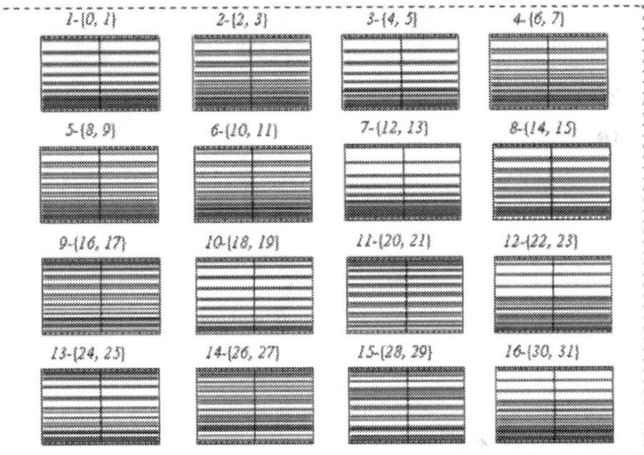

Figure 1b. The structure of subbarrier energy levels (with numbers n = 1–16) of electron transverse motion (γ = 20) along $\langle 110\rangle$ LiF.

The results of eigenvalues E^i calculation for electrons channeled along the $\langle 110\rangle$ LiF axes are presented in Fig.1a and Fig.1b. In fact, every level in Fig.1a is energy band. Red and blue lines on Fig.1b are the` odd and even numbers of quantum states corresponding to every energy band. In Fig.1b we denoted every energy level as n–$\{i_1,i_2\}$, where n is the number of the level, i_1, i_2 are the numbers of corresponding quantum states. That is all the levels are double degenerated.

Fig.2 presents the results of calculation of $|\Psi_i(\boldsymbol{r},t)|^2 = S^{-1}\phi_i^*(\boldsymbol{r}_\perp,\boldsymbol{k}_\perp)\phi_i(\boldsymbol{r}_\perp,\boldsymbol{k}_\perp)$ for the subbarrier quantum states (with numbers i = 0–31) of electrons with relativistic factor γ = 20 (energy 10.22 MeV) channeled along the axis $\langle 110\rangle$ LiF (S is the part of elementary cell cross–section per one axis [5], i.e. is different for L and F atoms). One can see that $|\Psi_i(\boldsymbol{r},t)|^2$ for the initial 22 quantum states (with numbers 0–21) coincide in pairs. For last 10 quantum states (with numbers 22–31) $|\Psi_i(\boldsymbol{r},t)|^2$ are differend – i.e. $|\Psi_i(\boldsymbol{r},t)|^2$ for quantum states of last 5 bands are varying within band.

4. DCR from the axially channeled electrons

The axial channeling of relativistic particles is accompanied by spontaneous radiation due to transitions between the energy levels $E^i(\boldsymbol{k}_\perp)$ and $E^f(\boldsymbol{k}_\perp)$ of the transverse motion. In the case of axial channeling, one can accept that photons arising under transitions can undergo diffraction on the crystallographic planes (diffraction planes) crossing the channeling axes (Fig. 3).

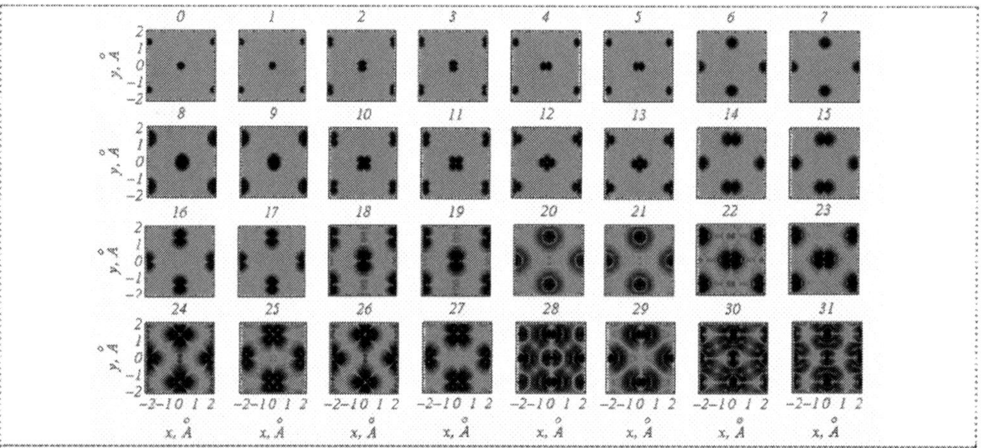

Figure 2. $|\Psi_i(\boldsymbol{r},t)|^2$ for subbarrier quantum states with numbers i = 0–31 of electron transverse motion.

Fig.3 shows the geometry of formation of axial diffracted channeling radiation (DCR). The momentum is conserved only in the direction parallel to the Z–axis, than the energy of the emitted photon can be written in the form

$$\hbar\omega = \hbar(\mathbf{gv} + \Omega_{if})(1 - \mathrm{v}_{\parallel}\sqrt{\varepsilon_0}\cos(2\theta_B)/c)^{-1}, \tag{8}$$

where $\hbar\Omega_{if} = E^f(\boldsymbol{k}_\perp) - E^i(\boldsymbol{k}_\perp)$, $\varepsilon_0 = 1 + \chi_0$ is the dielectric constant and χ_0 is electric susceptibility, $\theta_B = \arcsin(|\mathbf{vg}|/\mathrm{v}g)$ is Bragg angle and $\boldsymbol{g}$ is the reciprocal lattice vector corresponding to the diffraction planes, **v** is the velocity of electron along Z–axis.

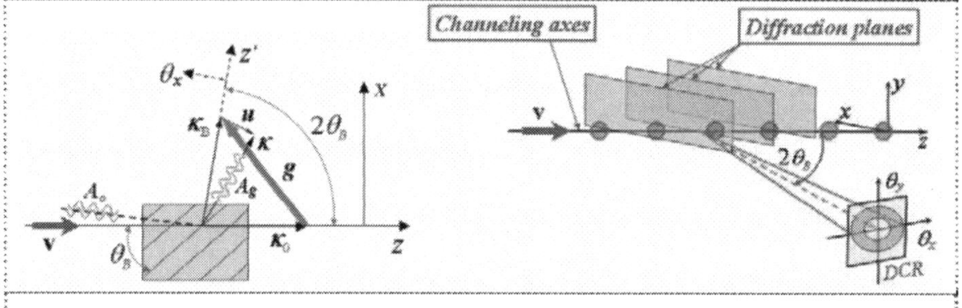

Figure 3. Formation of axial DCR from axially channeled electrons.

According to [2], the matrix element of DCR probability (transition probability of electron between energy levels $i \to f$ with photon emission) is written in the form:

$$M_{if} = e\langle\phi_f, p_z'|A\hat{p}|\phi_i, p_z\rangle/\gamma m_e c\,. \tag{9}$$

We had chosen Z–axis as channeling axis, so $\hat{\boldsymbol{p}}=(\hat{\boldsymbol{p}}_{\perp},\boldsymbol{p}'_z)$ is the momentum operator ($\boldsymbol{p}'_z$ is the electron longitudinal momentum after photon emission), where $\boldsymbol{p}'_z=m_e \mathrm{v}\gamma$, $\boldsymbol{A}=(\boldsymbol{A}_{\perp},\boldsymbol{A}_z)$ is the photon field that is given in a form of the Bloch wave:

$$\boldsymbol{A}=\sum_h \boldsymbol{A}_h \exp(\mathrm{i}\vec{\kappa}_h \boldsymbol{r})+c.c.=\sum_h\sum_\sigma A_{\sigma h}\vec{\varepsilon}_{\sigma h}\exp(\mathrm{i}\vec{\kappa}_h\boldsymbol{r})+c.c., \quad (10)$$

where $\boldsymbol{h}\in(0,\boldsymbol{g},...)$. If we set that the momentum of channeled electron along Z–axis is conserved, we can assign arising virtual photon both wave vector $\vec{\kappa}_0$ and the polarization vector $\vec{\varepsilon}_{\sigma h}$, $\sigma=\perp,\|$ is the polarization state.

Using Eq. (10), the matrix element M_{if} (for the σ–polarized photon) in two–wave approximation ($h=0,g$) may be decomposed into two parts:

$$M_\sigma^{if}=-e/\gamma mc\left(\langle\phi_f,\boldsymbol{p}'_z\left|A^*_{\sigma 0}\vec{\varepsilon}_{\sigma 0}\exp(-\mathrm{i}\vec{\kappa}_0\boldsymbol{r})\hat{\boldsymbol{p}}\right|\phi_i,\boldsymbol{p}_z\rangle+\langle\phi_f,\boldsymbol{p}'_z\left|A^*_{\sigma g}\exp(-\mathrm{i}\vec{\kappa}_g\boldsymbol{r})\vec{\varepsilon}_{\sigma g}\hat{\boldsymbol{p}}\right|\phi_i,\boldsymbol{p}_z\rangle\right). \quad (11)$$

According to [2], the first term in Eq (11) represents or the Čerenkov radiation (when $i=f$), or CR (when $i\neq f$), under the channeling condition. The second term in Eq. (11), according to [2], represents the emission of diffracted photon. When $i=f$ (the transition of the channeled electron is intraband) this term gives PXR under channeling conditions (PXRC). When $i\neq f$, this term describes the emission of diffracted photon due to the interband transition of electron between the transverse states – DCR.

According to Ref. [2], in two wave approximation

$$A_{\sigma g}=A_{\sigma 0}\big/2\sqrt{1+W_\sigma^2},\ W_\sigma=\left(R-\left(|\chi_g|P_\sigma\right)^2/R\right)\big/2|\chi_g|P_\sigma, \quad (12)$$

where W_σ is "resonance error" and

$$R=(\theta_x-\cot\theta_B\Omega_{if}/\omega_B)^2+\theta_y^2+\theta_{kin}^2-2\Omega_{if}/\omega_B,\ P_\sigma=\vec{\varepsilon}_{\sigma 0}\vec{\varepsilon}_{\sigma h},\ \theta_{kin}^2=\gamma^{-2}+|\chi_0|, \quad (13)$$

$\omega_B=c|\boldsymbol{g}|/(2\sin\theta_B)$ is the Bragg frequency, χ_g is the Fourier component of the electric susceptibility. Then, using the dipole approximation and relation $\langle f|\hat{\boldsymbol{p}}_{\perp}|i\rangle=-\mathrm{i}\gamma m_e\Omega_{if}\langle f|\boldsymbol{r}_{\perp}|i\rangle$ one can rewrite the second term of Eq. (11) in the following form:

$$2c\sqrt{1+W_\sigma^2}M_{-g\sigma}^{if}=\mathrm{i}eA^*_{\sigma 0}\left(\langle\phi_f,\boldsymbol{p}'_z\left|c\beta\varepsilon_{\sigma gz}\vec{\kappa}_{-g}(x\vec{\varepsilon}_{\|0}+y\vec{\varepsilon}_{\perp 0})+\Omega_{if}\vec{\varepsilon}_{\sigma g}(x\vec{\varepsilon}_{\|0}+y\vec{\varepsilon}_{\perp 0})\right|\phi_i,\boldsymbol{p}_z\rangle\right). \quad (14)$$

Here, $\vec{\kappa}_{-g}=\vec{\kappa}-\vec{g}=\vec{\kappa}_0+\vec{u}$ (Fig.3), $\varepsilon_{\sigma gz}$ is Z–component of $\vec{\varepsilon}_{\sigma g}$ and we have written the radius vector $\boldsymbol{r}_{\perp}$ through its components $\boldsymbol{r}_{\perp}=x\vec{\varepsilon}_{\|0}+y\vec{\varepsilon}_{\perp 0}$, where the polarization vectors are both $\vec{\varepsilon}_{\|0}$ and $\vec{\varepsilon}_{\perp 0}$ are lying in XY plane and $\vec{\varepsilon}_{\perp 0}$ is the polarization vector parallel to the diffraction plane, $\beta=\mathrm{v}/c$. Formally, Eq. (14) coincides with the formula obtained in Ref.[2] for DCR at planar channeling if X–component of radius vector $\boldsymbol{r}_{\perp}$ is zero $x=0$.

Then, taking into account the relations $P_{\|}\equiv\vec{\varepsilon}_{\|g}\vec{\varepsilon}_{\|0}$, $P_{\perp}\equiv\vec{\varepsilon}_{\perp g}\vec{\varepsilon}_{\perp 0}$, and definition of coordinates [1] $\theta_{x,y}=c(\vec{\kappa}-\vec{\kappa}_B)|_{x,y}\ \omega_B^{-1}=cu_{x,y}\omega_B^{-1}$, $\vec{\kappa}_B=\vec{\kappa}_0+\vec{g}$ (Fig.3), one can obtain the radiation probability for DCR $w_{if}\propto|M_{-g\sigma}^{if}|^2=|M_{-g\|}^{if}|^2+|M_{-g\perp}^{if}|^2$. Then number of photons emitted into elementary solid angle per unit length of the crystal:

$d^3N_{if}/d\theta_x d\theta_y dz=$

$$\frac{\alpha\omega_B^3\beta^4}{4\pi c^3\sin^2\theta_B}\left(\frac{\left(y_{if}\theta_x\theta_y+x_{if}\left(\theta_x^2-P_{\parallel}\Omega_{if}/\omega_B/\beta^2\right)\right)^2}{4\left(1+W_{\parallel}^2\right)}+\frac{\left(x_{if}\theta_x\theta_y+y_{if}\left(\theta_y^2-P_{\perp}\Omega_{if}/\omega_B/\beta^2\right)\right)^2}{4\left(1+W_{\perp}^2\right)}\right), \qquad (15)$$

where α is the fine structure constant, $x_{if}=\langle\phi_f|x|\phi_i\rangle$ and $y_{if}=\langle\phi_f|y|\phi_i\rangle$. It is clear that if $x_{if}=0$ (i.e. transverse motion exists only along *Y*–axis), the formula (15) coincides with the formula obtained in Ref.[2] for DCR at planar channeling.

5. Initial populations and matrix elements for axially channeled electrons

Using calculated 2D wave functions $\phi^i(\boldsymbol{r}_\perp,\boldsymbol{k}_\perp)$, we may start with analysis of axial channeling radiation and DCR.

First, it is necessary to know initial populations of transverse energy levels of channeling electrons:

$$P_i(\boldsymbol{k}_o,\boldsymbol{k}_\perp)=\left|\langle\phi^o(\boldsymbol{r}_\perp,\boldsymbol{k}_o)\|\phi^i(\boldsymbol{r}_\perp,\boldsymbol{k}_\perp)\rangle\left(d_x d_y\right)^{-1}\right|^2,\quad \phi^o(\boldsymbol{r}_\perp,\boldsymbol{k}_o)=\exp(-\mathrm{i}\boldsymbol{k}_o\boldsymbol{r}_\perp). \qquad (16)$$

Here $\phi^o(\boldsymbol{r}_\perp,\boldsymbol{k}_o)$ is the wave function for transverse motion of incident free electron with wave vector $\boldsymbol{k}=(\boldsymbol{k}_o,\boldsymbol{k}_\parallel)$, where $|\boldsymbol{k}_o|\cong|\boldsymbol{k}|\sin\theta_o\cong|\boldsymbol{k}|\theta_o$, with θ_o being the angle between the $\boldsymbol{p}=\hbar\boldsymbol{k}$ and channeling axis (*Z*–axis), d_x,d_y are the distances between the channeling axes in plane *XY*.

Fig.5 presents the results of calculation of the initial populations $P_i(\boldsymbol{k}_o,\boldsymbol{k}_\perp)$ in the case of electrons with γ = 20 (energy 10.22 MeV) channeled along $\langle 110\rangle$ axes of LiF. It is clearly seen that at axial channeling, as well as at planar channeling [3–4], the populations are strongly different for different bands, moreover the populations are nonzero for all bands. It is the effect of 2D periodicity of channeling axial problem. For example, if one considers axial channeling in one–string approximation, at $\theta_o=0$ the quantum states with nonzero angular momentum are not populated at all.

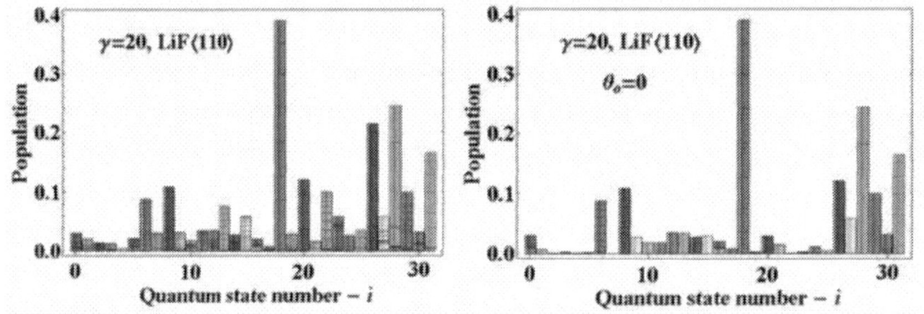

Figure 5. Initial population $P_i(\boldsymbol{k}_o,\boldsymbol{k}_\perp)$ of transverse energy bands. **Left:** Initial population for all initial angles θ_o in range $(0,2\theta_L)$ **Right:** Only for $\theta_o=0$. Here, θ_L is the Lindhard critical angle.

Second, for calculation of channeling radiation probability in dipole approximation, the transition matrix elements are required:

$$\boldsymbol{r}_{if}(\boldsymbol{k}_\perp^f,\boldsymbol{k}_\perp^i)=\langle\phi^f(\boldsymbol{r}_\perp,\boldsymbol{k}_\perp)|\boldsymbol{r}_\perp|\phi^i(\boldsymbol{r}_\perp,\boldsymbol{k}_\perp)\rangle. \qquad (18)$$

However, the calculation with Mathematica© of the matrix elements $r_{if}(\boldsymbol{k}_\perp^f,\boldsymbol{k}_\perp^i)$ according Eq. (18) takes enormous amount of time. We solved this problem in the following way: in Eq. (18) we divided the summation and integration and integrals are computed with special subroutine (written in C++).

Fig.6 shows the calculated matrix elements $|x_{if}(\boldsymbol{k}_\perp^f,\boldsymbol{k}_\perp^i)|^2$ and $|y_{if}(\boldsymbol{k}_\perp^f,\boldsymbol{k}_\perp^i)|^2$ for electrons with $\gamma = 20$ channeled along the $\langle 110 \rangle$ LiF axes. These results strongly differ from DCR at planar channeling [3-4] – the matrix elements $y_{if}(\boldsymbol{k}_\perp^f,\boldsymbol{k}_\perp^i)$ and $x_{if}(\boldsymbol{k}_\perp^f,\boldsymbol{k}_\perp^i)$ have nonzero values only for several interband transitions.

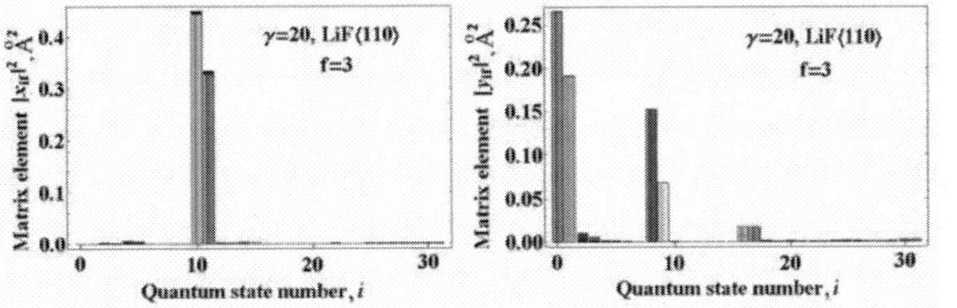

Figure 6. Calculated matrix elements $x_{if}(\boldsymbol{k}_\perp^f,\boldsymbol{k}_\perp^i)$ and $y_{if}(\boldsymbol{k}_\perp^f,\boldsymbol{k}_\perp^i)$ for $i \leftrightarrow 3$ transitions.

6. Angular distribution of DCR from axially channeled electrons

For calculations of the angular distribution of DCR at the Bragg angle from initial electron beam, we assume that the contribution to the DCR comes from those channeled electrons that are captured into subbarrier levels (the remaining electrons are involved in the formation of PXR), i.e.

$$d^3N_{if}/d\theta_x d\theta_y dz = \sum_f \left(\sum_i P_i(\boldsymbol{k}_o,\boldsymbol{k}_\perp) d^3N_{if}/d\theta_x d\theta_y dz\right). \quad (19)$$

Fig. 7 shows the ratio of calculated angular distribution of DCR to the maximal value of RXR (obtained using the standard formula, see e.g. [2]), in vicinity of the Bragg angle $\theta_B \cong 0.32$ rad for the electrons with $\gamma = 20$ channeled along $\langle 110 \rangle$ axes of LiF at incident angle $\theta_o = 0$.

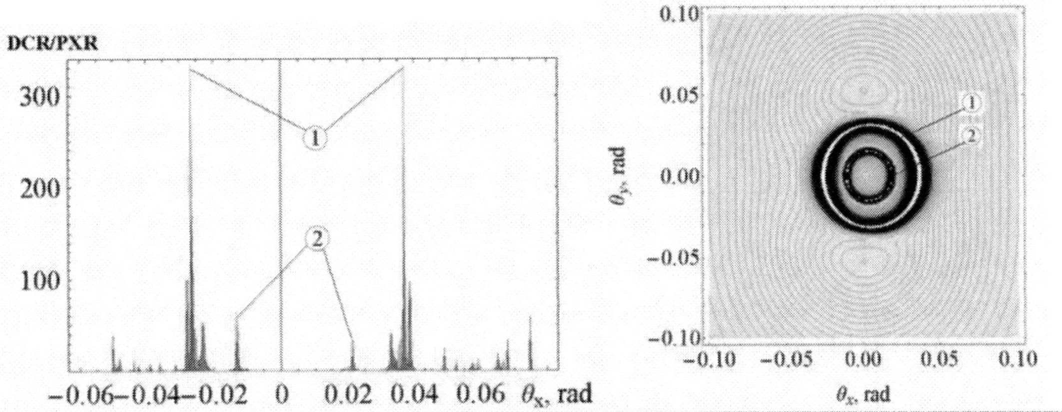

Figure 7. Left: the angular distribution of DCR versus the angle θ_x (at $\theta_y = 0$) for the electrons with $\gamma = 20$ and incident angle $\theta_0 = 0$ channeled along $\langle 110 \rangle$ LiF; the Bragg angle $\theta_B \cong 0.32$ rad. **Right:** contour diagram for angular distribution of DCR (only central (2) and maximal (1) peaks) and PXR. The photon energy $\hbar\omega_B = 2.174$ keV.

Finally, Figure 8 shows 3D fragment of the central and maximal peaks of DCR presented in Fig. 6. The peaks are ring–shaped and have complex doublet structure.

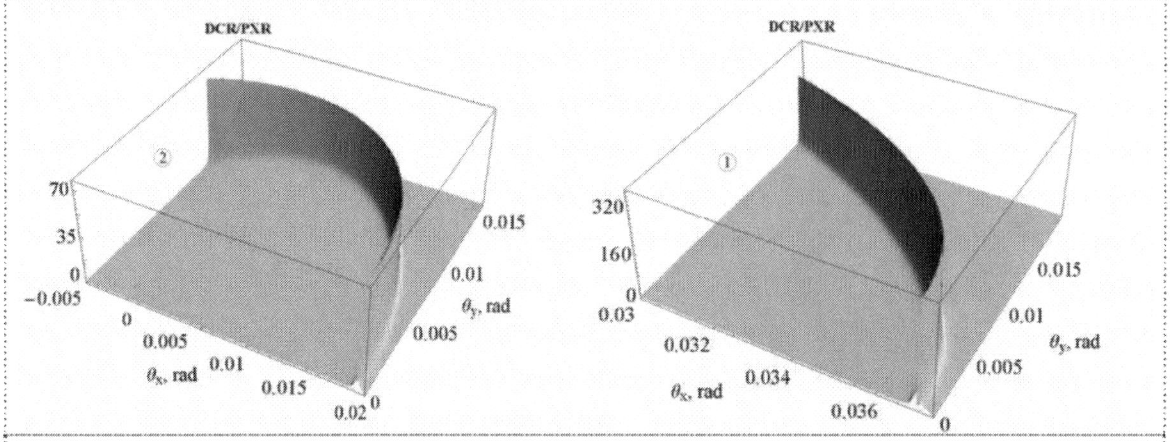

Figure 8. 3D fragments of the central (left) and maximal (rigth) peaks of the DCR angular distribution shown in Fig. 6.

7. Conclusions

The developed numerical method of solving 2D Schrödinger equation for axial channelling of electrons allowed for the first time calculations of angular distribution of DCR from axially channelled electrons. The calculations show an appearance of narrow brilliant DCR peaks in vicinity of the Bragg angle. That means, DCR amplitude exceeds sufficiently ordinary PXR amplitude and that DCR can be observed in experiments if the crystal thickness is less than dechanneling length.

The work is partially supported by Russian Science and Innovations Federal Agency under contract No 02.740.11.0238

8. Acknowledgments

The authors are grateful to Prof. H. Nitta for useful discussions.

References

[1] Baryshevsky V G, Feranchuk I D, Ulyanenkov A P 2005 *Parametric X–ray Radiation in Crystals* (Berlin: Springer–Verlag)
[2] Yabuki R, Nitta H, Ikeda T, Ohtsuki Y H 2001 *Physical Review* B **63** 174112
[3] Bogdanov O V, Korotchenko K B, Pivovarov Yu L 2007 *JETP Letters* **85** No. 11 555
[4] Korotchenko K B, Pivovarov Yu L Tukhfatullin T A 2008 *Nucl. Inst. & Meth. in Phys. Res.* **B266** 3753
[5] Bazylev V A, Zhevago N K 1987 *Radiation from fast particles in a matter and external fields* (Moscow: Nauka)
[6] Ashcroft N W, Mermin N D 1979 *Solid State Physics* (New York: Holt, Rinehart and Winston)
[7] Buxtont B F and Loveluck J E 1977 *J. Phys. C: Solid State Phys.* **10** 3941
[8] Klein R K, Kephart J O, Pantell R H, Park H, Berman B L, Swent R L, Datz S and Fearick R W 1985 *Phys.Rev.*B **31** 68
[9] Doyle P A and Turner P S 1967 *Acta Cryst.* A24, 390

Polarization bremsstrahlung in a backward direction for medium structure diagnostics.

N. Nasonov‡[1], P. Zhukova[1], V. Sergienko[2]

[1] Laboratory of Radiation Physics, Belgorod State University, 14 Studencheskaya st., 308007 Belgorod, Russia
[2] Lebedev Physical Institute RAS, Moscow, Russia

Abstract. Possibilities to elaborate the new energy dispersive methods of the diagnostics of a medium atomic structure based on the polarization bremsstrahlung from relativistic electrons moving in a sample are analyzed in the paper. The method of radial density of atoms measuring in a weakly ordered medium is proposed in the paper as well as the method of cluster size determination in small grained medium.

PACS numbers: 78.70.-g; 78.70.C_k

Submitted to: *J. Phys. B: At. Mol. Opt. Phys.*
Keywords: Polarization bremsstrahlung; Relativistic electrons; Radial density of atoms; Cluster size.

‡ Corresponding author: tel. +7 4722 315726; fax: +7 4722 311213;
e-mail: nnn@bsu.edu.ru

1. Introduction

When a fast charged particle moves through a dense medium it can emit electromagnetic waves due to the scattering of the particle Coulomb field by medium atoms. The distinctive property of the emission being discussed, known as the polarization bremsstrahlung (PB) [1, 2, 3], consists in the strong dependence of its characteristics on inter-atomic correlations in the target. This property opens the possibility to use PB for medium atomic structure diagnostics [4]. Since the spectrum of initial virtual photons associated with incident particle Coulomb field is wide enough, the spectral measurements of scattered photons are more convenient for the structure diagnostics on the base of PB. Such variant of X-ray diagnostics, known as EDXD (energy dispersive X-ray diffraction) [5, 6, 7], requires a well determined spectrum of primary photons. It should be noted in this connection that the use of PB for the diagnostics has an advantage consisting in the possibility to calculate exactly the spectrum of primary virtual photons. One further advantage of this approach connects with the possibility to elaborate the method of structure diagnostics with high space resolution. Indeed, the problem of a primary electron beam focusing on the sample by equipments of an ordinary magnetic optics is incomparably less than that for X-ray beam.

The suggestion to use PB for atomic structure diagnostics was verified by experiments, performed in the last few years. The measurements of the distances between atomic planes in polycrystalline targets have been realized experimentally on the base of PB from electrons with energies of about units MeV [8, 9, 10]. The possibility to determine texture in polycrystals by PB from 150 MeV electrons was demonstrated at the experiment [11]. The method of mosaic crystal diagnostics on the base of PB and the method of a nanocluster size measuring in small grained media have been recently proposed [12].

It should be noted that many crystallographjc planes in a polycrystalline target make the contribution to PB yield simultaneously in the form of Bragg reflexes. The number and positions of such reflexes in total spectrum depend strongly on the observation angle between directions of an incident electron beam and diffracted photon fluxes. It is significant that reflexes can overlap and hence the accuracy of measurements is determined by the spectral width of these reflexes. Generally the value of PB peak width $\Delta\omega$ is inverse proportional to Lorentz factor of an emitting electron γ, so that the energy resolution of the method under consideration increases with increasing γ. Unique possibility to increase the energy resolution in the process of polycrystal diagnostics is connected with coherent PB peaks emitted in a backward direction relative to the electron velocity [13]. It has been shown in [13], that $\Delta\omega$ for such anomalous peaks is proportional to γ^{-2} and so these peaks are of interest for diagnostics tasks.

It is significant that the analogous peaks occur in the process of coherent PB from relativistic electrons moving in a single crystal [14, 15] (coherent part of PB in a single crystal is identical to well known parametric X-ray radiation [16, 17, 18]). Experimental studies of parametric X-ray radiation peaks emitted to backward direction relative to an emitting electron velocity have demonstrated that the spectral width of these peaks ranges up to about tens MeV [15].

Our paper is devoted to the advancement of approach [13]. Possibilities to use backward PB peaks for the diagnostics of ordered in part media are the subject for our studies. New method of the diagnostics of short-range order in weakly ordered medium based on above PB peaks is elaborated in the next section of the paper. We discuss in section 3 the new method of cluster size measurement on the base of backward PB. Our conclusions are presented in section 4.

Relativistic system of units $\hbar = c = 1$ is used in the paper.

2. Relativistic electron PB to backward direction. Study of short-range order in weakly ordered medium

Let us consider the spectral-angular characteristics of PB photons emitted by relativistic electrons crossing a thin solid target starting from Maxwell equations for Fourier transform of excited electric field

$$(k^2 - \omega^2)\mathbf{E}_{\omega\mathbf{k}} - \mathbf{k}(\mathbf{k} \cdot \mathbf{E}_{\omega\mathbf{k}}) = 4\pi i\omega \mathbf{J}_{\omega\mathbf{k}} + \frac{i\omega e}{2\pi^2}\mathbf{V}\delta\left(\omega - \mathbf{k} \cdot \mathbf{V}\right), \tag{1}$$

Here the last item in the right side of (1) describes the current density of a fast electron, $\mathbf{J}$ is the induced current density of medium electrons, determined within the frame of dipole approximation [3] by the equations

$$4\pi i\omega \mathbf{J}_{\omega\mathbf{k}} = \omega^2 \int d^3k' G(\mathbf{k}' - \mathbf{k})\mathbf{E}_{\omega\mathbf{k}'}, \tag{2}$$
$$G = \frac{1}{2\pi^2}\frac{\alpha_0(\omega)}{1 + (\mathbf{k})' - \mathbf{k}^2\mathbf{R}^2}\sum_l e^{i(\mathbf{k}'-\mathbf{k})\mathbf{r}_l}, \quad \alpha_0(\omega) = \frac{e^2}{m}\sum_{n\geq 1}\frac{f_{n0}}{\omega_{n0}^2 - \omega^2},$$

where R is the screening radius in Fermi-Thomas atom model (the simplest model with exponential screening is used in our calculations), summation in (2) is performed over all atoms in the target, $\mathbf{r}_l$ is the coordinate of l-th atom, $\alpha_0(\omega)$ is the ordinary dipole polarizability of an atom.

The function G describes both refraction and scattering of photons in the target. It is convenient for the further analysis to separate G into averaged over coordinates of all atoms and accidental components

$$G = \bar{G} + \tilde{G},\ \bar{G} = \langle G \rangle = 4\pi n_0 \alpha_0(\omega)\delta(\mathbf{k}' - \mathbf{k}),\ 4\pi n_0\alpha_0(\omega) = \chi(\omega),\ \ 1 + \chi(\omega) = \varepsilon(\omega), \tag{3}$$

so that the scattering of a fast electron Coulomb field will be described by $\tilde{G}$ only. Here n_0 is the average density of atoms,$\chi(\omega)$ and $\varepsilon(\omega)$ are the dielectric susceptibility and permeability of the target respectively.

The solution of (1) should be determined by perturbation theory. Substituting (3) into (1) and representing the total field $\mathbf{E}_{\omega\mathbf{k}}$ as a sum of the electron Coulomb field $\mathbf{E}^{eq}_{\omega\mathbf{k}}$

and the emission field $\mathbf{E}^{rad}_{\omega\mathbf{k}}$ one can obtain in the first-order of $\tilde{G}$ the following expression for $\mathbf{E}^{rad}_{\omega\mathbf{k}}$

$$E^{rad}_{\omega\mathbf{k}} = \frac{i\omega^3 e}{2\pi^2}\frac{1}{k^2-\omega^2\varepsilon}\int \frac{d^3k'}{k'^2-\omega^2\varepsilon}\tilde{G}(\mathbf{k}'-\mathbf{k})\left(\mathbf{V}-\frac{\mathbf{k}'}{\omega\varepsilon}-\mathbf{k}\frac{\mathbf{k}\cdot\mathbf{V}-\frac{\mathbf{k}\cdot\mathbf{k}'}{\omega\varepsilon}}{\omega^2\varepsilon}\right)\delta(\omega-\mathbf{k}'\cdot\mathbf{V})(4)$$

Formula for the spectral-angular distribution of the number of emitted photons follows from (4) in the form

$$\omega\frac{dN}{d\omega d\Omega} = e^2\omega^6\int\frac{d^3k_1}{k_1^2+2\omega\sqrt{\varepsilon}\mathbf{n}\mathbf{k}_1}\frac{d^3k_2}{k_2^2+2\omega\sqrt{\varepsilon}\mathbf{n}\mathbf{k}_2}\langle\tilde{G}(\mathbf{k}_1)\tilde{G}^*(\mathbf{k}_2)\rangle\Big[\left(\mathbf{V}-\frac{\mathbf{k}_1}{\omega\varepsilon}\right)\left(\mathbf{V}-\frac{\mathbf{k}_2}{\omega\varepsilon}\right)-$$
$$\left(\mathbf{n}\mathbf{V}-\frac{\mathbf{n}\mathbf{k}_1}{\omega\varepsilon}\right)\left(\mathbf{n}\mathbf{V}-\frac{\mathbf{n}\mathbf{k}_2}{\omega\varepsilon}\right)\Big]\delta\left(\omega(1-\sqrt{\varepsilon}\mathbf{n}\mathbf{V})-\mathbf{k}_1\mathbf{V}\right)\delta\left(\omega(1-\sqrt{\varepsilon}\mathbf{n}\mathbf{V})-\mathbf{k}_2\mathbf{V}\right), \quad (5)$$

where all interatomic correlations of interest to us are contained in the correlator $\langle\tilde{G}(\mathbf{k}_1)\tilde{G}^*(\mathbf{k}_2)\rangle$.

To show the possibility to use PB for study of short-range order in solids one should calculate above correlator using two-particle density $f_2(\mathbf{r}_l,\mathbf{r}_m)$ which may be presented as a sum $f_2(\mathbf{r}_l,\mathbf{r}_m) = f_1(r_l)f_1(r_m) + g_2(|\mathbf{r}_l-\mathbf{r}_m|)$, where $f_1 = 1/V$ is one-particle distribution function for a homogeneous target, V is the volume of this target, g_2 is the two-particle correlation function. The desired radial density is expressed in term of the correlation function by the formula $n_0V^2n(r) = n(r) - n_0$ [19]. Based on above considerations one can represent (5) in the form

$$\omega\frac{dN}{d\omega d\Omega} = \omega\frac{dN_0}{d\omega d\Omega} + \omega\frac{d\Delta N}{d\omega d\Omega}. \quad (6)$$

Here the first item in the right side describes PB generated independently on each atom of the target

$$\omega\frac{dN_0}{d\omega d\Omega} \simeq \frac{2Z^2e^6}{\pi m^2}Ln_0\mid\frac{\omega^2}{\omega_p^2}\chi(\omega)\mid^2\frac{1}{(1+4\omega^2R^2)^2}\left(\ln\sqrt{\frac{1+4\omega^2R^2}{\omega^2R\rho^2}}-1\right),(7)$$

where Z is the number of electrons in a target atom, L is the thickness of the target, ω_p is the plasma frequency, $\rho^2 = \gamma^{-2} + \omega_p^2/\omega^2$, γ is Lorentz factor of an emitting fast electron.

The last item in (6) describes an addition to PB caused by interatomic correlations

$$\omega\frac{d\Delta N}{d\omega d\Omega} = \frac{2Z^2e^6}{m^2\omega}Ln_0\mid\frac{\omega^2}{\omega_p^2}\chi(\omega)\mid^2\int_1^\infty\frac{dx}{(1+4\omega^2R^2x^2)^2}\frac{x^2-1}{(x^2-1+\frac{1}{4}\rho^2)^2}\times \quad (8)$$
$$\int_0^\infty drr(n(r)-n_0)\sin(2\omega rx)$$

It is important to keep in mind that the results (7) and (8) have been obtained to fit PB photons emitted in a backward direction relative to an emitting electron velocity.

This is of no concern for PB on a separate atom (7), since the angular distribution of PB on an atom is close to isotropic one. On the other hand, it is of fundamental importance in the task of $\omega\frac{d\Delta N}{d\omega d\Omega}$ determination on the base of the result (8). Indeed, the connection between Fourier transform of radial density $n(r)$ and the quantity $\omega\frac{d\Delta N}{d\omega d\Omega}$ is integral, but in conditions of PB in a backward direction under consideration this connection can be approximated by local one due to extremely small value of the width of the range of momentum transfer, as it follows immediately from the coefficient $(x^2-1)/(x^2-1+\rho^2/4)^2$ consisting in a very sharp peak with the width of the order of $\rho^2/4 \approx 1/4\gamma^2 \ll 1$. Since the effective values of in the integral (8) is in the region of a distance between atoms l, the variable x in the function $\sin(2\omega rx)$ is approximately equal to unity with the proviso that

$$\frac{1}{4}\omega l\rho^2 \approx \frac{1}{4}\omega l\gamma^{-2} \ll 1 \tag{9}$$

As a result the formula (8) takes the form

$$\omega\frac{d\Delta N}{d\omega d\Omega} \approx A(\omega)\int_0^\infty drr(n(r)-n_0)\sin(2\omega r) \tag{10}$$

$$A(\omega) = \frac{2Z^2e^6}{m^2\omega}Ln_0 \mid \frac{\omega^2}{\omega_p^2}\chi(\omega) \mid^2 \frac{1}{(1+4\omega^2R^2)^2}\ln\frac{16}{\rho^2}$$

Formulae (6), (7) and (8) allow one to determine the function of radial density $n(r)$. Indeed, since PB spectral-angular distribution $\omega\frac{dN}{d\omega d\Omega}$ is defined experimentally and the coefficient $A(\omega)$ is calculated by the formula (10) as well as the distribution $\omega\frac{dN_0}{d\omega d\Omega}$ is calculated by (7) one can determine $n(r)$ by Fourier integral

$$n(r) - n_0 = \frac{1}{r}\frac{4}{\pi}\int_0^\infty d\omega \sin(2\omega r)\frac{\omega\frac{dN}{d\omega d\Omega} - \omega\frac{dN_0}{d\omega d\Omega}}{A(\omega)}, \tag{11}$$

analogous to that used in the classical Zernicke - Prins approach [20], based on angular measurements of scattered by a sample quasi-monochromatic X-ray beam.

It is of first importance that the approach being proposed does not constitute a conceptual disadvantage of Z-P method consisting in the availability of a finite top limit in the Fourier integral. The indicated limit causes substantial distortions in the function $n(r)$. The absence of above distortions within the frame of our approach is caused by EDXD technique and the possibility to determine the spectrum of initial virtual photons exactly.

3. Diagnostics of cluster size in small grained medium by anomalous PB peak

The general formula (5) is applicable for targets with arbitrary atomic structures. Let us use this formula for an analysis of the possibility to measure a cluster size in small grained medium on the base of PB. In order to calculate the correlator $\langle\tilde{G}(\mathbf{k}_1)\tilde{G}^*(\mathbf{k}_2)\rangle$ one should set the coordinates of atoms in the target $\mathbf{r}_n$ by the formula

$$\mathbf{r}_n = \mathbf{R}_n + \mathbf{r}_{nl} + \mathbf{u}_{nl}, \tag{12}$$

where $\mathbf{R}_n$ is the coordinate of n-th cluster, $\mathbf{r}_{nl}$ is the coordinate of l-th atom in n-th cluster, $\mathbf{u}_{nl}$ is the thermal displacement of this atom. Averaging over thermal displacements $\mathbf{u}_{nl}$, cluster sizes, forms and orientations leads to the following expression for the correlator:

$$\langle\tilde{G}(\mathbf{k}_1)\tilde{G}^*(\mathbf{k}_2)\rangle = \frac{2}{\pi}\frac{|\alpha_0(\omega)|^2}{(1+k_1^2R^2)^2}\Big[n_0\left(1-e^{-k_1^2u^2}\right)+ \tag{13}$$
$$n_S\left(\langle|S(\mathbf{k}_1)|^2\rangle + |\langle S(\mathbf{k}_1)\rangle|^2\int d^3r\,(n_c(r)-n_S)\,e^{-\mathbf{k}_1\mathbf{r}}\right)e^{-k_1^2u^2}\Big]\delta\left(\mathbf{k}_1-\mathbf{k}_2\right),$$

where u is the mean square amplitude of thermal vibrations of atoms, n_S is the density of clusters, $n_c(r)$ is the radial density of clusters, analogous to that in an amorphous medium, $\langle S(\mathbf{k})\rangle \equiv \langle S_l(\mathbf{k})\rangle$, $S_l(\mathbf{k}) = \sum_{n=1}^{N_l}\exp(i\mathbf{k}\mathbf{r}_{ln})$ is the structure factor of l-th cluster, containing N_l atoms.

Since $u \ll R$, the contribution of incoherent part of PB is strongly suppressed in (13) and hence PB spectral-angular distribution follows from (5) and (13) in the form

$$\omega\frac{dN^{coh}}{d\omega d\Omega} = \frac{Z^2e^6}{\pi^2m^2}Ln_S\left|\frac{\omega^2}{\omega_p^2}\chi(\omega)\right|^2\int\frac{d^3k}{(1+k^2R^2)^2}\frac{\left(\omega\mathbf{V}-\frac{\mathbf{k}}{\varepsilon}\right)^2-\left(\omega\mathbf{n}\mathbf{V}-\frac{\mathbf{n}\mathbf{k}}{\varepsilon}\right)^2}{(k^2+2\omega\sqrt{\varepsilon}\mathbf{n}\mathbf{k})^2}\times \tag{14}$$
$$\left[\langle|S(\mathbf{k})|^2\rangle + \frac{4\pi}{k}|\langle S(\mathbf{k})\rangle|^2\int_0^\infty drr(n_c(r)-n_S)\sin(kr)\right]e^{-k^2u^2}\delta\left(\omega(1-\sqrt{\varepsilon}\mathbf{n}\mathbf{V})-\mathbf{k}\mathbf{V}\right)$$

different from (6) by the availability of cluster structure factor $S(\mathbf{k})$, describing the coherent contribution of cluster atoms to PB yield. It is clear, that the quantity $\langle|S(\mathbf{k})|^2\rangle$ depends on the module of k only. In conditions of relatively small cluster sizes under consideration (measurement of cluster size of the order of 10 nm holds the greatest interest) the above function comprises a peak with a maximum at the point $k = 10$. Amplitude of this peak increases with increasing in cluster size, but its width decreases. Since the peak being discussed describes the cluster as an entire object, the microscopic arrangement of atoms in the cluster is of no importance to study of cluster size and hence we will use a very simple model of cubic lattice. Bearing the fundamental properties of the diagnostics method being studied in mind we neglect inter-cluster correlations in (14) and consider the cluster as a cube with N_S atoms arranged along its verge, so that the full number of atoms in the cluster is equal to N_S^3. Assume that a is the distance between neighboring atoms. As this takes place, the quantity $\langle|S(\mathbf{k})|^2\rangle$ is given by the formula

$$T(ka,N_S) = 1 + \frac{6}{N_S}\sum_{j=1}^{N}(N_S - j)\eta(j) + \frac{12}{N_S^2}\sum_{j=1}^{N}(N_S - j)\sum_{l=1}^{N}(N_S - l)\eta(\sqrt{j^2+l^2}) +$$

$$\frac{8}{N_S^3}\sum_{j=1}^{N}(N_S - j)\sum_{l=1}^{N}(N_S - l)\sum_{p=1}^{N}(N_S - p)\eta(\sqrt{j^2+l^2+p^2}) \quad (15)$$

$$\eta(\sqrt{j^2+l^2+p^2}) = \frac{\sin(ka\sqrt{j^2+l^2+p^2})}{ka\sqrt{j^2+l^2+p^2}}, \qquad \langle|S(\mathbf{k})|^2\rangle = N_S^3 T(ka,N_S)$$

The dependence $T(ka, N_S)$ presented in Fig.1 confirms completely above reasoning on the properties of cluster structure factor.

Let us consider now the basic idea of cluster size measuring on the base of coherent PB from relativistic electrons. First of all, it should be noted that the spectrum of coherent PB described by (14),(15) has a maximum. Indeed, the factor $|\frac{\omega^2}{\omega_p^2}\chi(\omega)|^2$ in (14) tends to zero with decreasing in ω within the small frequency range where the susceptibility $\chi(\omega)$ is approximately constant and hence PB spectrum is suppressed in small frequency range. On the other hand, coherent PB is suppressed in high frequency range, where the cluster structure factor decreases with increase in because of decreasing of coherent contribution of cluster atoms to the formation of PB yield. It is clear that the position of above maximum depends on the cluster size, determining the form of structure factor (see Fig.1). The performed analysis of the results (14) and (15) has shown that the frequency range where the maximum in question is realized depends strongly on the observation angle between an emitting electron velocity and the unit vector to the direction of emitted photon propagation. The most clear-cut dependence of above maximum position on the cluster size manifests for backward PB owing to extremely small width of the range of momentum transfer (in a similar way to that in the case of amorphous target), but the maximum is realized in small frequency range. A simple result follows from (14) and (15) in conditions under consideration

$$\omega\frac{dN^{coh}}{d\omega d\Omega} = \frac{2Z^2e^6}{\pi m^2}Ln_0|\frac{\omega^2}{\omega_p^2}\chi(\omega)|^2\int_1^{\infty}dx\frac{\exp(-4\omega^2u^2x^2)}{(1+4\omega^2R^2x^2)^2}xT(2\omega ax,N_S)\frac{x^2-1}{(x^2-1+\rho^2/4)^2} \approx$$

$$\frac{2Z^2e^6}{\pi m^2}Ln_0\ln\left(\frac{4}{\rho}\right)|\frac{\omega^2}{\omega_p^2}\chi(\omega)|^2T(2\omega a,N_S) \quad (16)$$

where the last formula must be used with care because its accuracy is about of 15-20 per cent.

It is well to bear in mind that the calculated term in the right side of the equation (16) which is compared with measured left side of this equation is not convenient for cluster size determination because of strong distortions contributed by the dispersion of dielectric susceptibility $\chi(\omega)$. Considerably more convenient for comparison with data function is defined as $Q(\omega a, N_S) = \omega^2a^2T(2\omega a, N_S)$.

The function $Q(\omega a, N_S)$ should be compared with experimental results by the formula

$$Q(\omega a, N_S) = \frac{\pi m^2}{2Z^2 e^6} \frac{\omega_p^2 a^2}{Ln_0 \ln(4/\rho)} \frac{\omega_p^2}{\omega^2} \frac{1}{|\chi(\omega)|^2} \omega \frac{dN^{coh}}{d\omega d\Omega} \tag{17}$$

The task of cluster size determination is solved by the selection of the parameter N_S equivalent to cluster size for good agreement between calculated function $Q(\omega)$ and measured quantity, given by the right side of the equation (17). It is clear, that more realistic results may by obtained within the frame of more general model of the target as an ensemble of clusters with different forms and sizes, but such analysis is beyond the scope of this paper.

High sensitivity of the function $Q(\omega)$ to cluster size is demonstrated by curves presented in Fig.2. An important point is that the desired cluster size is determined by the position of maximum in PB spectrum, which can by measured with accuracy substantially better than that for traditional measuring of the width of PB coherent peak.

Conclusions

The performed analysis shows the possibility to use collective effects in PB from relativistic electrons moving in a dense media for the target atomic structure diagnostics. PB in a backward direction is particularly appealing for the diagnostics because of unique small width of the range of momentum transfer in the emission process.

A new EDXD method of the diagnostics of radial density function for atoms in a weakly ordered medium on the base of relativistic electron PB in a backward direction is proposed. The desired function is reconstructed by the equation (11). The field of integration in this equation is not bounded from overhead and the spectrum of initial virtual photons associated with incident electron Coulomb field is well defined, which is why the desired function $n(r)$ is not distorted in distinction to Zernicke-Prins approach.

PB may be used for the measurement of cluster sizes in a small grained medium. The proposed method is based on one-to-one correspondence between the cluster size and the position of coherent maximum in PB spectrum caused by the scattering of incident electron Coulomb field by the cluster as an entire object. An important point is that PB in a backward direction is found to be proportional to average square of module of cluster structure factor. This circumstance allows one to determine the cluster size on the base of simple equation (18).

A conceptual sketch of the method being proposed is considered in this paper. The inclusion of a spread of cluster sizes and forms as well as inter-cluster correlations will allow one to realize substantially more realistic model for interpretations of observational data.

Acknowledgments

This work was accomplished in the context of both FCP of Russian Ministry of Education and Science and RFBR (grant:09-02-97528).

References

[1] Amusia M et al. 1992 *Polarization Bremsstrahlung from Particles and Atoms* (New York: Plenum Press)

[2] Korol A et al. 1997 *J.Phys.B:At.Mol.Opt.Phys.* **30** 1105-1150.

[3] Korol A and Lyalin A and Solov'ev A 2004 *Polarization Bremsstrahlung* (St. Petersburg: St.PSPU publ) (In Russian).

[4] Nasonov N 1998 *Nucl.Instr.Meth. B* **145** 19-24.

[5] Coppens P., Cox D., Vieg E. and Robinson I.K. 1992 *Synchrotron Radiation Crystallogrphy.* (Academic Press. London, San Diego, New York, Boston, Sydney, Tokio, Toronto). 316p.

[6] Buras B. and Gerward L. 1995 *Techniques for X-rays In: International Tables for Crystallography* (Ed.by A.J.C. Wilson. IUCr: Dordrecht, Boston, London: Kluver Academic Publlishers) 84-87.

[7] Clarck S.M. 2002 *Thirty years of energy dispersive powder diffraction* Crystallography Reviews. **8** 57-92.

[8] Blazhevich S et al. 1999 *Phys.Lett. A* **254** 230.

[9] Astapenko V., Kubankin A., Nasonov N. et al. 2006 *JETP Lett.* **84** 281-284.

[10] Gostishchev N., Kubankin A., Nasonov N. et. al 2008 *Technical Physics Lett.* **34** 78-82.

[11] Takabayashi Y et al. 2006 *Nucl.Instr.Meth. B* **243** 453.

[12] Astapenko V and Gostishchev N and Zhukova P et al 2008 *Bulletin of the Russian Academy of Science:Physics* **6** 926-929.

[13] Astapenko V and Nasonov N and Zhukova P 2007 *ZJ. Phys. B: At. Mol. Opt. Phys.* **40** 1-10.

[14] Brenzinger K-H and Limburg B and Backe H et al. 1997 *Phys. Rev. Lett.* **79** 2462-2465.

[15] Backe H and Ay C and Clawiter N et al. 2003 *Proc. Symp. Channeling-Bent Crystals-Radiation Processes* 41-58.

[16] Baryshevsky V and Feranchuk I 1971 *Sov.Phys.JETP* **61** 944-948.

[17] Garibian G and Yang S 1971 *Sov. Phys. JEPT* **61** 930-943.

[18] Baryshevsky V and Feranchuk I and Ulyanenkov A 2005*Parametric X-ray Radiation in Crystals. Theory, Experiment and Applications* (Springer. Berlin) 176 p.

[19] James R. 1950 *Optical principles of the diffraction of X-rays* (Ed. L. Bragg. London:The crystalline state) **2**.

[20] Zernicke F and Prins A 1927 *Zs. f. Phys.* **41** 184.

Figure captions

Figure 1.

Square of the module of cluster structure factor versus the momentum transfer, calculated for different cluster size.

Figure 2.

The position of the maximum of PB spectral peak versus the cluster size.

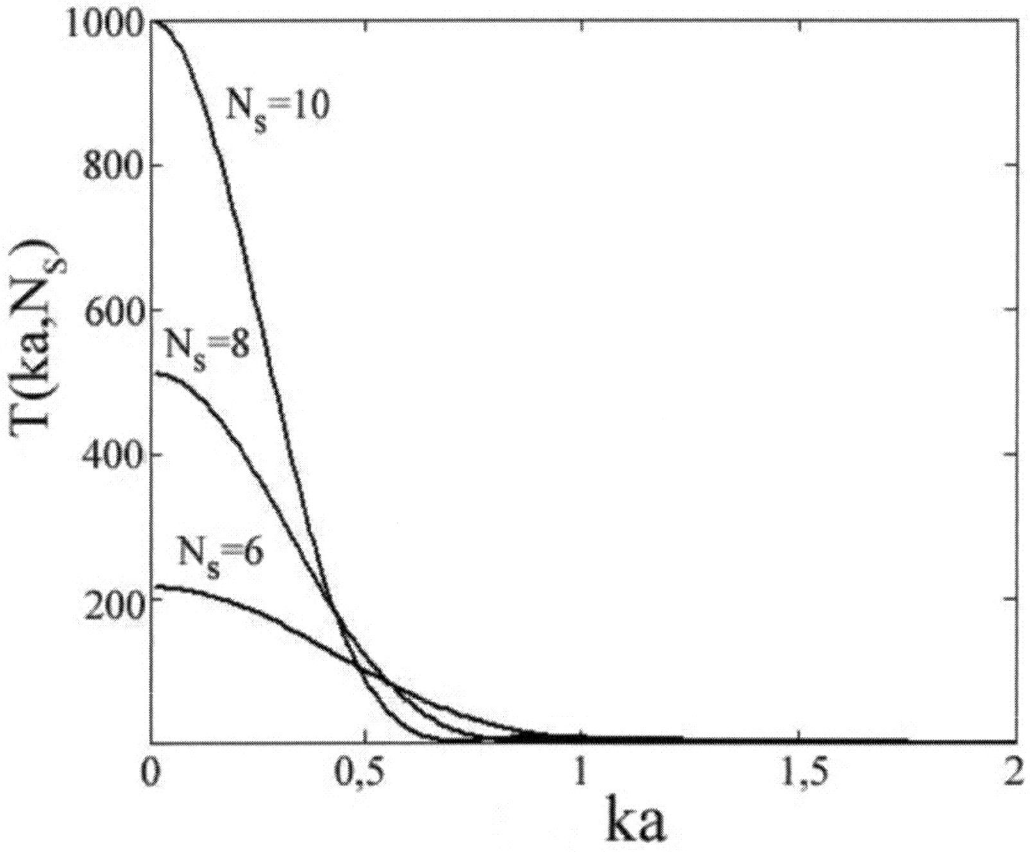

Fig.1

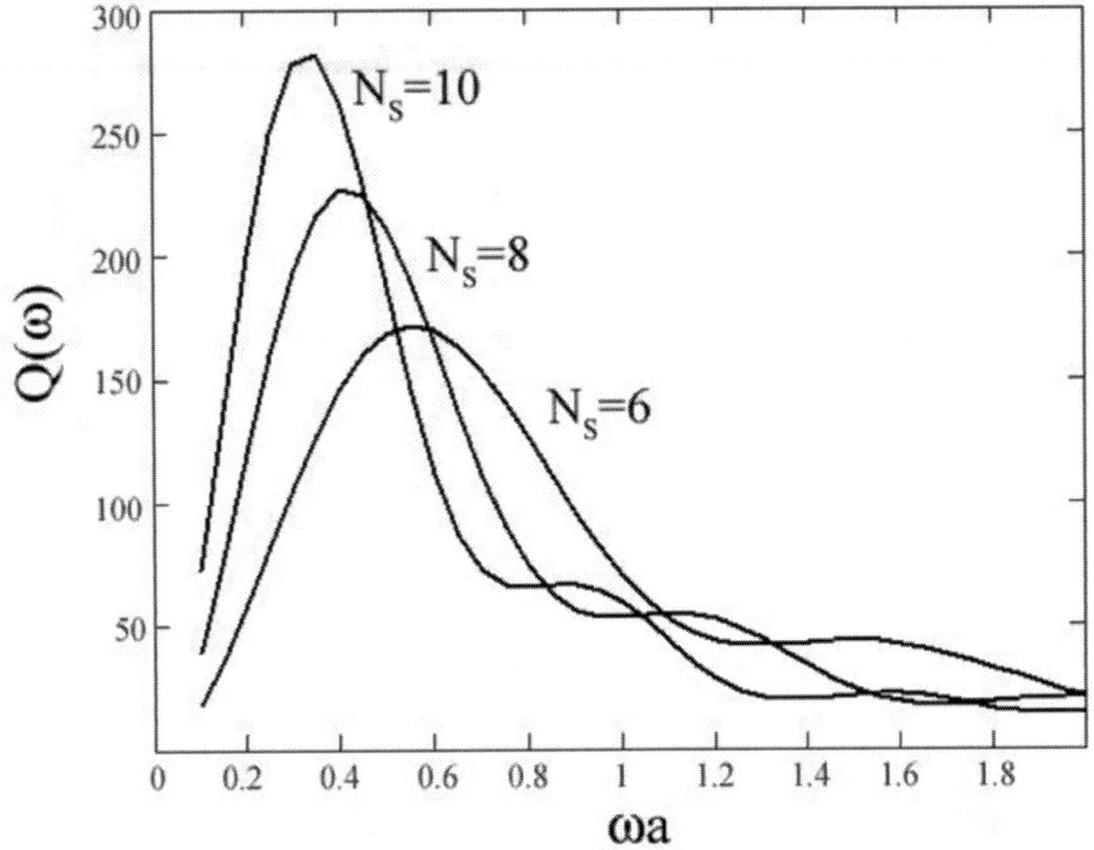
300
250
200
150
100
50
0
0.2
0.4
0.6
0.8
1
1.2
1.4
1.6
1.8
$Q(\omega)$
ωa
N_S=10
N_S=8
N_S=6

Fig.2

First observation of parametric X-ray radiation enhancement for grazing incident electrons

A.N. Eliseev[1], V.A. Khablo[1], A.S. Kubankin[2]*, N.N. Nasonov[2], R.M. Nazhmudinov[2], V.I. Sergienko[1], A.V. Subbotin[2]

[1] Lebedev Physical Institute RAS, 53 Leninskiy prospekt, Moscow, 119991, Russia.

[2] Belgorod State University, 14 Studencheskaya st., Belgorod, 308007, Russia.

*e-mail: kubankin@bsu.edu.ru

Absract. The results of first observation of PXR enhancement realizing in the scheme of grazing incidence of emitting electrons onto the surface of crystalline target in conditions of asymmetric diffraction of electron Coulomb field are presented. The measured enhancement for 7MeV electrons and Si crystal was found to be nearly 2.7. The experimental result is in agreement with theoretical predictions.

1. Introduction.

Parametric X-ray radiation (PXR) appears due to the Bragg diffraction of a fast charged particle Coulomb field on a system of atomic planes in a crystal [1-5]. The spectrum of PXR comprises a set of narrow reflexes. The emission mechanism being discussed is of interest for creating of effective source of quasimonochromatic tunable X-rays, but photoabsorption bounds the yield of PXR and the brightness of such sources is not enough for many applications. Because of this, the search of possibilities to suppress an influence of a photoabsorption on PXR generation is one of the main directions of explorations in the field of creating of sources alternative to classical synchrotron.

The main idea of the suppression of photoabsorption consists in increase of the effective generation path L_{el} of emitting electrons in a target (photons generated on L_{el} are free to escape from an absorbing target). Two approaches to the problem of L_{el} increasing have been proposed [4,6]. The first of them has been realized in the experiment [4] wherein PXR was observed for the first time. An electron beam directed at butt-end of the target moved inside the target parallel to its surface. As this took place, generated photons escaped from the target traveling through above surface and hence the path of emitted photons in the target did not depend on the electron path L_{el} which was bounded by multiple scattering only. An essential disadvantage of the method being discussed consists in the limitation on the effective transverse size of an electron beam, which must be less than a photoabsorption length L_{ab}. It should be noted that the position of reflecting plane in the experiment [4] was not parallel to a crystal surface, so that the geometry of asymmetric scattering was realized in this experiment (such geometry provided a take-off of PXR photons perpendicularly to a crystal surface).

Another way is based on the grazing incidence of an electron beam onto the crystal surface in conditions of asymmetric scattering geometry [6]. In line with [6] the grazing geometry allows to increase PXR yield tens times. The aim of this paper consists in the experimental verification of the effect [6]. The paper is organized as following. The results of calculations of optimal conditions for verification of the effect of PXR enhancement in grazing incidence geometry are presented in the next Section. The Section 3 is devoted to the description of both experimental setup and the obtained data. Our conclusions are presented in Section 4.

2. PXR enhancement in conditions of grazing incidence of emitting electrons onto the crystal surface.

PXR has been searched in many theoretical and experimental works. Traditionally PXR is considered in symmetrical geometry of diffraction, when the crystallographic plane coincides with the surface of the target and radiation propagates near to mirror direction relative to surface of the target. The used geometry of PXR process is presented in Fig.1. The electron beam propagates along the direction $\mathbf{e}_1$, its orientation relative to the surface of the target is controlled by angle α. The angle between reflecting crystallographic plane R and the crystal surface is designated as β. The direction of $\mathbf{e}_1$ coincides with the mirror direction for incident electron beam relative to the reflecting crystallographic plane R, specified by the reciprocal lattice vector $\mathbf{g}$. The angular distribution of PXR is concentrated near to $\mathbf{e}_2$ and the variable $\Theta_{\|}$ describes the distribution of emitted photons in incidence plane, specified by the axis of an electron beam $\mathbf{e}_1$ and the vector $\mathbf{g}$. The traditional geometry is realized when $\beta = 0$. The yield of PXR is limited by photoabsorption. In the traditional case PXR yield is collected from the length of trajectory of emitting particle L_{el} approximately equal to the photoabsorption length L_{ab} for generated PXR signal.

The possibility to increase PXR yield, described in [6], is based on PXR propagation under large angle relative to the velocity of emitting electron. The relation between L_{el} and L_{ab} may be changed substantially in asymmetrical geometry of diffraction. This relation may be presented in simple form

$$\frac{L_{el}}{L_{ab}} = \frac{\sin(2\beta + \alpha)}{\sin(\alpha)} \quad (1)$$

Presented formula shows the ratio between PXR yield for asymmetrical and symmetrical geometries. Two conditions should be fulfilled for bright manifestation of the effect of PXR enhancement. The first of them is grazing incidence of emitting electrons on the surface of the target ($\alpha \to 0$). This condition is difficult to fulfill for real electron beam that has a finite transverse size (the linear size of the target increases when the incidence angle tends to zero). In conditions under consideration the incidence angle is limited by the value asin(d / L) (d – diameter of the beam, L – the linear size of the target in incidence plane), otherwise a part of the beam does not interact with the target.

The second condition consists in PXR propagation normally to the surface of the target. An influence of the photoabsorption is minimal in conditions under consideration.

PXR spectral-angular characteristics in asymmetric geometry may be calculated on the basis of kinematical diffraction theory. A simple formula obtained in [6]

$$N(\alpha,\beta)=\frac{2e^2\omega_g^4}{g^4}\left(\ln\left(1+\gamma^2\Theta_d^2\right)-\frac{\gamma^2\Theta_d^2}{1+\gamma^2\Theta_d^2}\right)\times\frac{\sin(2\beta+\alpha)\sin^2(\beta+\alpha)\left(1+\cos^2(2\beta+2\alpha)\right)}{\sin(\alpha)\chi_0''\left(\frac{g}{2\sin(\beta+\alpha)}\right)}, \quad (2)$$

$$\omega_g^2=\omega_0^2F(g)S(\mathbf{g})e^{-\frac{1}{2}g^2u_T^2},$$

differs from traditional one by coefficient (1) only. Here ω is the energy of emitted photon, ω_0 is the plasma frequency, $F(g)$ is the formfactor of an atom, $S(\mathbf{g})$ is the structure factor of an elementary cell, u_T is the mean-square amplitude of thermal vibrations of atoms of the target, γ is Lorentz factor of the emitting electron, Θ_d is collimator angular size, $\chi_0''(\omega)$ is the imaginary part of dielectric susceptibility of the target.

In real experiment PXR enhancement is limited by the transverse size of the beam. In order for all beam electrons to interact with the target, grazing angle must exceed the relation d/L ($d \ll L$). As a result the maximum possible enhancement is determined by d/L. Assuming the beam size d to be of the order of 0.1 cm and the length of the target L to be of about 1cm one can calculate PXR enhancement as a function of β. Such dependence calculated on the base of (1) is presented in Fig.2.

3. Experimental setup and results of measuring.

For the experimental verification of the discussed effect of PXR enhancement in grazing incidence geometry it is necessary to measure both the yield of PXR in conditions of normal incidence and that in conditions of grazing incidence. The comparison of PXR yields in traditional and grazing cases should be performed with minimal difference in experimental parameters for both cases. It is very convenient to compare the yields when the difference between the measured PXR yields lies in replacement of the target only. In conditions under consirderation the energy of PXR photons and the collimator size are the same in both cases.

The experimental setup was created on base of 7MeV microtron of P.N. Lebedev Physical Institute RAS. The scheme of the experiment is presented in Fig.3. The electron beam 2, generated by microtron 1, is formed by carbon collimators 3. The magnets 4 clean the beam from background photons appearing in the microtron and collimators (collimators decrease intensity of the original beam 10^3 -10^4 times). The magnetic lenses 5 and corrector 6 shape the beam and direct it onto the target 7. The target was placed into a goniometer with three degrees of freedom of rotation and the possibility of taking it out of the electron beam. The target was controlled in horizontal plane with the step of rotation $7.1\cdot10^{-5}$Rad and in vertical plane with the step of rotation $4\cdot10^{-5}$Rad. The beam intensity, location and angular divergence were controlled by proportional chamber 11 and Faraday cup 12. The photon signal from the target 7 was collimated by collimator 8 and detected by energy-dispersive X-ray p.i.n. detector 9 with resolution 152eV (Amptek XR-100CR: 6mm^2 x 500μm, Be window 0.3mil). The above intensive background from the microtron and collimators 3, was reduced by lead protection 13 and noise shielding. The vacuum of the system was better then 10^{-4}torr.

The calculation of PXR in grazing incidence geometry shows a possibility to obtain the enhancement of PXR yield using two perfect silicon crystals with different orientation of the crystallographic plane towards the surface of the crystal. For traditional case the target with surface coinciding with crystallographic plane (111) ($\beta = 0$) was chosen. For the grazing case the plane of the surface of the target coincided with (100) crystallographic plane ($\beta = 0.955$ Rad.). The thickness of targets was 450mk for traditional case and 200mk for grazing case. One of the most important problems was choosing the observation angle determining the energy of detected PXR. The energy of PXR should be in area where the background has a smooth spectral distribution. The measurement of spectral distribution of the background without the target showed the smooth distribution with peaks of characteristic X-ray lines of the elements of target's vacuum chamber (Al, Ti, Cr, Fe and Ni). The background of the target became apparent as characteristic radiation of K-lines. The influence of multiple scattering of electrons in the target on PXR properties increases in grazing geometry, it may distort the PXR spectrum. Thus, we chose the energy range of PXR behind of Si K-edge where the influence of the multiple scattering is minimal because the photoabsorption is maximal and the background has no peaks. In conditions undr consideration the observation angle was 40 degrees (angle between axes of electron beam and X-ray detector). The orientation dependences for traditional and grazing cases, calculated on base of (2) with account of test conditions are presented in Fig.4.

The radiating area of target's surface increases when incidence angle is decreased. It changes the observation angle of the detector and hence the detected spectrum becomes wider than for a point source. In our case the diameter of the beam was 5mm and the long size of the beam trace on the target in grazing geometry was about 19mm ($\alpha = 0.267$ Rad). The relation of the long size of the beam trace on the target to the distance between the detector and the target should be much more, than the characteristic size of PXR angular distribution $\sqrt{\gamma^{-2} + \omega_0^2 / \omega^2} = 68$ mRad. In this case the line spreading is negligible. The detector was placed on a distance 320cm from the target (the above relation 0.006). The size and position of collimator 8 were so chosen that the target was fully available for the detector in traditional and grazing cases. The collimating angle was $5.86 \cdot 10^{-7}$ster.

For the correct comparison of PXR yields in traditional and grazing cases, the background should be subtracted from the total detected signal (the background increases in grazing case) and the measuring of each point should be normalized on the beam current. The background was smooth. Its level did not exceed 20% of the signal in area of PXR manifestation. The normalizing and comparison were performed on the base of characteristic K-line yield of Si, measured in traditional and grazing cases for identical current. The results of PXR measurement in traditional and grazing geometry are presented in Fig.5. The fitting of the experimental points by the theoretical curve, calculated in kinematical approach for PXR, is presented by solid line. The position of maximum of orientation distribution is 41mRad. It well coincides with the theoretical prediction 34mRad. The difference may be explained by the influence of multiple scattering and initial angular spread of electron beam.

4. Conclusions

Presented experimental results show the possibility to increase PXR yield on the base of grazing incidence of emitting electrons onto the surface of crystalline target in conditions of asymmetric diffraction of the electron Coulomb field. The enhancement has been measured for 7MeV electrons interacting with Si crystal. The comparison of the yields for traditional and grazing cases shows enhancement near 3 times. The obtained data is in agreement with theoretical predictions.

Acknowledgements
One of the authors (A.K.) is grateful for financial support (grant of the President of RF for young scientists, MK-4199.2009.2).

References

[1] M.L. Ter-Mikaelian, High Energy Electromagnetic Processes in Condensed Media, (Wiley, NewYork, 1972).
[2] G.M. Garibian, C. Yang, Sov. Phys. JETP, 34, (1972), 495.
[3] V.G Baryshevsky, I.D. Feranchuk, Sov. Phys. JETP, 34, (1972), 283.
[4] A.N. Didenko, B.N. Kalinin, Pak S.D. et al // Phys. Lett. A, 110 (1985) 177-179.
[5] V. Baryshevsky, I. Feranchuk, A. Ulyanenkov. Parametric X-ray Radiation in Crystals. Theory, Experiment and Applications. Berlin: Springer, 2005, 176 p.
[6] N. Nasonov, P. Zhukova, M. Piestrup, H. Park // NIM, B 251, (2006), 96.

Figures captions

Fig.1. Geometry of PXR process.

Fig.2. PXR enhancement in condition of grazing incidence and asymmetric diffraction geometry.

Fig.3. The experimental setup.

Fig.4. The orientation dependences of PXR. 1 - traditional case; 2 - grazing geometry.

Fig.5. Orientation dependences of PXR yield for Si crystal. ○ – traditional case; □ – grazing case.

Active quartz monochromator

A.S. Gogolev, Yu.A. Popov, A.R. Wagner, A.P. Potylitsyn

Tomsk Polytechnic University, 30, Lenin Avenue, Tomsk, 634050, Russia

E-mail: alextpuftf@tpu.ru

Abstract. The experimental results of the active X-ray monochromator research based on quartz crystal with ultrasonic waves are reported in the paper. At frequency of f = 5.46 MHz the intensity increasing of diffracted Lβ1 line tungsten beam by 100% was detected. The tension amplitude of variable electric field causing acoustic excitement was equal to 10 V. The relationship between external impact parameters and diffracted beam intensity were studied.

1. Introduction

Many experiments demonstrate the advantages of using monochromatic X-ray radiation of high intensity for different applications in science, technology and medicine (see, for example [1]). As a rule, in such tasks monochromatization bremsstrahlung or synchrotron radiation (SR) by crystal monochromator is used. SR beam parameters received in electronic storage is undoubtedly exceed parameters of other monochromatic sources. However, due to the high cost of such facilities it remains unavailable for public use, for example, in medical diagnosis. On the other hand X-ray tubes are commercially available and easy to use but the intensity of the monochromatized beam for most tasks is below the required values [2].

In the works [3, 4] the authors demonstrated the possibility to increase the intensity of the diffracted beam when on the piezo-electric crystal monochromator to influence by acoustic or non-uniform temperature field. The authors of [5-7] have shown the possibility to control spatiotemporal characteristics of the diffracted radiation by means of surface acoustic waves. As an alternative way to increase the intensity, listed in the cited works, in the article [8] the piezoquasimosaic effect is proposed to use.

In this paper we investigate the possibility of creating active elements of the X-ray optics based on piezo-electric crystals with the external deformation fields.

2. Experiment

The experimental setup is shown in Fig. 1. As X-ray source a pulsed X-ray tube with a tungsten anode RAP 160-5 was used. All measurements were performed with the same parameters of the tube, the voltage and average current were 60 kV and 1 mA, respectively. X-ray machine was placed in a lead house with a wall thickness of 5 cm. Radiation was formed by collimator with diameter 3 mm located at a distance of 90 mm from the output window of X-ray tube, and then fall onto a quartz crystal mounted in a remotely controlled goniometer at a distance of 215 mm from the collimator.

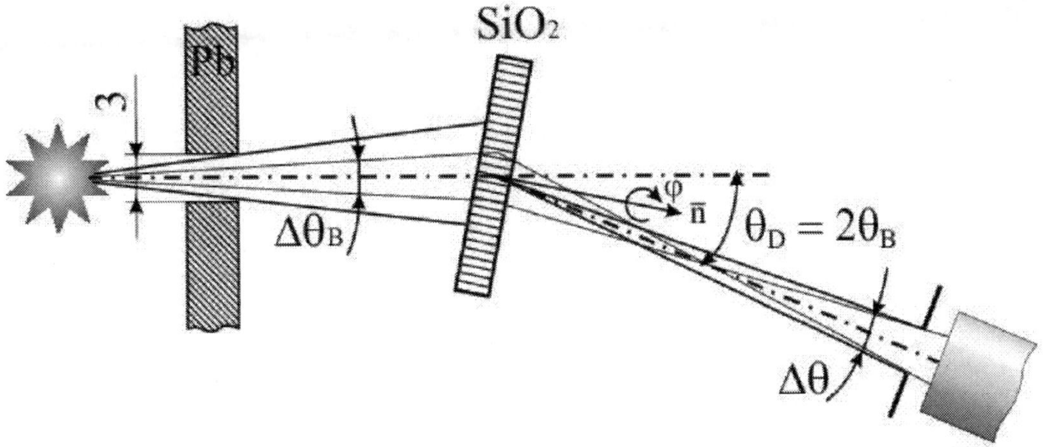

Figure 1. Scheme of the experiment. θ_B - Bragg angle, θ_D - the angle of observation, n - direction normal to the surface of the crystal.

Goniometer has three translational and three rotational freedom degrees with an accuracy of orientation not worse than 0.5 mm and 10 μrad, respectively. It is possible to install the crystal at the Bragg angle for any families of reflecting atomic planes, so that a detector located in the horizontal plane could register only beams propagating in this plane. In the experiment the X-cut quartz crystal with diameter of 15 mm and thickness of 0.5 mm with a silver sputtering was used. The reflection from atomic planes oriented perpendicular to the large surface of the sample (10-11) with interplanar distance d = 3.3429 Å is studied. Radiation was registered in the Laue geometry by scintillation detector (NaI) working in current mode at a distance of 620 mm from the crystal monochromator. The diameter of the sensitive volume was equal to 30 mm with a beryllium output window thickness of 20 μm which provides to register x-ray radiation in the range from 8.5 to 50 keV with the efficiency of not less than 60%. Slit collimator with a 30×3 mm before detector was placed for decrease background radiation. $L_{\beta 1}$-line of tungsten with an energy of 9,671 keV for which the Bragg angle is $\theta_B = 11.09°$ for quartz reflection (10-11) was observed. For the ultrasound excitation in the quartz a generator of sinusoidal electrical pulses with smooth changes of amplitude from 0 to 10 V and frequencies from 1 Hz to 25 MHz was used.

To find the reflex (10-11) the angles of θ and φ were scanned, and then orientation angles corresponding to the maximum yield of the diffracted radiation were selected.

Fig. 2 shows the study results of the diffracted beam intensity dependence on the excitation frequency. The piezo-electric resonance frequency is determined by the formula $f_n = nc/(2l)$, where n - odd integers, l - thickness of the crystal, c - ultrasonic waves velocity in quartz by about 5700 m/s [9]. As seen from Fig. 2 at frequencies of 5.46 and 16.42 MHz, corresponding to the first and second piezo-electric resonance, the increasing of diffracted beam intensity was observed, whereas at a frequency of 10.92 MHz increasing the intensity of the radiation was not observed. The FWHM of the resonances was 28 kHz in both cases.

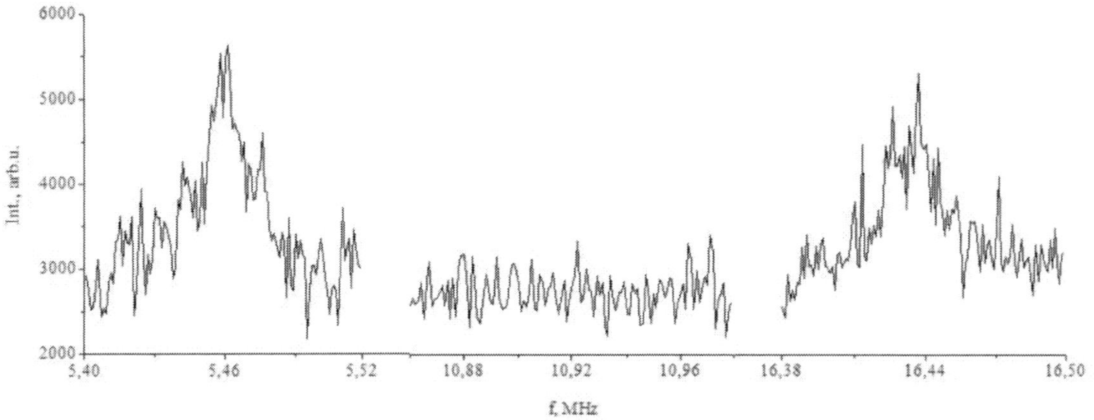

Figure 2. The diffracted beam intensity dependence on the frequency of external excitation, the oscillations amplitude is 10 V.

The following figure shows the study results of the diffracted beam intensity dependence on the amplitude of the external electromagnetic influence on the crystal-monochromator. The radiation intensity increases linearly in the diffraction direction to increasing voltage electromagnetic oscillations.

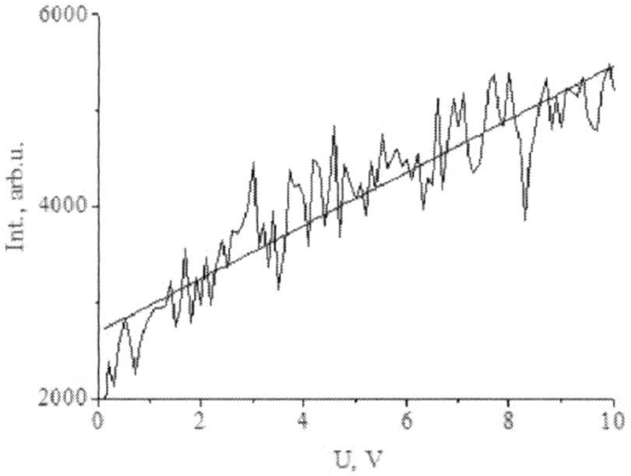

Figure 3. The diffracted beam intensity dependence on the external impact.

Fig. 4 shows the orientational dependence of the diffracted radiation yield at a voltage U = 10 V for the first resonance and the absence of external field. On the fig. 4 zero corresponds to the angle orientation of the crystal 11.09° relative to the initial direction of radiation. The figure shows that the diffracted radiation intensity is more than without external influence in two times, with the width of the rocking curves differing slightly. Rocking curve width is determined primarily with the beam divergence and an angular capture detector $\Delta\theta = 4.9$ mrad which corresponds to the deviation from the Bragg angle $\Delta\theta_B = 2.5$ mrad.

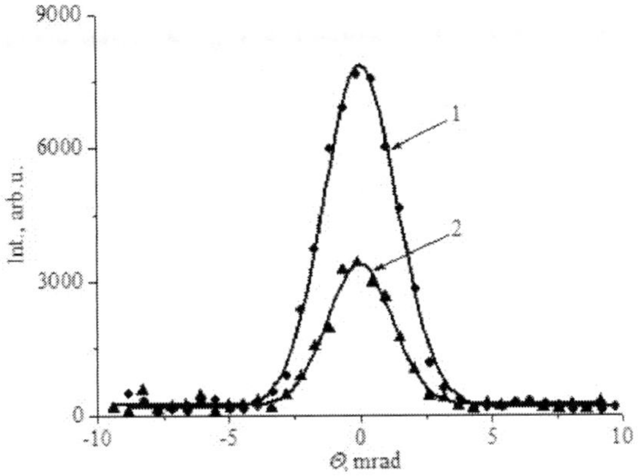

Figure 4. Rocking curves, the dots represent experimental values: 1) with the ultrasound frequency of 5.46 MHz in the crystal, the FWHM = (3.30 ± 0.06) mrad; 2) without external influence on the crystal, the FWHM = (2.92 ± 0.09) mrad.

3. Conclusion

Experiments showed the possibility of creating active X-ray monochromators based on quartz crystals with ultrasonic vibrations. With the help of external influence on the crystal monochromator the control change and increase the intensity of the diffracted X-ray beam were carried out.

Based on the effects of the deformation fields influence on the X-rays scattering can be constructed the new elements of X-ray optics. In particular, active monochromators with acoustic waves will increase the luminosity of the monochromatic sources and transport of intense monochromatic X-ray beams without loss. In [4] the possibility of strengthening the diffracted radiation on the order of magnitude when increasing the voltage field was indicated.

The authors express their gratitude to the staff of the Institute of Applied Problems of Physics NAN RA (Yerevan, Armenia), A.R. Mkrtchyan, A.G. Mkrtchyan, V.R. Kocharian for numerous stimulating discussions and for the opportunity to conduct research on quartz monochromator. This work was partly supported by AVTSP RNPVSH № 2.1.2/3013 and GK № P1202.

References

[1] H. Wiedemann, M. Balfay, R. Carr et al., NIMA **347**, 515 (1994)
[2] V. Arkadiev, H. Brauninger, W. Burker et al., NIMA **455**, 589 (2000)
[3] A.R. Mkrtchyan, M.A. Navsardyan, V.K. Mirzoyan, Pisma v ZhTF **8**, 677 (1982) (in Russian)
[4] A.R. Mkrtchyan et al., Pisma v ZhTF **8**, 1181 (1983) (in Russian)
[5] H. Cerva, W. Graeff, Phys. Stat. Sol. **(a) 82**, 35 (1984)
[6] L.A. Kocharian, P.P. Sukiasyan, A.S. Borhazyan et al., Izv. Akad. Nauk Arm. SSR Fiz **21**, 317 (1986) (in Russian)
[7] E.G. Lapin, V.M. Samsonov, G.P. Solodov et al., JETP **73**, 1016 (1997) (in Russian)
[8] Remi Tucoulou, Francois de Bergevin, Olivier Mathon, and Dimitry Roshchupkin, Phys. Rev. B **64**, 134108 (2001)
[9] Krasilnikov V.A. Introduction to acoustics: Textbook. - M.: Izd-vo MGU, 1992. - 152 p. (in Russian)

Possibilities for measurement of nano-crystallites size with use of parametric X-ray radiation

A V Shchagin

Kharkov Institute of Physics and Technology, Kharkov 61108, Ukraine
e-mail address: shchagin@kipt.kharkov.ua

Abstract Spectral and angular properties of the parametric X-ray radiation (PXR) of relativistic particles emitted in the backward direction from a thin crystalline layer and polycrystal are considered. The shapes of the natural spectral peak of the PXR from a crystalline ball and spectral distribution of PXR from a polycrystal are estimated. It is shown that the natural spectral peak width determines the PXR spectral peak width emitted in backward direction from crystalline layer and grains with size of nanometer range. The method for measurements of crystalline layer thickness and crystalline grains size of nanometer range with use of spectral peak width of the PXR emitted in the backward direction is discussed.

1. Introduction

The first research of the width of the spectral peak of parametric X-ray radiation (PXR) emitted by relativistic charged particle moving in a singlecrystal has been performed at Kharkov [1]. It was found that the width is determined mainly by the experimental angular resolution and the natural width due to the crystal thickness is negligible in forward hemisphere. In research [2,3] it was demonstrated that the PXR spectral peak width from a singlecrystal dramatically reduces at observation in the backward direction relative the incident particle velocity. The multiple scattering effect on the spectral peak width was analyzed in [4]. It was shown in [5] that normalized spectral peak width of focused PXR emitted from a bent singlecrystal by channeling particles can reach order of value about 10^{-8}. PXR from a polycrystal was first treated by Nasonov [6] and observed at Moscow [7]. Intense PXR from textured polycrystal was first studied at Hiroshima [8,9]. In these experiments, the PXR emitted by relativistic electrons in crystalline grains of a polycrystal has been identified. A possibility for application of the PXR natural spectral peak width for measurements of crystalline grains size was noted in [10]. Theoretical research of PXR from a polycrystal has been performed in [11]. In present paper we consider some features of the spectral peak of PXR emitted in the backward direction relative to the vector of incident particles velocity and possibilities for its applications for measurements of crystalline layer thickness and grains size in nanometer range. The consideration is performed in the frameworks of the Ter-Mikaelian theory of the PXR [12,13].

2. PXR yield in the backward direction

The general formula for absolute differential yield of PXR within the framework of the Ter-Mikaelian theory has been confirmed in a number of experiments with singlecrystals, see e.g. reviews [13-15]

$$\frac{dN}{d\Omega}=\frac{e^2L\left|\chi_{\vec{g}}(\omega)\right|^2 k}{2\pi\hbar\varepsilon_0^3\xi\left(V\xi^{-1}-\vec{V}\vec{\Omega}\right)}\cdot\left|\frac{\vec{k}\times\vec{k}_0}{\left(\vec{k}_\perp-\vec{g}_\perp\right)^2+k^2\gamma_{eff}^{-2}}\right|^2, \tag{1}$$

where dN is the number of quanta emitted in the solid angle $d\Omega$ by the particle with charge e passed crystal of effective thickness L; $\chi_{\vec{g}}(\omega)$ is the Fourier component of dielectric susceptibility; $\hbar$ is the Plank constant divided by 2π; ε_0 is the average permittivity; $\vec{k}=k\vec{\Omega}$ is the radiation wave vector; $k=\omega\sqrt{\varepsilon_0}/c$; ω is the radiation frequency; $\vec{\Omega}$ is the unit vector for the radiation direction; $\vec{\xi}=\sqrt{\varepsilon_0}\vec{V}/c$ is the particle velocity $\vec{V}$ in terms of the phase wave velocity in the crystal $c/\sqrt{\varepsilon_0}$; $V=|\vec{V}|$; $\xi=|\vec{\xi}|$; $\xi^{-1}=1+\gamma_{eff}^{-2}/2$; c is the light velocity; $\vec{k}_0=\vec{k}\xi+\vec{g}$; $\gamma_{eff}=\left[\gamma^{-2}+|\chi_0|\right]^{-1/2}$ is the effective relativistic factor of the particle in the crystal with taking into account of the density effect [16]; γ is the relativistic factor of the particle; $\chi_0=\chi_0(\omega)$ is the average dielectric susceptibility; $\vec{g}$ is the reciprocal lattice vector of the crystallographic planes; $g=2\pi/a$; a is the distance between crystallographic planes, $\vec{k}_\perp$ and $\vec{g}_\perp$ are the components of vectors $\vec{k}$ and $\vec{g}$ perpendicular to $\vec{V}$; $k=\dfrac{-\vec{g}\cdot\vec{V}}{V\xi^{-1}-\vec{V}\cdot\vec{\Omega}}$. The effective thickness L of the target in the shape of crystalline slab with allowance of the radiation attenuation has form $L=T_e\left|\dfrac{\vec{t}\cdot\vec{\Omega}}{\vec{t}\cdot\vec{v}}\right|\left[1-\exp\left(-\dfrac{T}{T_e\left|\vec{t}\cdot\vec{\Omega}\right|}\right)\right]$, where $T_e=T_e(\omega)$ is the e-fold attenuation length of the radiation in the target; T the slab thickness; $\vec{t}$ is the unit vector perpendicular to the slab surface; $\vec{v}=\vec{V}/V$. In transparent crystal $L=\dfrac{T}{|\vec{t}\cdot\vec{v}|}$.

The PXR differential yield in approximation of small angles $\delta,\alpha<<1$ in the vicinity of the PXR reflection center was derived from equation (1) with use of spherical coordinates in [15]

$$\frac{dN}{d\Omega}=\frac{z^2 n_p\left|\chi_{\vec{g}}(cg)\right|^2}{137\varepsilon_0^{\frac{1}{2}}\xi^2\sin^4\phi}\cdot\frac{\left[\xi^{-1}-\cos(\theta_c-\delta)\right]^2\left(\alpha^2+\delta^2\cos^2\theta_c\right)}{\left(\delta^2+\alpha^2+\gamma_{eff}^{-2}\right)^2}, \tag{2}$$

where $n_p=\dfrac{L}{a}\sin\phi$ is the number of crystallographic planes taking part in production of PXR, z is the particle's charge in units of the electron charge, ϕ is the Bragg angle between the crystallographic planes and vector $\vec{V}$, $\chi_{\vec{g}}(cg)=inv(\omega)$, $\theta_c=2\phi$ is the polar angle of the reflection center, δ,α are coordinates of the observation direction in the plane of small angles that is perpendicular to the direction of the reflection center, δ is in the reaction plane containing vectors $\vec{g},\vec{V}$ and α is perpendicular to the reaction plane. Formula (2) differs from the well known Feranchuk-Ivashin formula [17] e.g. by the factor in square brackets that provides proper description of experimentally observed asymmetry in the PXR reflection [1,16].

In order to consider properties of PXR emitted by electrons ($z=1$) in the backward direction one can use spherical coordinates from [15] and derive specific formula for PXR differential yield in the backward direction in approximation of small angles $\tau,\delta,\rho<<1$ in the plane perpendicular to the incident particle velocity vector $\vec{V}$

$$\left(\frac{dN}{d\Omega}\right)_{bw}=\frac{4n\left|\chi_{\vec{g}}(cg)\right|^2}{137\xi^2\sqrt{\varepsilon_0}}\cdot\frac{\rho^2}{\left[\rho^2+\gamma_{eff}^{-2}\right]^2}, \tag{3}$$

where

$$\rho^2=\left|\vec{\delta}-2\vec{\tau}\right|^2=\delta^2+4\tau^2-4\delta\tau\cos\varphi, \tag{4}$$

ρ is the angle between the observation direction $\vec{\Omega}$ and the PXR reflection center direction, $\vec{\delta}$ is the angle between the observation direction $\vec{\Omega}$ and vector $-\vec{V}$, $\delta=\pi-\theta$, $\vec{\tau}$ is the angle between vectors $\vec{g}$ and $\vec{V}$, $\tau=\frac{\pi}{2}-\phi$, $2\vec{\tau}$ is the angle between the PXR reflection center and vector $-\vec{V}$, $2\tau=\pi-\theta_c$, φ is the angle between the reaction plane containing vectors $\vec{g},\vec{V}$ and the observation plane containing vectors $\vec{\Omega},\vec{V}$. The number of crystallographic planes producing the PXR in the backward direction in a thin transparent crystal slab or layer of thickness T along the trajectory is $n\approx\frac{Tg}{2\pi}=\frac{T}{a}$.

3. PXR and Bragg frequencies in the backward direction

The general formula for PXR frequency ω within the Ter-Mikaelian theory has been confirmed experimentally in [1,16]

$$\omega=\frac{c\vec{g}\vec{V}}{c-\sqrt{\varepsilon}\vec{V}\vec{\Omega}}=\frac{cg\cos\tau}{\sqrt{\varepsilon}\left(\xi^{-1}-\cos\theta\right)}, \tag{5}$$

where θ is the polar angle of the observation direction $\vec{\Omega}$. With use of the second-order Taylor's formula one can find expression for deviation of the PXR frequency in the vicinity of the observation angle θ and the angle of crystal alignment τ

$$\frac{d\omega}{\omega}=-\frac{\sin\theta}{\xi^{-1}-\cos\theta}d\theta-\tan\tau\cdot d\tau-\frac{d\tau^2}{2}+\frac{1+\sin^2\theta-\xi^{-1}\cos\theta}{2\left(\xi^{-1}-\cos\theta\right)^2}d\theta^2+\frac{\tan\tau\sin\theta}{\left(\xi^{-1}-\cos\theta\right)^2}d\tau d\theta. \tag{6}$$

The first term in equation (6) is dominating for singlecrystal of fixed alignment τ and at θ not in the vicinity of $\theta=0,\pi$. In this case we obtain

$$\frac{\Delta\omega}{\omega}\approx-\frac{\Delta\theta}{\tan(\theta/2)}. \tag{7}$$

Formula (7) for PXR spectral peak width $\Delta\omega$ at experimental angular resolution $\Delta\theta$ has been experimentally confirmed for PXR reflection emitted in forward hemisphere in [1]. In the vicinity of the exact backward direction at $\theta=\pi$, $\tau=0$ three terms in equation (6) vanish and $d\omega$ dramatically reduces. In such case the formula for normalized PXR frequency in backward direction is

$$\frac{\omega-\omega_\pi}{\omega_\pi}=\frac{\delta^2}{4}-\frac{\tau^2}{2}, \tag{8}$$

where

$$\omega_\pi=\frac{cg}{\sqrt{\varepsilon}\left(\xi^{-1}+1\right)} \tag{9}$$

is the frequency at $\tau,\delta=0$. It is interesting to note that the frequency (8) increases at declination of the observation direction from exact backward direction $\theta=\pi$, and decreases at declination of the crystal alignment from the one when the reciprocal lattice vector $\vec{g}$ is parallel to $\vec{V}$. The yield of PXR at $\tau,\delta=0$ is zero, see equation (3).

Consider also the difference between the Bragg frequency in the observation direction $\omega_B=-\frac{cg^2}{2\sqrt{\varepsilon_0}(\vec{\Omega}\vec{g})}$ and PXR frequency. The formula for the normalized difference in the vicinity of the PXR reflection center at θ not in the vicinity of 0 or π has been derived with use of spherical coordinates in approximation of small angles in [18]. It is a function of square of the angle between the reflection center and observation direction $\rho^2=\alpha^2+\delta^2$,

$$\frac{\omega_B - \omega}{\omega} = \frac{\gamma_{eff}^{-2} + \rho^2}{4\sin^2\phi}. \quad (10)$$

Using spherical coordinates from [18], one can find specific formula for the difference in the backward direction. It is similar to one shown above

$$\frac{\omega_B - \omega}{\omega} = \frac{\gamma_{eff}^{-2} + \rho^2}{4} \quad (11)$$

but the square of the angle between the reflection center and the observation direction is determined by equation (4). The Bragg frequency in the backward direction $\omega_B = \frac{cg}{\sqrt{\varepsilon_0}\left(2-\left|\vec{\delta}-\vec{\tau}\right|^2\right)}$ always exceeds the PXR frequency, the absolute difference is no less then $\omega_B - \omega \geq \frac{cg}{8\sqrt{\varepsilon_0}}\gamma_{eff}^{-2}$.

The transition radiation, bremsstrahlung, channeling radiation of frequency ω_B can be diffracted in the backward direction. The frequency of any kind of diffracted radiation and the PXR frequency are separated by the frequency interval (11).

4. Natural spectral peak of the PXR emitted in the backward direction from a crystalline slab

From classical point of view, the PXR is the wavetrain of electromagnetic oscillations. The number of oscillations n in the wavetrain is determined by the effective number of crystallographic planes crossed by the particle and taking part in production of the PXR. In present paper we deal with PXR emitted in the backward direction from a crystal or crystallite that are thin enough to neglect both multiple scattering of electrons and attenuation of radiation. The PXR in backward direction is produced at crystallographic planes aligned almost perpendicularly the particle trajectory. The number of oscillations n of permanent amplitude in the wavetrain from a thin crystal of thickness along the particle trajectory T is $n = \frac{Tg}{2\pi}$. One can obtain spectral distribution of the radiation intensity $I(s)$ of the wavetrain with use of Fourier transformation. It is described by the well known function

$$I(s) \sim \pi^2 n^2 \frac{\sin^2 s}{s^2}, \quad (12)$$

where $s = \frac{\Delta\omega\pi n}{\omega_\pi}$. The full width at half of maximum (FWHM) of the central peak in (12) is

$$\left(\frac{\Delta\omega}{\omega_\pi}\right)_{nat} = 0.89 n^{-1}. \quad (13)$$

Therefore one can determine the crystal slab thickness as $T = 0.89 \cdot a \cdot \left(\frac{\Delta\omega}{\omega_\pi}\right)_{nat}^{-1}$. Normalized spectral distribution (12) is shown in figure 1.

5. Spectral peak of the PXR emitted in the backward direction from a crystalline ball

Crystallites in a polycrystal can have different forms. Calculations of X-rays scattering on a crystalline ball are described in [19]. Let us estimate spectral properties of the PXR emitted in the backward direction by relativistic particles crossing a ball-shaped crystal of radius R. We will suppose that the radius of the cylinder $\lambda\gamma_{eff}$ (where $\lambda \approx 2a$, see (9)) around of the particle trajectory where the PXR is produced is less then the crystallite ball radius R, $\lambda\gamma_{eff} < R$. Also, we will believe that noticeable yield of the PXR in backward direction can provide the crystallographic planes that are almost perpendicular to the particle's trajectory only.

The number of crystallographic planes n crossed by the particle is a function of the impact parameter r, $n(r)=n_D\sqrt{1-\left(\frac{r}{R}\right)^2}$, where $n_D=\frac{2R}{a}$, $0\le r\le R$. Integrating (12) over the ball cross section $\int_0^R \frac{\sin^2\left[\frac{\Delta\omega}{\omega}\pi n(r)\right]}{\left(\frac{\Delta\omega}{\omega}\right)^2}\frac{2\pi r dr}{\pi R^2}$ one can find spectral distribution of the PXR emitted in the backward direction from a crystalline ball

$$I_b(s)\sim 2\pi^2 n_D^2\int_0^1 \frac{\sin^2\left(s\sqrt{1-x}\right)}{s^2}dx, \quad (14)$$

where $s=\frac{\Delta\omega\pi n_D}{\omega_\pi}$. Factor 2 is included for normalization at $s=0$. The full width at half of maximum (FWHM) of the central peak in the distribution (14) is

$$\left(\frac{\Delta\omega}{\omega_\pi}\right)_{B-nat}=1.11\cdot n_D^{-1}, \quad (15)$$

where n_D is the diameter of the ball in units of the interplanar spacing. Thus, the natural spectral peak width from the crystalline ball exceeds in 1.25 times the one from the slab (13) of thickness equal to the ball diameter. One can determine the crystalline ball diameter $D=1.11\cdot a\cdot\left(\frac{\Delta\omega}{\omega_\pi}\right)_{B-nat}^{-1}$ after measurement of the natural spectral peak width $\left(\frac{\Delta\omega}{\omega_\pi}\right)_{B-nat}$ of the PXR emitted in the backward direction. Note that natural spectral peak shapes (12, 14) are independent of incident particle energy. Normalized spectral distribution (14) is shown in figure 1.

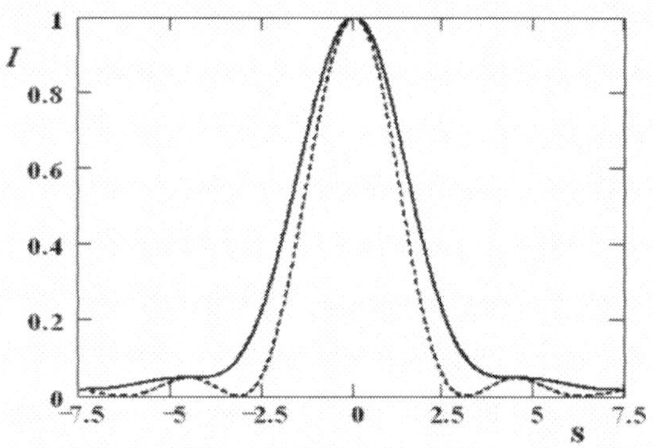

Figure 1. Comparison of spectral distribution of PXR emitted in the backward direction from a crystalline slab and crystalline ball. The distribution from the slab is calculated by equation (12) and shown by dashed line. The distribution from the ball is calculated by equation (14) and shown by solid line. Thickness of the slab is equal to the diameter of the ball, $n=n_D$. Normalized spectral distributions are shown as a functions of $s=\frac{\Delta\omega\pi n_D}{\omega_\pi}$.

6. Influence of finite experimental angular resolution

We will suppose that an X-ray spectrometric detector with circular aperture is installed at observation angle 180° and experimental angular resolution is restricted by value $\delta<\delta_{\max}$. The value $\delta_{\max}$ is determined mainly by finite sizes of the beam spot at the target and the aperture of the detector, the divergence of the incident particle beam, and also by multiple scattering of electrons in the target. Considering (8) one can see that such restriction leads to the restriction of the spectral range observed by the detector due to finite angular resolution. Radiation is absent at frequencies out of the range

$$-\frac{\tau^2}{2} < \frac{\omega-\omega_\pi}{\omega_\pi} < -\frac{\tau^2}{2} + \frac{\delta_{max}^2}{4} \qquad (16)$$

of full width $\delta_{max}^2/4$. At $\delta_{max} = \sigma\gamma_{eff}^{-1}$ the full width of the range due to finite angular resolution is

$$\left(\frac{\Delta\omega}{\omega_\pi}\right)_{ang} = \frac{\sigma^2\gamma_{eff}^{-2}}{4}. \qquad (17)$$

For example, at $\sigma = 0.2$ the full width of the spectral range is $\left(\frac{\Delta\omega}{\omega_\pi}\right)_{ang} = 10^{-2}\gamma_{eff}^{-2}$.

The spectral distribution within the range (the spectral peak shape) can have different forms depending of the alignment of $\vec{\tau}$. The frequency $\omega_{max} = \omega_\pi\left(1+\sigma^2\gamma_{eff}^{-2}/4\right)$ is the maximum PXR frequency that can be observed from a polycrystal by the detector installed in the backward direction.

7. PXR from polycrystal in the backward direction

Let us consider spectral properties of the PXR emitted in the backward direction at $\theta = \pi$ from a thin polycrystal with the reciprocal lattice vector $\vec{g}$ of grains randomly distributed in solid angle 4π. The PXR yield is proportional to normalized number of the reciprocal lattice vectors $\frac{2\pi\tau d\tau}{4\pi} = \frac{d\omega}{2\omega_\pi}$, see (8,9). With this expression and $\rho^2 = 2\tau^2$, one can derive from equation (3) with use of equations (4,8)

$$\frac{dN}{d\Omega d\omega} = \frac{4n\left|\chi_{\vec{g}}(cg)\right|^2}{137\xi^2\sqrt{\varepsilon_0}} \cdot \frac{1}{2\omega_\pi} \cdot \frac{-8x}{(1-8x)^2}\gamma_{eff}^2, \qquad (18)$$

where $x = \frac{\omega-\omega_\pi}{\omega_\pi}\gamma_{eff}^2$, $x<0$. Formula (18) describes spectral distribution of the intensity of the PXR emitted at $\theta = \pi$ by relativistic particles from a thin polycrystal. The distribution has the shape of the asymmetrical spectral peak at $\omega < \omega_\pi$, $x<0$ shown in figure 2. The maximum of the peak is at $x = -\frac{1}{8}$, $\frac{\omega-\omega_\pi}{\omega_\pi} = -\frac{\gamma_{eff}^{-2}}{8}$. The similar maximum was obtained in equation (18) in work [11]. The full width at half of maximum (FWHM) of the peak (18) is

$$\Delta x = \frac{1}{\sqrt{2}}, \quad \left(\frac{\Delta\omega}{\omega_\pi}\right)_{polycr} = \frac{\gamma_{eff}^{-2}}{\sqrt{2}}. \qquad (19)$$

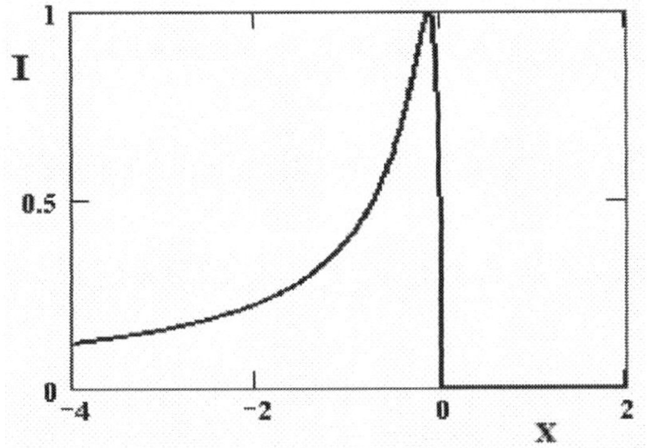

Figure 2. The shape of the spectral peak of the PXR emitted in the backward direction from a polycrystal at $\theta = \pi$. The spectral distribution is calculated by equation (18). Normalized spectral distribution is shown as a function of $x = \frac{\omega-\omega_\pi}{\omega_\pi}\gamma_{eff}^2$.

8. Measurement of the crystalline layer thickness

There are two main reasons for broadening of the spectral peak of the PXR emitted in the backward direction from a crystalline slab or layer: i. because of finite thickness of the slab or layer (13) and ii. because of finite experimental angular resolution (17). The total spectral peak width $\left(\frac{\Delta\omega}{\omega_\pi}\right)_{tot}$ can be estimated as

$$\left(\frac{\Delta\omega}{\omega_\pi}\right)_{tot} = \sqrt{\left(\frac{\Delta\omega}{\omega_\pi}\right)^2_{ang} + \left(\frac{\Delta\omega}{\omega_\pi}\right)^2_{nat}} . \tag{20}$$

Measurements of the slab or layer thickness is possible at condition

$$\left(\frac{\Delta\omega}{\omega_\pi}\right)_{nat} > \left(\frac{\Delta\omega}{\omega_\pi}\right)_{ang} . \tag{21}$$

From equations (13,17,21) one can find that broadening because of the slab thickness dominates at

$$T < \frac{3.6a}{\sigma^2}\gamma^2_{eff} . \tag{22}$$

For example, at $\sigma = 0.2$, typical inter-planar distance in a crystal $a = 3\cdot 10^{-8} cm$, and $\gamma_{eff} = 20$ that corresponds to electron beam energy about 10 MeV the broadening because of the slab thickness can dominate up to crystal thickness about $T \sim 10\mu m$ at normalized spectral peak width up to $3\cdot 10^{-5}$. But real possibilities for measurements of so narrow spectral peak width from a thick crystal are restricted by energy resolution of X-ray spectrometers and increasing of δ in so thick target due to multiple electron scattering. Besides, there exist more traditional methods for measurements of thickness of thick crystals. However, generation of PXR in thick singlecrystal is interesting for creation of X-ray source with high spectral density [2-4].

The method seems much more convenient for measurements of crystalline layers of nanometer range. At well executed inequality (21), the crystalline layer or slab thickness may be estimated as

$$T = 0.89\cdot a\cdot\left(\frac{\Delta\omega}{\omega_\pi}\right)^{-1}_{exp} . \tag{23}$$

For example, at the layer thickness in the range 3-300 nm and $a = 3\cdot 10^{-8} cm$ the normalized PXR spectral peak width measured experimentally $\left(\frac{\Delta\omega}{\omega}\right)_{exp}$ should be in the range of about $0.89\cdot 10^{-1} - 0.89\cdot 10^{-3}$ respectively. At the layer thickness 60 nm the widths are $\left(\frac{\Delta\omega}{\omega_\pi}\right)_{nat} = 2.5\cdot 10^{-3}$, $\left(\frac{\Delta\omega}{\omega_\pi}\right)_{ang} = 2.5\cdot 10^{-5}$.

9. Measurement of the crystalline grains size in a polycrystal

Discuss some details concerned to determination of the crystalline grains size in a thin polycrystal by the method of measurement of the PXR spectral peak width. There are three main sources for broadening of spectral peak of the PXR emitted in the backward direction from grains of a thin polycrystal: i. because of grains (balls) size (15) ii. because of finite experimental angular resolution (17) and iii. because of arbitrary alignments of grains (19). The effect of the multiple electron scattering can be neglected in a thin target. The total spectral peak width $\left(\frac{\Delta\omega}{\omega_\pi}\right)_{tot}$ can be estimated as

$$\left(\frac{\Delta\omega}{\omega_\pi}\right)_{tot} \approx \sqrt{\left(\frac{\Delta\omega}{\omega_\pi}\right)^2_{ang} + \left(\frac{\Delta\omega}{\omega_\pi}\right)^2_{B-nat} + \left(\frac{\Delta\omega}{\omega_\pi}\right)^2_{polycr}} \tag{24}$$

Measurements of the grains size is possible at condition

$$\left(\frac{\Delta\omega}{\omega_\pi}\right)^2_{B-nat} > \left(\frac{\Delta\omega}{\omega_\pi}\right)^2_{ang} + \left(\frac{\Delta\omega}{\omega_\pi}\right)^2_{polycr} . \tag{25}$$

From equations (17,19) one can find that $\left(\frac{\Delta\omega}{\omega_\pi}\right)_{polycr} >> \left(\frac{\Delta\omega}{\omega_\pi}\right)_{ang}$ in popular experimental case $\sigma < 1$.

In such case it is possible to neglect term $\left(\frac{\Delta\omega}{\omega_\pi}\right)_{ang}$ in (24,25). Then from inequality (25) we obtain the condition for the grain size $G = 2R$ that can be measured

$$G < 1.11\sqrt{2}a\gamma^2_{eff} . \tag{26}$$

At this condition the natural spectral peak gives dominating contribution to the width of emitted spectral peak. At well executed inequality (25) the size of grains may be estimated as

$$G = 1.11 \cdot a \cdot \left(\frac{\Delta\omega}{\omega_\pi}\right)^{-1}_{\exp} , \tag{27}$$

where $\left(\frac{\Delta\omega}{\omega_\pi}\right)_{\exp}$ is the experimentally measured FWHM of the PXR peak. For example, at $a = 3 \cdot 10^{-8} cm$ and $\gamma_{eff} = 20$ that corresponds to electron beam energy about 10 MeV the broadening because of the crystallites size dominates at $G < 190\, nm$. At grains of size in the range 30-150 nm the experimentally observed PXR spectral peak width $\left(\frac{\Delta\omega}{\omega}\right)_{\exp}$ should be in the range of about $1.11 \cdot 10^{-2} - 2.22 \cdot 10^{-3}$ respectively. At the grains size $G = 60$ nm and $\sigma = 0.2$ the widths are $\left(\frac{\Delta\omega}{\omega_\pi}\right)_{B-nat} = 5.5 \cdot 10^{-3}$, $\left(\frac{\Delta\omega}{\omega_\pi}\right)_{polycr} = 1.8 \cdot 10^{-3}$, $\left(\frac{\Delta\omega}{\omega_\pi}\right)_{ang} = 2.5 \cdot 10^{-5}$.

10. Results and discussion

The main result of this paper is: There is the real possibility for measurements of the crystal layer thickness and crystalline grains size in polycrystal in nanometer range with use of the natural spectral peak width of the PXR emitted in the backward direction. Besides, note possibilities to study by PXR other objects of nanometer size range e.g. fullerene-like structures.

The influence of the density effect is taken into account in present consideration due to use in calculations the effective relativistic factor γ_{eff}. The influence of diffracted transition radiation should be insignificant at electron beam energy $E_e < mc^2 \frac{\omega_\pi}{\omega_p}$. The yield of the PXR from a polycrystal (18) has to be multiplied by the factor equal to doubled number of the same type crystallographic planes in crystallites. Concrete grains shape and distribution of grains by sizes in polycrystal has to be taken into account. Also, the plane effect [20] can increase the PXR yield and width in backward direction especially at low electron beam energy. Besides, diffraction of channeling radiation is possible in backward direction at channeling of electrons in crystallographic axes and planes as well as diffraction of bremsstrahlung at frequencies $\omega \geq \omega_\pi \left(1 + \gamma^{-2}_{eff}/4\right)$ exceeding the maximum PXR frequency ω_π.

Experimental research at an electron accelerator with energy of about ten(s) MeV is desirable to check possibilities of the method. The electron beam should excite PXR in a thin crystalline or polycrystalline layer. Textured polycrystal can be used as a target to increase the PXR yield. Target should be installed in a goniometer to provide optimal alignment of the target relative to the electron beam. A crystal-diffraction X-ray spectrometer can be used for measurements of the PXR spectra at observation angle $\theta = 180^{\circ}$.

11. References

1. Shchagin A V, Pristupa V I and Khizhnyak N A 1990 A fine structure of parametric X-ray radiation from relativistic electrons in a crystal *Phys. Lett.* **A 148** 485-488
2. Brenzinger K-H, Limburg B, Backe H, Dambach S, Euteneuer H, Hagenbuck F, Herberg C, Kaiser K H, Kettig O, Kube G, Lauth W, Schope H and Walcher Th 1997 How narrow is the linewidth of parametric X-ray radiation? *Phys. Rev. Lett.* **79** 2462-2465
3. Backe H., Kube G. and Lauth W. 2002 On the line shape of backward emitted parametric X-radiation *Electron-Photon Interaction in Dense Media: Proc. NATO Advanced Research Workgroup (Nor-Hamberd, Yerevan, Armenia, 25–29 June 2001) (NATO Science Series, Series II: Mathematics, Physics and Chemistry* vol **49**) ed H. Wiedemann (Dordrecht / Boston / London Kluwer Academic Publishers) pp 153–182
4. Shul'ga N F and Tabrizi M 2002 Multiple scattering effect on the line width of parametric X-ray relativistic electron radiation in a crystal *JETP Letters* **76** 279-282
5. Shchagin A V Focusing of parametric X ray radiation 2004 *JETP Letters* **80** 469-473.
6. Nasonov N N 1998 Collective effects in the polarization bremsstrahlung of relativistic electrons in condensed media *Nucl. Instr. and Meth.* **B 145** 19-24
7. Blazhevich S, Chepurnov A, Grishin V, Ishkhanov B, Nasonov N, Petukhov V and Shvedunov V 1999 Polarization bremsstrahlung of relativistic electrons in aluminium *Phys.Lett* **A 254** 230-232
8. Endo I, Iseki D, Ohnishi T, Moriyoshi C and Shchagin A V 2004 On the origin of mysterious X-ray spectral peaks observed at the REFER electron ring *Nucl. Instr. and Meth.* **B 217** 666-670
9. Takabayashi Y, Endo I, Weda K, Moriyoshi C and Shchagin A V 2006 Observation of intense PXR from textured polycrystal *Nucl. Instr. and Meth.* **B 243** 453-456
10. Shchagin A V 2007 Diffraction in forward direction of parametric X-ray radiation from relativistic particles of moderate energy *Proceedings of SPIE* ed S B Dabagov **6634** 663418
11. Astapenko V A, Nasonov N N, Zhukova P N 2007 Anomalous peak in the spectrum of polarizational bremsstrahlung from relativistic electrons moving through solid target *Journal of Physics B: Atomic, Molecular & Optical Physics* **40** 1-10.
12. Ter-Mikaelian M L 1969 *High-Energy Electromagnetic Processes in Condensed Media* in Russian: (Edition of Armenian Academy of Science, Yerevan, 1969), in English: (Wiley-Interscience, New York, 1972)
13. Ter-Mikhaelyan M L 2001 Electromagnetic radiative processes in periodic media at high energies *Physics Uspekhi* **44** 571-596
14. Shchagin A V and Maruyama X K 1997 Parametric X-rays *Accelerator-Based Atomic Physics Techniques and Applications* ed S M Shafroth and J C Austin (AIP Press, New York) chapter 9 pp 279-307
15. Shchagin A V 2001 Current status of parametric X-ray radiation research *Radiation physics and Chemistry* **61** 283-291
16. Shchagin A V and Khizhnyak N A 1996 Differential properties of parametric X-ray radiation from a thin crystal *Nucl. Instr. and Meth.* **B 119** 115-122
17. Feranchuk I D and Ivashin A V 1985 Theoretical investigation of the parametric X-ray features *J. Phys. (Paris)* **45** 1981-1986
18. Shchagin A V 2008 The difference between frequency of parametric x-ray radiation and Bragg frequency *The Journal of Kharkov National University* **823** *physical series "Nuclei, Particles, Fields"* **3/39** 91-96 [in Russian]
19. Landau L D, Lifshits E M 1992 *Electrodynamics of continuous media* **VIII** Moscow, Nauka
20. Shchagin A V 1999 Parametric X-rays at the right angle to the particle beam *Phys. Lett.* **A 262** 383-388

Longitudinal beam profile monitor at CTF3 based on Coherent Diffraction Radiation

M. Micheler[1], R. Ainsworth[1], G.A. Blair[1], G. Boorman[1], R. Corsini[2], P. Karataev[1], T. Lefevre[2], K. Lekomtsev[1]

[1]John Adams Institute at Royal Holloway, University of London, Egham, Surrey, TW20 0EX, United Kingdom

[2]European Organization for Nuclear Research (CERN), CH-1211 Genve 23, Switzerland

E-mail: `micheler@pp.rhul.ac.uk`

Abstract. A setup for the investigation of Coherent Diffraction Radiation (CDR) from a conducting screen as a tool for non-invasive longitudinal electron beam profile diagnostics has been designed and installed in the Combiner Ring Measurement (CRM) line of the CLIC Test Facility (CTF3, CERN). In parallel, the investigation of Coherent Synchrotron Radiation (CSR) generated mostly in the last bending magnet of the combiner ring is foreseen. The first measurements of CDR and CSR are obtained. In this paper we present the status of the experiment and future plans.

1. Introduction

The longitudinal electron bunch profile is one of the parameters which need to be monitored closely. The optimisation and detailed control of the longitudinal electron distribution in the bunch is crucial for the maximisation of the luminosity in future linear colliders. Close monitoring can prevent luminosity losses due to the hour-glass effect if the bunches are too long, and losses due to the pinch effect if the bunches are too short [1].

The monitoring of the longitudinal bunch profile will also be very important for the Compact Linear Collider (CLIC) [2]. For an optimal performance of the CLIC drive beam, the longitudinal beam profile must be controlled after it has been stretched for injection into the combiner rings and after it has been extracted and compressed. The power extraction in the so-called Power Extraction and Transfer Structures (PETS) also depends on the longitudinal beam profile, a good knowledge of which is highly desirable. According to the last CLIC Instrumentation Workshop more than 200 bunch length measurement stations are required for optimal performance of the machine [3].

Coherent radiation is a widely used tool for monitoring the longitudinal bunch profile. Coherent radiation occurs at wavelengths comparable to, or longer than, the bunch length, when all electrons in the bunch irradiate more or less in phase. The intensity of coherent radiation is proportional to the square of the beam current. The spectral distribution of the coherent radiation contains the information about the electron distribution in the bunch [4, 5]. Coherent Diffraction Radiation (CDR) was suggested as a mechanism for coherent radiation generation due to its non-invasive nature [6, 7, 8].

2. Theoretical background

CDR arises when a charged particle beam passes by in the vicinity of a target. In this case the effect of the beam interaction with the target material is minimal and a smaller perturbation to the beam is produced compared with other diagnostics, such as coherent transition radiation (CTR).

When considering Diffraction Radiation (DR) theory the analogy between the processes of radiation and light scattering is used. The electromagnetic field of a uniformly moving particle is represented as a sum of pseudo-photons. Hence, the processes of radiation are reduced to those of the scattering of pseudo-photons off a metal screen [9, 10].

Assuming an electron is traveling along the positive z-direction and the vacuum-metal interface is in the x-y-plane, the DR field is simply a superposition of the real photons created on the target surface and can be described by

$$E^l_{x,y} = \frac{1}{4\pi^2} \iint E^i_{x,y}(x_s, y_s) \frac{e^{i\varphi}}{r} dy_s dx_s \tag{1}$$

where $E^l_{x,y}$ is the amplitude of the x- and y-polarisation components of DR, respectively, and the integration is performed over the target surface. $E^i_{x,y}$ is the amplitude originating from an elementary radiation source on the target at a position (x_s, y_s), φ is the phase advance of the photons and r is the distance from an arbitrary point on the target surface to the observation point with Cartesian coordinates ξ and η.

The amplitude $E^i_{x,y}$ one needs to substitute into Eq. 1 is just [9, 10]

$$E^i_{x,y}(x_s, y_s) = \frac{iek}{\pi\gamma} \begin{pmatrix} \cos\psi_s \\ \sin\psi_s \end{pmatrix} K_1\left(\frac{k}{\gamma}\rho_s\right) \tag{2}$$

where $\rho_s = \sqrt{x_s^2 + y_s^2}$. $k = 2\pi/\lambda$ is the radiation wave vector, λ is the radiation wavelength, γ is the charged particle Lorentz-factor, K_1 is the first order McDonald function, and e is the electron charge. A natural unit system is also used, where $h = m_e = c = 1$ [10].

From a geometrical discussion, $\frac{e^{i\varphi}}{|\vec{r}|}$ can be represented as [10]

$$\frac{e^{i\varphi}}{|\vec{r}|} = \frac{e^{ika}}{a} \exp\left[\frac{ik}{2a}\left(x_s^2 + y_s^2\right) - \frac{ik}{a}\left(x_s\xi + y_s\eta\right)\right] \tag{3}$$

Substituting Eq. (3) and Eq. (2) into Eq. (1) one obtains equations for the correct amplitudes and the DR spatial-spectral distribution can be calculated using

$$\frac{d^2W^{DR}}{d\omega du} = 4\pi^2k^2\left[\left|E_x^{DR}\right|^2 + \left|E_y^{DR}\right|^2\right] \tag{4}$$

where E_x^{DR} and E_y^{DR} are the x- and y-polarisation components of DR, and $du = d\xi d\eta$ is the detector aperture element.

Simulation studies are performed where the target is placed to one side of the electron beam with impact parameter h. The single electron spectrum $S(\omega)$ for a given target size, particle energy, impact parameter, and detector aperture can be determined by integrating Eq. 4 over the detector aperture.

The integration limits used for this integration are the dimensions of the detector opening along the x and y axis, i.e. $\xi_{detector} = 46\,mm$ and $\eta_{detector} = 35\,mm$. The distance between the detector and the target, $a = 2\,m$, and the target size was taken into account, which in the case at the CLIC Test Facility (CTF3) is just $40 \times 40\,mm$, which is a projection of the target surface onto the plane perpendicular to the beam trajectory.

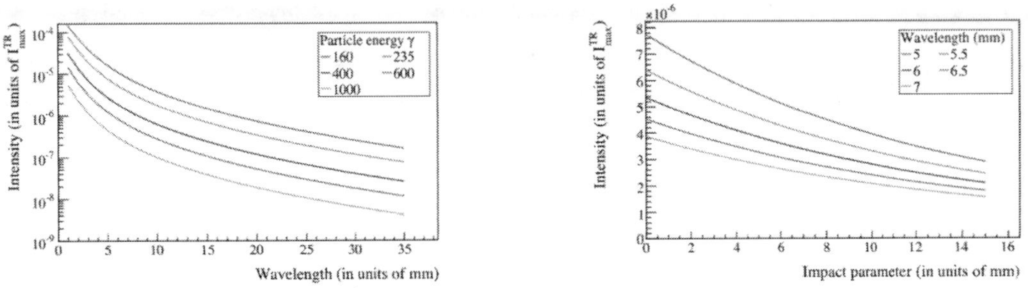

(a) DR spectra for different particle beam energies

(b) DR intensity variation with impact parameter for different observation wavelengths.

Figure 1. DR simulations for a $40 \times 40\,mm$ target and a detector at a distance $a = 2\,m$ ($I^{TR}_{max} = \frac{\gamma^2\lambda}{2\pi}$)

A plot of the DR spectra for different particle beam energies and an impact parameter of $h = 20\,mm$ can be seen in Fig. 1(a). Another investigation is the change of intensity with impact parameter for fixed wavelength, which is shown in Fig. 1(b). It is interesting to note that the intensity for a $15\,mm$ impact parameter only decreases by a factor of 2 compared to a zero impact parameter. This means that one does not have to worry about the signal levels when keeping the target at a distance from the electron beam. Therefore a working point at $h = 15\,mm$ is perfectly acceptable for accelerator operation. The spectra calculated by this model will be used for data analysis at a later stage.

Kramers-Kronig relation will be used to derive the longitudinal particle distribution in a bunch [11] from the experimentally obtained form factor

$$\rho^2(k) = \frac{S(k)}{N_e^2 S_e(k)} \tag{5}$$

where $S(k)$ is the experimentally measured coherent diffraction radiation spectrum, N_e is the number of particles in a bunch, $S_e(k)$ is the single electron spectrum calculated in Section 2 and $k = \frac{2\pi}{\lambda}$ is the wavenumber, where λ is the wavelength of emitted radiation.

Since $\rho(k)$ is a Fourier transform of the longitudinal particle distribution in a bunch, the normalised bunch distribution function can be determined as

$$S(z) = \frac{1}{\pi}\int_0^\infty dk\rho(k)\cos\left(\psi(k) - zk\right) \tag{6}$$

where z is the longitudinal coordinate. The phase factor $\psi(k)$ and the form factor amplitude $\rho(k)$ in Eq. (6) are related by Kramers-Kronig relation so that if the form factor $\rho^2(k)$ is measured at all wave numbers, then the phase factor can be obtained as [11, 12]

$$\psi(k) = -\frac{2c}{\pi}\int_0^\infty dx \frac{\ln(\rho(x)/\rho(k))}{x^2 - k^2} \tag{7}$$

where x is the integration variable in the wave number domain.

Realistically, measuring of $\rho^2(k)$ over the entire wave number interval is practically impossible, therefore it is necessary to extrapolate. At first we replace the integration by summation over the chosen wave number domain, as in the experiment we will have a number of discrete spectral points. We suggested the longitudinal charge distribution, $S_{ideal}(z) = \frac{exp(-z^2/2\sigma_1^2)}{4\sqrt{2\pi}\sigma_1} + \frac{3exp(-(z-z_0)^2/2\sigma_2^2)}{4\sqrt{2\pi}\sigma_2}$, where z_0=1.2 mm, σ_1=0.3 mm, and σ_2=0.45 mm. The calculated form factor $\rho^2_{calc}(k)$ is presented in Fig.2(a). The data area confined within the two vertical lines was assumed to be a given data set (from 1.3 to 4.6 mm^{-1}). For the interpolation between the form factor data points the following function was applied:

$$\rho^2_{int}(k) = \frac{\sum_{n=0}^{N} \rho^2(k_n)\, exp\left(-\frac{(k_n-k)^2}{2\sigma^2}\right)}{\sum_{n=0}^{N} exp\left(-\frac{(k_n-k)^2}{2\sigma^2}\right)} \tag{8}$$

where $\rho^2(k_n)$ is the form factor data and σ is the smoothing parameter for the reconstruction presented in Fig.2. To avoid significant smoothing of the data, σ was chosen to be $\frac{\Delta k_n}{3}$, where $\Delta k_n = k_n - k_{n-1}$.

For the low wave number extrapolation the following extrapolation function was used [13]:

$$\rho^2_{low}(k) = \rho^2_{int}(k_0) exp(-ak^2 + bk + c) \tag{9}$$

with $a = (\ln \rho^2_{int}(k_0) - k_0 \frac{s}{\rho^2_{int}(k_0)}) \frac{1}{k_0^2}$, $b = \frac{s}{\rho^2_{int}(k_0)} + 2ak_0$, $c = -\ln \rho^2_{int}(k_0)$, where $\rho^2_{int}(k_0)$ is the interpolation function value corresponding to the lowest wave number, k_0 is the lowest wave number and s is the slope derived from the interpolation function: $s = \frac{\rho^2_{int}(k_4) - \rho^2_{int}(k_0)}{k_4 - k_0}$.

The following function was used to extrapolate towards large wavenumbers:

$$\rho^2_{large}(k) = exp(-\beta k^2 + \gamma k + \delta) \tag{10}$$

where β, γ, δ are chosen to smoothly join the larger wave numbers. ρ^2_{large} should match the data at the largest wave number and both the first and second derivatives of ρ^2_{large} should match the first and second derivatives of ρ^2_{int} at the largest wave number.

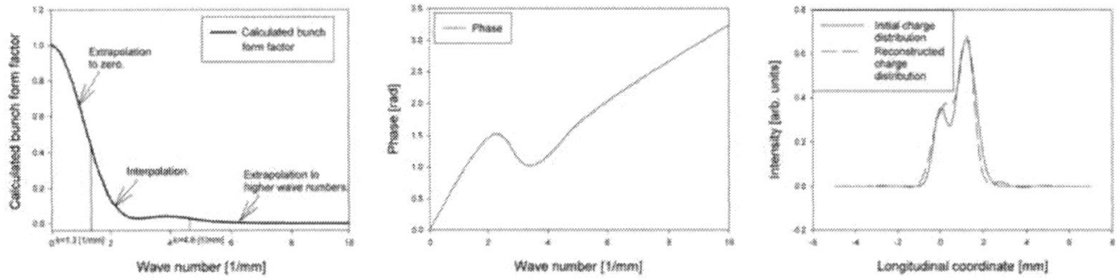

(a) (b) (c)

Figure 2. (a) Calculated bunch form factor, when interpolation and extrapolation functions were applied. (b) Phase reconstruction. (c) Initial and reconstructed longitudinal charge distributions.

The bunch profile and the phase reconstruction are presented in Fig.2(c) and Fig. 2(b), respectively, when extrapolation procedures to lower and larger wave numbers, as shown in Fig.2(a), were applied. A comparison of the reconstructed bunch profile with the initial bunch profile provides information about the accuracy of the extrapolation methods. As seen from

Fig.2(c), the reconstructed bunch length and the amplitude agree fairly well with the original distribution, even though the method is not completely sensitive to the left non-dominant peak of the initial pulse.

It is worth noting, that a sufficiently large spectral detector coverage is very important when we want to apply extrapolation. If the spectral range is too short, especially towards larger wave numbers, the method does not reconstruct the minimal phase accurately enough to be able to detail the trailing structures that follow the dominant peak.

3. Experimental setup

The CDR setup is installed in CTF3 [2]. CTF3 aims to demonstrate the feasibility of the CLIC two-beam acceleration scheme and other necessary components including non-invasive bunch length measurements. CTF3, as seen in Fig. 3(a), consists of a linac producing a 125 MeV

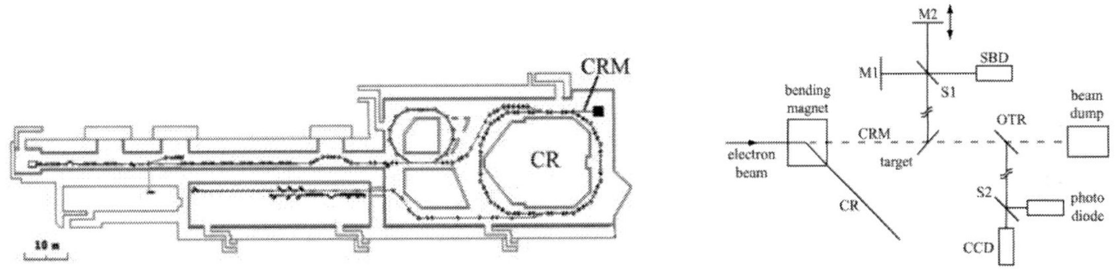

(a) Schematic view of CTF3 with the combiner ring (CR) located at the right.

(b) Schematic drawing of the CRM line including the CDR setup, an OTR screen, and the beam dump

Figure 3. CDR setup at CTF3

electron beam, a Delay Loop (DL), a Combiner Ring (CR) and the CLIC Experimental Area (CLEX) where measurements are carried out to prove the feasibility of CLIC. The DL and CR are used to interleave and combine the electron beam to produce a high current electron beam with a bunch sequence frequency of 12 GHz. For the CDR setup, the DL after the linac is bypassed and the beam is injected into the CR. After the first bending magnet in the CR, an extended straight section can be found, which is the CR Measurements (CRM) line. A schematic layout of the CRM line with a section of the CR can be seen in Fig. 3(b).

Two ultra-high vacuum (UHV) six-way crosses form our vacuum vessel. The inner diameter of the six-way crosses is 95.7 mm. The target is a 60 mm$\times$ 40 mm $\times$ 300 μm silicon wafer coated with aluminium and is placed to one side of the electron beam with impact parameter h. The target is attached to the shaft of a 4D UHV manipulator, which is mounted on top of the downstream cross and provides precise remote control of the rotational and vertical translation axis. The manipulator is equipped with stepper motors which provide a single step precision, relating to a 0.004° and 1μm rotational and translational precision, respectively. The motors are driven and monitored with a *BALDOR NextMove e100* motion controller via custom cable connections. The vacuum output window is a fused silica vacuum window, which is transmissive for mm and sub-mm radiation, with a viewing diameter of 40 mm. The adjacent viewport is a standard Kodial window with a viewing diameter of 38 mm and is only used for alignment purposes. Both viewports are attached to the six-way cross through 150-to-70 mm OD adapter flanges.

The radiation originating from the target is translated vertically with a periscope to avoid backgrounds from the horizontal particle beam plane. On an optical table the Michelson interferometer is installed and the reference working point is at a height of 5 inch (127 mm) above the table surface.

The mirrors (M1 and M2 in Fig. 3(b)) are broadband aluminium coated mirrors with a diameter of 4 inch (101.6 mm). Two axes of each mirror holder can be controlled with fine adjustment screws, which can potentially be motorised.

Mirror M2 is mounted on top of a translation stage. The translation stage has a travel range of 150 mm, a resolution of 0.1 μm, a minimum incremental motion of 0.3 μm, and a uni-directional repeatability of 1 μm. The translation stage is interfaced with a controller which can be accessed via a RS232 serial connection.

In order to align the interferometer an optical laser alignment procedure is used. A HeNe laser is stably mounted on the adjacent side of the beam line and directed into the six-way cross through the Kodial viewport. The laser then resembles the path of radiation originating from the target. The mirrors are then adjusted locally to obtain circular fringes produced by the interferometer. Even a rough alignment with an optical laser will be nearly perfect for millimetre waves.

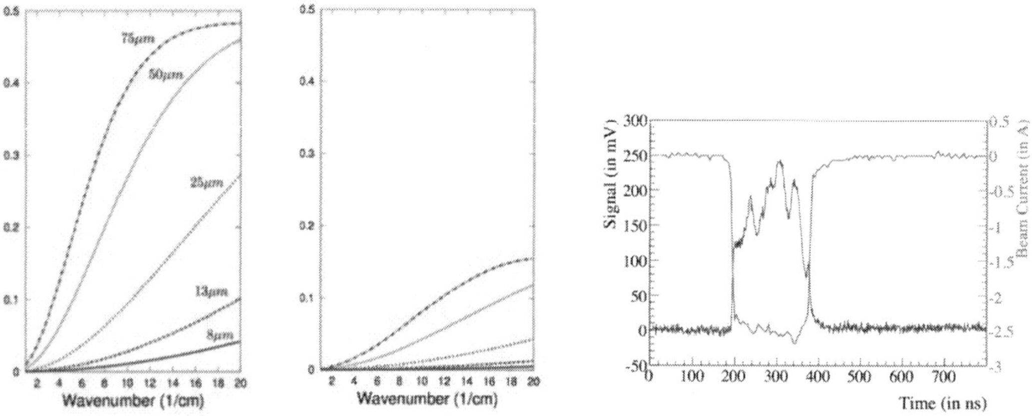

(a) Kapton R_s polarisation

(b) Kapton R_p polarisation

(c) Example CDR signal with corresponding beam current reading from a nearby BPM for a 200 ns bunch train

Figure 4. Comparison of efficiencies for different thickness for Kapton films (a,b) and an example CDR signal (c)

A 50 μm Kapton film was used as a splitter. The efficiencies of the splitter are represented in Fig. 4(a) and 4(b). From the studies it was concluded that the best compromise between splitter efficiency and linearity for commercially available films was for a 50 μm Kapton film [14, 15].

The detectors are based on ultra fast Schottky barrier diodes (SBD) (with response time typically around 250 ps). All SBD detectors used are polarisation sensitive and one can only measure one polarisation at a time. High quality RF cables with a bandwidth of 10 GHz are used to transport the signal. The detector used at the moment is sensitive in a wavelength region from 2.14 mm to 3.33 mm (90 - 140 GHz) with an average sensitivity of around 1000 mV/mW. An example signal of DR is shown in Fig. 4(c). It can be seen that for a fairly constant beam current the DR intensity varies significantly. It suggests that the longitudinal electron distribution throughout the train is non-uniform and the detector can be used for machine tuning.

The data acquisition is performed with a 10-bit *Acqiris DC282* digitiser. The DC282 offers synchronous four channel sampling at up to 2 GS/s, or interleaved dual- or single-channel sampling at up to 4 and 8 GS/s respectively. The input for an external trigger provides a precise synchronisation to the electron gun trigger. The internal acquisition memory of the digitiser is 256 kSample/channel and is large enough to theoretically store around 100 bunch trains of $1.4\mu s$ length.

4. Experimental results

With the CDR setup in the CRM line, DR and Synchrotron Radiation (SR) can be measured. With the bending magnet turned on and the electron beam circulating in the combiner ring, SR can be observed. The target is therefore lowered and used as a mirror to direct SR into the detector. For DR measurements the magnet is simply turned off. After the CDR setup and the OTR screen, the beam is terminated in the CRM beam dump.

According to theory (cf. Fig. 1(b)) the CDR intensity must decrease monotonically as a function of impact parameter. The impact parameter dependance of the signal for two different beam trajectories was investigated. Fig. 5(a) shows the dependance for the beam on the CRM reference orbit, i.e. the beam is centred in the beam pipe, and Fig. 5(b) illustrates the dependance for the beam lowered by 7 mm in the CRM line by a vertical corrector in the CR.

As there are two different screens available in the OTR station behind the CDR setup - a semitransparent silica screen and high-reflective aluminised silica screen - the influence of the two different screens was also studied. For the nominal beam orbit, upon changing the OTR screens behind the setup a change in CDR intensity was observed. The different intensities can be explained by reflections of transition radiation within the UHV hardware as outlined in Fig. 5(c). Transition radiation emitted by the two OTR screens and the aluminium UHV flange of the beam dump contribute to the CDR signal.

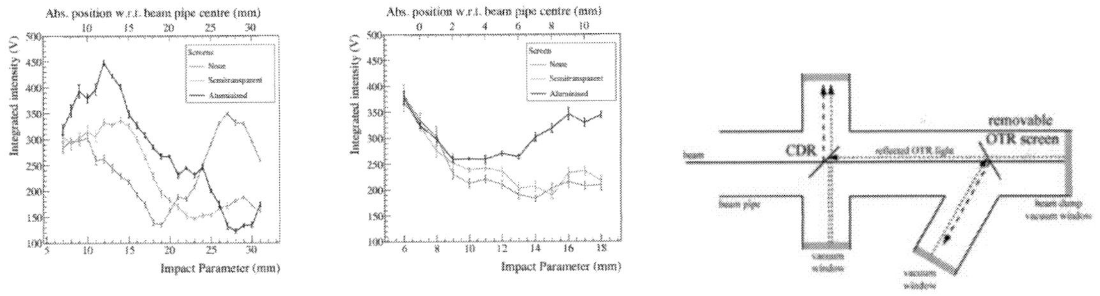

(a) Impact parameter dependance for nominal beam operation in the CRM line

(b) Impact parameter dependance for a lowered beam in the CRM line

(c) Schematic drawing explaining the origin of the backgrounds

Figure 5. Dependence of the CDR signals on the impact parameter and OTR screens behind for different beam positions in the CRM line.

For low impact parameters and an OTR screen inserted, the contribution to the CDR signal is dominated by the OTR screens. For large impact parameters the contribution is dominated by the beam dump UHV flange when none of the OTR screens are inserted.

For the beam lowered in the CRM line, the impact parameter dependance is shown in Fig. 5(b). When no screen or the semitransparent screen is inserted the signal agrees fairly well with the expected impact parameter dependance. However, the aluminised silica screen shows a contribution to the signal for large impact parameters, which can be explained by the vertical position of the CDR target. For small impact parameters the target is lowered further in the beam line and the reflected TR is simply blocked which manifests itself in the convergence of the signals and a monotonic signal increase for decreasing impact parameters.

To be able to block the reflected background completely we are planning to design and install an off-centre adapter flange, as discussed in Section 5. This upgrade is very important to continue the CDR measurements.

SR measurements were also performed and therefore the target edge is 1 mm below the beam pipe centre. As expected, the CSR distribution is a single peak distribution, as seen in Fig. 6(a). The maximum intensity found during the rotation scan was the mirror reflection direction and the angle of mirror reflection was used as the working point.

The first interferogram is shown in Fig.6(b). The target was placed at an orientation angle of 34.0° and a vertical position of $0\,mm$. The CSR signal is integrated over a small part of the bunch train - typically $50\,ns$ - as the longitudinal profile is fluctuating quite significantly throughout the train. For each translation stage position, the arithmetic mean of the integrated CSR intensity is taken over 10 bunch trains. It was chosen to average over 10 bunch trains as this is a good compromise between statistics and the stability of the electron beam.

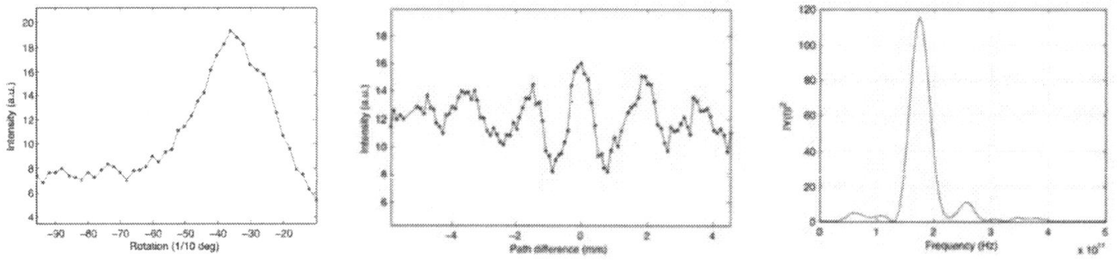

(a) Rotation dependance (b) Interferogram (c) Spectrum

Figure 6. a) CSR Interferogram and b) corresponding spectrum

From the interferogram in Fig. 6(b), the corresponding spectrum can be found performing a discrete Fourier transform. As the interferogram is expected to be symmetric about the zero path difference a discrete Cosine transform was performed. The resulting spectrum is shown in Fig. 6(c). One may see that the CSR spectrum is rather narrow. To be able to perform Kramers-Kronig analysis we shall perform measurements using a few SBD detectors in the near future.

5. Conclusion and outlook

In this paper we have presented the status of the CDR longitudinal beam profile monitor development in CTF3 at CERN. We have designed and installed the vacuum hardware, the Michelson interferometer system, and we have also developed the data acquisition and hardware control software. The first CSR and CDR signals were observed using an ultrafast SBD detector. For a fairly flat charge distribution throughout the train, we observed a signal variation which suggests that there is a bunch-by-bunch longitudinal profile fluctuation in the train.

The first CSR interferometric measurements have been performed with the system and the first CSR spectra are obtained. As described above, the spectra need to be normalised with respect to various spectral hardware dependencies and the single electron radiation spectrum. Thereafter the longitudinal bunch profile will be reconstructed using Kramers-Kronig relation. The experience gathered from parasitic synchrotron radiation studies will be transferred to CDR interferometric measurements which are scheduled after an upgrade.

Unfortunately, a significant background in the CRM line did not allow us to perform proper CDR interferometric measurements. To exclude the background the system will be modified. This can be achieved by installing an off-centre adapter flange. As the study of the background suggests, a vertical off-set of this inner bore by +15 mm would allow for the backgrounds to be minimised while still being able to align the interferometer with a laser through the adjacent viewport. Such an off-centre flange has been designed and manufactured at the CERN workshop, and is due for installation in the near future.

In January/February 2010 we intend to install a second target in order to block the upstream background radiation.

References

[1] A. Chao, M. Tigner, *Handbook of accelerator physics and engineering*, World Scientific Publishing, 1999

[2] G. Geschonke et al.,*CTF3 Design Report*,CERN/PS 2002-08 (2002)

[3] T. Lefevre, Review of Instrumentation needs and Critical items, *http://indico.cern.ch/conferenceDisplay.py ?confId=55244*, 2009

[4] K. Ishi, et al., Phys. Rev. A 43, 5597 (1991)

[5] Y. Shibata, et al., Phys. Rev. E 49, 785 (1994)

[6] A.H. Lumpkin, N.S. Sereno, D.W. Rule, Nucl. Instr. and Meth. A 475 (2001), pp. 470-475

[7] M. Castellano, et al., Phys. Rev. E 63, 056501 (2001)

[8] B. Feng, et al., Nucl. Instr. and Meth. A 475 (2001), pp. 492-497

[9] M. L. Ter-Mikaelian, *High-Energy Electromagnetic Processes in Condensed Media*, Wiley-Interscience, 1972

[10] P. Karataev, Physics Letters A 345, 428 (2005)

[11] R. Lai, A.J. Sievers, Phys. Rev. E 50, R3342 (1994)

[12] R. Lai, A.J. Sievers, Phys. Rev. E 50, R4294 (1994)

[13] V. Blackmore, *Determination of the Time Profile of Picosecond-Long Electron Bunches through the use of Coherent Smith-Purcell Radiation*, PhD Thesis, Univeristy of Oxford, 2008

[14] D. A. Naylor, R. T. Boreiko, and T. A. Clark, Appl. Opt., 17(7):10551058, 1978

[15] D.R. Smith, E.V. Loewenstein, Appl. Opt., 14(6):1335 1341, 1975

Generation of induced Smith-Purcell radiation in the absence of resonator

Dmitry N. Klochkov[1†], Alexander I. Artemyev[2], Koryun B. Oganesyan[3], Yuri V. Rostovtsev[4], Marlan O. Scully[4], Chin-Kun Hu[1,5‡]

[1]Institute of Physics, Academia Sinica, Nankang, Taipei 1152, Taiwan
[2]General Physics Institute RAS, Vavilova 38, 119991 Moscow, Russia
[3]Yerevan Physics Institute, 375036, Armenia, Yerevan, 2 Alikhanian Br.
[4]Department of Physics and Institute for Quantum Studies, Texas A&M University, College Station, Texas 77843-4242, USA
[5]Center for Nonlinear and Complex Systems and Department of Physics, Chung-Yuang Christian University, Chungli 32023, Taiwan

E-mail: † `dimitri.klochkoff@gmail.com`

E-mail: ‡ `huck@phys.sinica.edu.tw`

Abstract. The simplest model of the magnetized infinitely thin electron beam is considered. For the grating, which has depth of grooves as a small parameter, the dispersion equation of the Smith-Purcell instability was obtained. It was found that the condition of the Thompson or the Raman regimes of excitation does not depend on beam current but depends on the height of the beam above grating surface. The growth rate of instability in both cases is proportional to the square root of the electron beam current.

The spontaneous Smith-Purcell (SP) radiation is emitted with an electron passes close to the matal surface of a grating [1]. Last years there has been a renewed interest to the induced intense SP radiation produced by relativistic electron beam. The experiment at Dartmouth College [2, 3], where radiation in the SP system was observed, has stimulated new investigations concerning the SP FEL as an open slow wave structure [4, 5, 6, 7, 8, 9].

Several theories have been proposed to describe the operation of a Smith-Purcell FEL. In particular, Schaechter and Ron [10] proposed a theory based on the interaction of an electron-beam with a wave traveling along the grating. The interaction is found to amplify waves incident on the grating and reflected by it, with a gain that depends on the reflection matrix of the grating. They have found that the gain is proportional to the cube root of the electron-beam current, which agrees with the behavior of Cherenkov free-electron lasers and other slow-wave devices. More recently, Kim and Song [4] have proposed a theory in which they assume that the electrons interact with a wave that travels along the surface of the grating. Assuming that at least one Fourier component of the traveling wave is radiative, and they have found that the gain is proportional to the square root of the electron-beam current $\sim I_b^{1/2}$.

A different result has been obtained in Ref. [5]; in this paper a rectangular grating is considered, assuming that the entire space above the grating is filled by a uniform electron beam. The authors have established the dispersion relation and found the dispersion law $\omega(\mathbf{k})$.

It turns out that the dispersion equation allows only evanescent solutions and the operating point of a SP FEL is fixed by the intersection of the dispersion curve with the beam line. The corresponding small signal gain follows the $\sim I_b^{1/3}$ law.

Hence, there is no general consent on the mechanisms of the SP FEL and the small signal gain obtained by different approaches. Taking also into account that there is currently substantial interest in the implementation of a SP FEL, it is of interest to reexamine the theory of SP FEL.

One can select two theoretical target settings for SP radiation. First of them is a problem of generation of SP-radiation. In this case there is no incoming from infinity wave. SP system generates outgoing waves. Another problem is an amplification or attenuation of incoming waves by SP-system. One can expect that these different problem have different solutions.

In this work we consider only the problem of generation of wave s by SP-system, when the incoming waves are absent. In order that mathematical problems do not cover physical picture, we consider the simplest model of the beam.

For the small signal gain we consider the linear over the field amplitude solutions of system of Maxwell's and beam's equations. We take a Cartesian coordinate system in a such manner that the electron beam propagates to the positive x direction in the vacuum over the grating (parallel to the surface of a grating) as it is shown in Fig. 1. For simplicity, we consider a infinitely in z direction sheet of the beam with thickness Δ_b and the non-disturbed density n_b stabilized by external strong magnetic field. We assume the system to be translationally invariant in the z direction. This means that all physical values depend on coordinates x and y only, and $E_z = 0$.

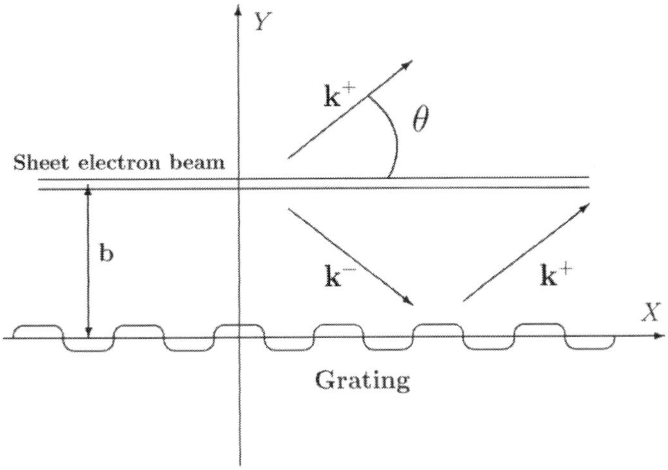

Figure 1. Schematic of an SP-FEL using a sheet electron beam. The z-axis is a perpendicular to the list. The sheet electron beam is in the plane $y = b$. Period of the grating is d, amplitude of the grating is h.

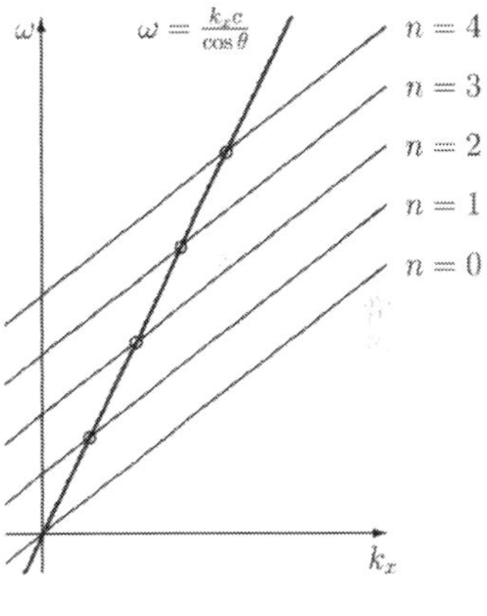

Figure 2. The dispersion relations for the electromagnetic wave $\omega = k_x c/\cos\theta$ and the beam waves with $\omega_n = (k_x + n\varkappa)u$ are shown in $k_x - \omega$ plane. The intersection points determine the spectrum of frequency (22).

We start from the Maxwell's equations

$$\nabla \times \mathbf{E} = -\frac{1}{c}\frac{\partial \mathbf{B}}{\partial t}, \qquad \mathrm{div}\mathbf{B} = 0, \tag{1}$$

$$\nabla \times \mathbf{B} = \frac{1}{c}\frac{\partial \mathbf{E}}{\partial t} + \frac{4\pi}{c}\mathbf{j}, \qquad \mathrm{div}\mathbf{E} = 4\pi\rho. \tag{2}$$

and the hydrodynamical equations for the beam, namely, equation of motion

$$\frac{d\mathbf{p}}{dt} = \left(\frac{\partial}{\partial t} + \mathbf{v}\nabla\right)\mathbf{p} = e\mathbf{E} + e\frac{\mathbf{v}}{c} \times \mathbf{B} \tag{3}$$

and continues equation

$$\frac{\partial \rho}{\partial t} + \mathrm{div}\mathbf{j} = 0, \tag{4}$$

where $\mathbf{j} = \rho\mathbf{v} = en\mathbf{v}$. An initial velocity of electrons is $\mathbf{u} = (u, 0, 0)$, and $\gamma_0 = (1 - u^2/c^2)^{1/2}$ is a relativistic factor.

In our case there are only three non-zero components E_x, E_y and B_z of electromagnetic field. In order to describe the field in vacuum we introduce the potential P and determine the field components with

$$E_x = \frac{\partial P}{\partial y}, \qquad E_y = -\frac{\partial P}{\partial x}, \qquad B_z = \frac{1}{c}\frac{\partial P}{\partial t}. \tag{5}$$

To describe the electromagnetic field in the vacuum space ($\rho = 0$ and $\mathbf{j} = 0$ in Eqs.(1),(2)), we take advantage of Floquets theorem and expand the fields above ($y > b$) and below ($y < b$) beam level in the form :

$$P = \begin{cases} \sum\limits_n P_n e^{-i\omega t + i\mathbf{q}_n^+\mathbf{r}}, & y > b \\ \sum\limits_n (P_n^+ e^{i\mathbf{q}_n^+\mathbf{r}} + P_n^- e^{i\mathbf{q}_n^-\mathbf{r}})e^{-i\omega t}, & y < b. \end{cases} \tag{6}$$

Here the wave vectors are $\mathbf{q}_n^{\pm} = (q_{nx}, \pm q_{ny}) = (k_x + n\varkappa, \ \pm q_{ny})$, and $\varkappa = 2\pi/d$ is a wave number of the grating, d is a grating period. The dispersion relation for electromagnetic wave in vacuum space is $\omega^2 = (\mathbf{q}_n^{\pm})^2 c^2$.

Using the boundary conditions on the sheet beam

$$E_x(b+0) = E_x(b-0), \qquad E_y(b+0) - E_y(b-0) = 4\pi\tilde{\rho}. \tag{7}$$

we can express the partial amplitudes P_n^+ and P_n^- in terms of the partial amplitudes P_n of the outgoing waves. Here the surface density of disturbed beam charge $\tilde{\rho}$ is obtained from the linearized equations for the beam.

The solutions, obtained above, define the amplitudes of the field, but do not determine its frequency. In order to get the frequency we have to consider the boundary condition on the surface of the grating. We consider the simplest model of the grating, when the equation of metal surface is defined as

$$y(x) = h\sin(\varkappa x) = h\sin\left(\frac{2\pi}{d}x\right). \tag{8}$$

Tangential vector to the surface is $\mathbf{r}_0 = (1, h\varkappa\cos(\varkappa x))$. Boundary condition $(\mathbf{r}_0\mathbf{E}(x, y(x))) = 0$ takes the form

$$E_x(x, y(x)) + h\varkappa\cos(\varkappa x)E_y(x, y(x)) = 0. \tag{9}$$

Here E_x and E_y are given by (5). Substituting the solutions of Maxwell equations (5), (6) the in boundary condition (9), multiplying by $\exp(-ip\varkappa x)$, where $p = 0; \pm1; \pm2; \pm3, ...,$ and integrating then over coordinate x within the interval $(0; d)$ we obtain the infinite system of algebraic equation respectively the partial amplitudes P_n:

$$\sum_n \left(q_{ny} + (n-p)\varkappa\frac{q_{nx}}{q_{ny}}\right)\left(J_{n-p}(-q_{ny}h) + K(n)\left[J_{n-p}(-q_{ny}h) - J_{n-p}(q_{ny}h)e^{2iq_{ny}b}\right]\right)P_n = 0. \tag{10}$$

Here $J_\alpha(x)$ is a Bessel function.

In order to cut the infinite system we assume that the depth of the grating groove h is a moderate. Since the Bessel function on small variable x drops as $J_\alpha(x) \approx (x/2)^\alpha/\alpha!$ with α increasing, the assumption that h is a small parameter permits us to consider 3-waves approximation. We suppose that there are only three non-zero partial amplitudes, namely, P_{n-1}, P_n and P_{n+1}. Other amplitudes are assumed to be zero. Then we get three algebraic equations, matrix of which is

$$\begin{Vmatrix} q_{n-1y}\alpha_{n-1} & \beta_n^+ & \gamma_{n+1}^+ \\ \beta_{n-1}^- & q_{ny}\alpha_n & \beta_{n+1}^+ \\ \gamma_{n-1}^- & \beta_n^- & q_{n+1y}\alpha_{n+1} \end{Vmatrix}. \tag{11}$$

Here coefficients are

$$\begin{aligned} \alpha_n &= 1 + K(n)(1 - e^{2iq_{ny}b}), \\ \beta_n^\pm &= \mp\frac{h}{2}(q_{ny}^2 \pm \varkappa q_{nx})\left[1 + K(n)(1 + e^{2iq_{ny}b})\right], \\ \gamma_n^\pm &= \frac{h^2}{8} q_{ny}(q_{ny}^2 \pm 2\varkappa q_{nx})\alpha_n. \end{aligned} \tag{12}$$

Here we introduce the notation

$$K(n) = \frac{i}{2}\Delta_b\omega_b^2\gamma_0^{-3}\frac{q_{ny}}{(\omega - (\mathbf{q}_n^+\mathbf{u}))^2}. \tag{13}$$

The nontrivial solution of infinite system of equations (11) takes place if its determinant equals zero. The latter is dispersion equation describing both the spectrum of frequencies and the growth rates of the induced Smith-Purcell radiation. Neglecting the small terms proportional to h^4 and h with higher powers, we obtain the simple dispersion equation:

$$D(\omega, \mathbf{k}) = q_{n-1y}q_{ny}q_{n+1y}\alpha_{n-1}\alpha_n\alpha_{n+1} - q_{n-1y}\alpha_{n-1}\beta_n^-\beta_{n+1}^+ - \beta_{n-1}^-\beta_n^+q_{n+1y}\alpha_{n+1} = 0. \tag{14}$$

What is zero-order approximation for dispersion equation that is the question. In the absence of the beam, when $\omega_b = 0$, we find $P_n^- = 0$ and $P_n^+ = P_n$, that means $\mathbf{E}_n^- = 0$ and $\mathbf{E}_n^+ = \mathbf{E}_n$. Substituting this amplitudes in the boundary condition we find that $\mathbf{E}_n = 0$. Since all amplitudes equal zero, the determinant of the system is not equal to zero. Indeed, substituting $\omega_b = 0$ in equation (14) we get

$$\begin{aligned} D(\omega, \mathbf{k}, \omega_b = 0) &= q_{n-1\,y}q_{ny}q_{n+1\,y} \\ &+ \frac{h^2}{4}\left\{q_{n-1\,y}(q_{ny}^2 - \varkappa q_{nx})(q_{n+1y}^2 + \varkappa q_{n+1x}) + q_{n+1\,y}(q_{n-1y}^2 - \varkappa q_{n-1x})(q_{ny}^2 + \varkappa q_{nx})\right\} \neq 0. \end{aligned} \tag{15}$$

The latter has a simple physical explanation. In the absence of the beam, the electromagnetic field in the system is absent too, because the grating does not emit the waves. Indeed, the grating plate does not shine in a dark room!

Therefore we can not use condition $\omega_b = 0$ as a zero-order approximation to solve dispersion equation. Instead we use as zero-order approximation condition $h = 0$, which gives

$$D_0(\omega, \mathbf{k}; h = 0) = q_{n-1y}q_{ny}q_{n+1y}\alpha_{n-1}\alpha_n\alpha_{n+1} = 0. \tag{16}$$

In general case for an arbitrary angle θ the product of $q_{n-1y}q_{ny}q_{n+1y}$ does not equal zero, therefore we can write

$$D_0(\omega, \mathbf{k}; h = 0) = \alpha_{n-1}\alpha_n\alpha_{n+1} = 0. \tag{17}$$

Without loss generality we assume that the n-mode of electromagnetic spectrum excites in the system. The latter means that $\alpha_n = 0$, which gives equation

$$(\omega - (k_x + n\varkappa)u)^2 + \frac{i}{2}\Delta_b\omega_b^2\gamma_0^{-3}q_{ny}\left(1 - e^{2iq_{ny}b}\right) = 0. \tag{18}$$

We seek solution of the equation in the form

$$\omega = \omega_n \pm \Omega_{bn}, \tag{19}$$

where ω_n satisfies the equation (18) in the limit $\omega_b \to 0$:

$$\omega_n = (k_x + n\varkappa)u, \tag{20}$$

which is the dispersion relation for longitudinal beam waves. Taking into account the dispersion relation for electromagnetic wave in vacuum

$$k_x = \frac{\omega_n}{c}\cos\theta, \tag{21}$$

we find the frequency of n-mode of spectrum for the limit $\omega_b \to 0$ as the intersection point in the $k_x - \omega$ diagram shown in the Fig. 2 :

$$\omega_n = \frac{n\varkappa u}{1 - \beta\cos\theta}, \tag{22}$$

where $\beta = u/c$ is a dimensionless velocity of the electron, θ is an angle between vector $\mathbf{k}^+$ and positive direction of x-axis, or say another, the angle of observation. Please *do not mix up* the n-mode in spectrum of radiation with the n-mode in q_x-spectrum of surface wave E_n. With help of Eq. (22) we find that

$$q_{ny} = i\frac{\omega_n}{u\gamma_0}, \tag{23}$$

so the n-mode of surface wave E_n does not radiate, because it damps in the positive direction of y-axis.

Considering the presence of beam as perturbation and substituting the value q_{ny} given by Eq. (23) in Eq. (18) we find from Eq.(18) the beam frequency Ω_{bn}:

$$\Omega_{bn} = \omega_b\gamma_0^{-2}\sqrt{\frac{\Delta_b\omega_n}{2u}\left(1 - e^{-2\frac{\omega_n b}{u\gamma_0}}\right)}. \tag{24}$$

Here we give the physical interpretation of solutions obtained. The condition $h = 0$ means that there is a flat mirror in plane $y = 0$. The fluctuations of charge and current densities of the electron beam produce fluctuations of electromagnetic field, the plane wave of which reflect from the mirror. The resonant condition (22) is a crossing point of dispersion lines for the beam wave and reflect electromagnetic plane wave. The boundary condition gives discrete spectrum, modes of which contain two coupled frequencies: "high" frequency with $\omega = \omega_n + \Omega_{bn}$ and "low" frequency with $\omega = \omega_n - \Omega_{bn}$. The beam frequency Ω_{bn} depends on the height of beam above surface $y = 0$ and has asymptotes:

$$\Omega_{bn} = \begin{cases} \omega_b\gamma_0^{-2}\sqrt{\dfrac{\Delta_b\omega_n}{2u}}, & b \to \infty \\ \omega_b\gamma_0^{-5/2}\dfrac{\omega_n}{u}\sqrt{\Delta_b b}, & b \to 0. \end{cases} \tag{25}$$

Now we consider the presence of the grooves of mirror ($h \neq 0$) as perturbation and will seek the solution of dispersion equation near synchronism point (19) in the form:

$$\omega = \omega_n \pm \Omega_{bn} + \delta\omega. \tag{26}$$

Substituting solution (26) in Eq. (14) and neglecting small terms under realistic condition $\{\omega_b, \Omega_{bn}, |\delta\omega|\} \ll \varkappa u$ we obtain dispersion equation respectively small shift of frequency:

$$\delta\omega^2 \pm 2\Omega_{bn}\delta\omega = \frac{h^2}{4}\frac{\Delta_b\omega_b^2}{\gamma_0^5}\left(\frac{\omega_n}{u}\right)^3 e^{-2\frac{\omega_n b}{u\gamma_0}}\Big(X_n(\theta,u) + iY_n(\theta,u)\Big). \tag{27}$$

For $\omega = \omega_n$ the wave number q_{n+1y} is always an imaginary value. We rewrite q_{n+1y} as $q_{n+1y} = ig_{n+1}$, where $g_{n+1} = \frac{\omega_n}{nu}\sqrt{(n+1-\beta\cos\theta)^2 - n^2\beta^2}$. The wave number q_{n-1y} can be as a real as an imaginary for $\omega = \omega_n$. Under condition:

$$\frac{1-\beta\cos\theta}{1+\beta} < n < \frac{1-\beta\cos\theta}{1-\beta} \tag{28}$$

the wave number q_{n-1y} is a real and then the coefficients X_n and Y_n are equal to

$$\begin{aligned} X_n &= \left(1 + \frac{\varkappa u\gamma_0^2}{\omega_n}\right)\left(\frac{\varkappa}{g_{n+1}}\left(1 + \frac{\varkappa u}{\omega_n}\right) - \frac{g_{n+1}u}{\omega_n}\right), \\ Y_n &= \left(1 - \frac{\varkappa u\gamma_0^2}{\omega_n}\right)\left(\frac{q_{n-1y}u}{\omega_n} - \left(1 - \frac{\varkappa u}{\omega_n}\right)\frac{\varkappa}{q_{n-1y}}\right). \end{aligned} \tag{29}$$

Another case realizes under condition

$$n \notin \left(\frac{1-\beta\cos\theta}{1+\beta}, \quad \frac{1-\beta\cos\theta}{1-\beta}\right), \tag{30}$$

when the wave number $q_{n-1y} = ig_{n-1}$ is an imaginary and then the coefficients X_n and Y_n are equal to

$$\begin{aligned} X_n &= \left(1 + \frac{\varkappa u\gamma_0^2}{\omega_n}\right)\left(\frac{\varkappa}{g_{n+1}}\left(1 + \frac{\varkappa u}{\omega_n}\right) - \frac{g_{n+1}u}{\omega_n}\right) - \left(1 - \frac{\varkappa u\gamma_0^2}{\omega_n}\right)\left(\frac{g_{n-1}u}{\omega_n} + \left(1 - \frac{\varkappa u}{\omega_n}\right)\frac{\varkappa}{g_{n-1}}\right), \\ Y_n &= 0. \end{aligned} \tag{31}$$

Here $g_{n-1} = \frac{\omega_n}{nu}\sqrt{(n-1+\beta\cos\theta)^2 - n^2\beta^2}$.

The square equation (27) has a simple solutions. Here we consider two interesting cases. At first, we consider the Thompson (single-particle approximation) type of excitation, when the electron beam radiates as a single particle or a bunch. In this case the beam waves can be neglected and condition of generation is $|\delta\omega| \gg \Omega_{bn}$. Under the condition we can omit the second term in left-hand side of dispersion equation and obtain the solution:

$$\delta\omega_n = \frac{h\omega_n}{2u}\sqrt{\frac{\Delta_b\omega_n}{u}\frac{\omega_b}{\gamma_0^{5/2}}}\, e^{-\frac{\omega_n b}{u\gamma_0}}\left(X_n + iY_n\right)^{1/2}. \tag{32}$$

The growth rate of instability of n-mode (22) ($\delta\omega = \delta\omega_n' + i\delta\omega_n''$) in this case is

$$\delta\omega_n'' = \pm\frac{h\omega_n}{2u}\frac{\omega_b}{\gamma_0^{5/2}}\sqrt{\frac{\Delta_b\omega_n}{u}}\, e^{-\frac{\omega_n b}{u\gamma_0}}\left(X_n^2 + Y_n^2\right)^{1/4}\sin\left|\frac{\psi}{2}\right|, \tag{33}$$

where angle ψ is defined with $\psi = \arccos\left(X_n/\sqrt{X_n^2+Y_n^2}\right)$. For sufficiently large value of the relativistic factor γ_0, when the inequality $\gamma_0 \gg \omega_n b/u$ holds, the growth rate of instability is proportional to square root of inverse γ_0: $\delta\omega'' \sim \gamma_0^{-1/2}$. The condition of the Thompson regime excitation does not depend on beam current and has form:

$$\frac{h\omega_n}{u}\frac{|X_n+iY_n|}{\sqrt{2\gamma_0\left(e^{2\frac{\omega_n b}{u\gamma_0}}-1\right)}} \gg 1. \tag{34}$$

This inequality holds for $b \to 0$ and for the big value of grooves depth h of grating.

Another case is the Raman (or collective) regime of generation, when influence of beam waves is appreciable and $|\delta\omega| \ll \Omega_{bn}$. For the Raman regime we obtain

$$\delta\omega = \pm\frac{h^2\omega_n^2}{4u^2}\frac{\omega_b}{\gamma_0^3}\sqrt{\frac{\Delta_b\omega_n}{2u}}\frac{e^{-2\frac{\omega_n b}{u\gamma_0}}}{\sqrt{1-e^{-2\frac{\omega_n b}{u\gamma_0}}}}(X_n+iY_n). \tag{35}$$

The growth rate of instability is

$$\delta\omega_n'' = \pm\frac{h^2\omega_n^2}{4u^2}\frac{\omega_b}{\gamma_0^3}\sqrt{\frac{\Delta_b\omega_n}{2u}}\frac{e^{-2\frac{\omega_n b}{u\gamma_0}}}{\sqrt{1-e^{-2\frac{\omega_n b}{u\gamma_0}}}}Y_n(\theta,u). \tag{36}$$

If $Y_n(\theta,u) > 0$, then the hight-frequency branch with $\omega = \omega_n + \Omega_{bn}$ excites. In the opposite case the low-frequency waves with $\omega = \omega_n - \Omega_{bn}$ generates. For sufficiently large value of the relativistic factor γ_0, when the inequality $\gamma_0 \gg \omega_n b/u$ holds, the growth rate of instability is proportional to inverse γ_0: $\delta\omega'' \sim \gamma_0^{-1}$. Condition for the Raman type of excitation is

$$\frac{h^2\omega_n^2}{4u^2\gamma_0}\frac{|Y_n|}{e^{2\frac{\omega_n b}{u\gamma_0}}-1} \ll 1. \tag{37}$$

It does not depend on Langmuir beam frequency (or beam current), but it depends on the beam height b above the grating.

The mechanism of light radiation is following. The surface wave with mode number n acting on the beam electrons, perturbs the trajectories of latter. Electrons, oscillating near equilibrium point $\mathbf{r} = \mathbf{r}_0 + \mathbf{u}t$, emit electromagnetic wave. Because, the amplitude of oscillation is proportional to the magnitude of surface wave on the height h of beam, the value of growth rate depends on the beam height above the grating surface.

When $q_{n-1y} = 0$ the 3-waves approximation does not work. It can be when $n \approx (1-\beta\cos\theta)(1\pm\beta)$, for example, for $n = 1$ and $\theta = 0$. In this case we have to include additional evanescent modes and consider 5-waves approximation, namely, we assume that there are only five non-zero amplitudes: E_{n-2x}, E_{n-1x}, E_{nx}, E_{n+1x} and E_{n+2x}. Other amplitudes are assumed to equal zero. Taking into account that h is a small parameter and neglecting small terms as above, we obtain the following dispersion equation:

$$\begin{aligned} D(\omega,\mathbf{k}) = \big(&q_{n-2y}\alpha_{n-2}q_{n-1y}\alpha_{n-1}q_{n+1y}\alpha_{n+1}q_{n+2y}\alpha_{n+2} - q_{n-2y}\alpha_{n-2}q_{n-1y}\alpha_{n-1}\beta_{n+1}^-\beta_{n+2}^+ \\ &- \beta_{n-2}^-\beta_{n-1}^+q_{n+1y}\alpha_{n+1}q_{n+2y}\alpha_{n+2}\big)q_{ny}\alpha_n - q_{n-2y}\alpha_{n-2}\beta_{n-1}^-\beta_n^+q_{n+1y}\alpha_{n+1}q_{n+2y}\alpha_{n+2} \\ &- q_{n-2y}\alpha_{n-2}q_{n-1y}\alpha_{n-1}\beta_n^-\beta_{n+1}^+q_{n+2y}\alpha_{n+2} = 0. \end{aligned} \tag{38}$$

Considering the solution of Eq. (38) near the synchronism point (26), we can rewrite the equation respectively $\delta\omega$:

$$\delta\omega^2 \pm 2\Omega_{bn}\delta\omega + \frac{i}{2}h\frac{\Delta_b\omega_b^2\omega_n^2}{u^2\gamma_0^5}\frac{A}{B}e^{-2\frac{\omega_n b}{u\gamma_0}} = 0, \tag{39}$$

where

$$A = q_{n-2y}q_{n+2y}\left\{\left(1+\frac{\varkappa u\gamma_0^2}{\omega_n}\right)q_{n-1y}\beta_{n+1}^{+} - \left(1-\frac{\varkappa u\gamma_0^2}{\omega_n}\right)q_{n+1y}\beta_{n-1}^{-}\right\},$$
$$B = q_{n-2y}q_{n-1y}q_{n+1y}q_{n+2y} - q_{n-2y}q_{n-1y}\beta_{n+1}^{-}\beta_{n+2}^{+} - \beta_{n-2}^{-}\beta_{n-1}^{+}q_{n+1y}q_{n+2y}. \tag{40}$$

For the case $q_{n-1y} = 0$ the dispersion equation reduces tom the form:

$$\delta\omega^2 \pm 2\Omega_{bn}\delta\omega + i\frac{\Delta_b\omega_b^2\omega_n^2}{u^2\gamma_0^5}\frac{q_{n-2y}e^{-2\frac{\omega_n b}{u\gamma_0}}}{q_{n-2y}^2 + \varkappa q_{n-2x}} = 0. \tag{41}$$

Here $q_{n-2y}^2 = 2\varkappa^2\left(\frac{n}{1-\beta\cos\theta} - 1.5\right)$ is a positive value. For the Thompson regime of excitation we find the growth rate:

$$\delta\omega_n'' = \pm\frac{\omega_b\omega_n}{u\gamma_0^{5/2}}\sqrt{\frac{\Delta_b}{2}\frac{q_{n-2y}}{q_{n-2y}^2 + \varkappa q_{n-2x}}}e^{-\frac{\omega_n b}{u\gamma_0}}, \tag{42}$$

which does not depend on small parameter h.

For the Raman regime of generation we find the growth rate:

$$\delta\omega_n'' = \pm\sqrt{\frac{\Delta_b\omega_n}{2u}}\frac{\omega_b\omega_n}{u\gamma_0^3}\frac{q_{n-2y}}{q_{n-2y}^2 + \varkappa q_{n-2x}}\frac{e^{-2\frac{\omega_n b}{u\gamma_0}}}{\sqrt{1-e^{-2\frac{\omega_n b}{u\gamma_0}}}}. \tag{43}$$

In this case the growth rate does not depend on small groove depth h too. Therefore, we have a peak of the growth rate under condition $q_{n-1y} = 0$.

We have found that zero-order approximation for solution of dispersion equation of SP-generation corresponds to the mirror boundary case, when the electron beam propagates above plane metal surface (mirror). The growth rates for both the Thompson and the Raman types of waves excitation are proportional to the Langmuir frequency ω_b or square root of beam current $I_b^{1/2}$. The conditions of excitations for both Thompson and Raman types do not depend on Langmuir beam frequency (or beam current), but depend on the beam height b above the grating. It is essential distinction between the Cherenkov and SP instabilities.

This work was supported by the National Science Council of the Republic of China (Taiwan) under Grant Nos. NSC 96-2911-M-001-003-MY3 & 98-2911-I-001-028 (for Chin-Kun Hu), and National Center for Theoretical Sciences in Taiwan and supported by the International Science and Technology Center, Moscow (project A-1602). We gratefully acknowledge the support from the Defense Advanced Research Projects, the Office of Naval Research, the NSF grant EEC-0540832 (MIRTHE ERC), and the Robert A. Welch Foundation (Grant #A1261).

[1] S.J. Smith and E.M. Purcell, Phys. Rev. **92**, 1069 (1953).
[2] J. Urata, M. Goldstein, M. F. Kimmitt, A. Naumov, C. Platt, and J. E. Walsh, Phys. Rev. Lett. **80**, 516 (1998).
[3] A. Bakhtyari, J. E.Walsh, and J. H. Brownell, Phys. Rev. E **65**, 066503 (2002).
[4] K.-J. Kim and S.-B. Song, Nucl. Instrum. Methods Phys. Res., Sect. A **475**, 158 (2001).
[5] H. L. Andrews and C. A. Brau, Phys. Rev. ST Accel. Beams **7**, 070701 (2004).
[6] H.L. Andrews, C. H. Boulware, C. A. Brau, and J. D. Jarvis, Phys. Rev. ST Accel. Beams **8**, 050703 (2005).
[7] H.L. Andrews, C. H. Boulware, C. A. Brau, J. T. Donohue, J. Gardelle, and J. D. Jarvis, New J. Phys. **8**, 289 (2006).
[8] Vinit Kumar and Kwange-Je Kim, Phys. Rev. E **73**, 026501 (2006).
[9] G. F. Mkrtchian, Phys. Rev. ST Accel. Beams **10**, 080701 (2007).
[10] L. Schaechter and A. Ron, Phys. Rev. A 40, 876 (1989).
[11] P. M. Van den berg, Appl. Sci. Res. **24**, 261 (1971).

Smith-Purcell radiation from a chain of spheres

K V Lekomtsev*[1], M N Strikhanov[2] and A A Tishchenko[2]

[1]John Adams Institute at Royal Holloway University of London, Egham Hill, Egham, Surrey, TW20 0EX, UK
[2]National Research Nuclear University "MePhi" , Moscow, Kashirskoe sh. 31, 115409, Russia

*E-mail: K.V.Lekomtsev@rhul.ac.uk

Abstract. Smith-Purcell and diffraction radiation were investigated. These types of radiation appear when a charged particle moves close to a conducting target. Spectral and angular distribution of diffraction radiation from the non-periodic chain of spheres is obtained analytically; local field effects are discussed. Analytical expression for the distribution of Smith-Purcell radiation from the periodic chain of spheres is obtained as well. For the first time it has been shown, that Smith-Purcell radiation for such a system is distributed over the cone. The results are investigated for the particles of different sizes, dielectric and metal, and for both ultrarelativistic and nonrelativistic cases.

1. Introduction

Smith-Purcell radiation arises when charged particles fly above a grating at the distance less than $\beta\gamma\lambda$, where $\beta = v/c$, v is the velocity of charged particles, $\gamma = 1/\sqrt{1-\beta^2}$ is the Lorentz factor and λ is a wavelength. This effect for the first time was predicted by I.M. Frank [1] and was experimentally observed by Smith and Purcell in optical wavelength range, using 300 keV electrons [2]. Of late years, close attention has been paid to investigation of Smith-Purcell radiation from one and two-dimensional systems, that consists of small spherical particles [3]-[11]. In [3] radiation emission probability and electron energy loss spectra for finite and infinite strings of Al and silica spheres were obtained by solving Maxwell's equations in the frames of a multipole expansion approach. The results obtained in Ref. [3] are general enough, but they are expressed in terms of spherical harmonic expansion, which makes the formulas rather complicated. Besides, nonrelativistic case only was analysed in that work.

Another way to describe electromagnetic processes in matter consists in calculation of the local field that acts on a single particle in a physical system. This work is devoted to development of theoretical basis of interaction between a moving charge and a chain of spherical particles characterized by arbitrary dielectric function. Spherical particles are assumed to be of different sizes, dielectric or metal. As dipole approximation had been considered, wavelength of emitted radiation must be much greater than a size of particles: $\lambda_{rad} >> R$, where R - is sphere diameter. Electron energy losses are assumed to be negligible to its kinetic energy: $E_{los} << E_{kin}$.

The local field theory for the chain of spheres was developed. Local field effects relate to physical interaction between scattering particles and also proved to lead to a sharp increase of the radiation intensity at some frequencies [6], [12].

2. Local field effects in diffraction radiation from the non-periodic chain of spheres.

Charge travelling above a system of spherical particles generates the polarisation current density $\mathbf{j}_{mic}(\mathbf{r}',\omega)$, defined at frequency ω and at $\mathbf{r}'$. It defines the radiation distribution with respect to an angle and a frequency [14]:

$$d^2E(\mathbf{n},\omega)=\frac{\omega^2 d\omega d\Omega}{c^3}\left|\int d^3r'\exp\left(-i\mathbf{n}\mathbf{r}'\frac{\omega}{c}\right)\left[\mathbf{n},\mathbf{j}_{mic}\left(\mathbf{r}',\omega\right)\right]\right|^2. \quad (1.1)$$

In dipole approximation:

$$\mathbf{j}_{mic}(\mathbf{r}',\omega)=-i\omega\alpha(\omega)\sum_b \mathbf{E}_{mic}(\mathbf{R}_b,\omega)\delta(\mathbf{r}'-\mathbf{R}_b)=-i\omega\sum_b \mathbf{d}(\mathbf{R}_b,\omega)\delta(\mathbf{r}'-\mathbf{R}_b), \quad (1.2)$$

where $\mathbf{E}_{mic}(\mathbf{R}_b,\omega)$ is exact microscopic field acting on each spherical particle in the chain positioned at $\mathbf{R}_b$, $\mathbf{d}(\mathbf{R}_b,\omega)$ is the particle dipole moment, $\alpha(\omega)$ is the polarizability of a spherical particle and summation is performed over all particles located at $\mathbf{R}_b$.

Using microscopic Maxwell's equations exact microscopic field can be obtained as:

$$\mathbf{E}_{mic}(\mathbf{r},\omega)=\mathbf{E}_0(\mathbf{r},\omega)-\frac{1}{2\pi^2}\int\frac{d^3l}{l^2-\frac{\omega^2}{c^2}}\sum_b\left[\mathbf{l}\left[\mathbf{l},\mathbf{d}(\mathbf{R}_b,\omega)\right]\right]\exp\left(i\mathbf{l}(\mathbf{r}-\mathbf{R}_b)\right), \quad (1.3)$$

where $\mathbf{E}_0(\mathbf{r},\omega)$ is the field of external origin or electron self-field, $\mathbf{l}$ is the Fourier transform variable.

The main idea of the local field theory consists in replacement of exact microscopic field $\mathbf{E}_{mic}(\mathbf{R}_b,\omega)$ by some effective field $\mathbf{E}_{loc}(\mathbf{R}_b,\omega)$ called local field. Hence:

$$\mathbf{E}_{loc}(\mathbf{r},\omega)=\mathbf{E}_0(\mathbf{r},\omega)-\frac{1}{2\pi^2}\int\frac{d^3l}{l^2-\frac{\omega^2}{c^2}}\sum_b\left[\mathbf{l}\left[\mathbf{l},\mathbf{d}(\mathbf{R}_b,\omega)\right]\right]\exp\left(i\mathbf{l}(\mathbf{r}-\mathbf{R}_b)\right). \quad (1.4)$$

The value of the local field can be obtained as a result of averaging over the distribution of all the rest particles relative to the certain a-th particle. It is assumed, that $f(\mathbf{R}_{ab})$ or $g(\mathbf{R}_{ab})=Nf(\mathbf{R}_{ab})$ probability density of such a distribution, where $\mathbf{R}_{ab}=\mathbf{R}_a-\mathbf{R}_b$.

From (1.4) we can derive the expression for the dipole moment of the spherical particle located at $\mathbf{R}_a$:

$$\mathbf{d}(\mathbf{R}_a,\omega)=\alpha(\omega)\mathbf{E}_0(\mathbf{R}_a,\omega)-\frac{\alpha(\omega)}{2\pi^2}\sum_b\int\frac{d^3l}{l^2-\frac{\omega^2}{c^2}}\left[\mathbf{l}\left[\mathbf{l},\mathbf{d}(\mathbf{R}_b,\omega)\right]\right]\exp\left(i\mathbf{l}(\mathbf{R}_a-\mathbf{R}_b)\right). \quad (1.5)$$

Averaging of (1.5) over $\mathbf{R}_b$ and applying Fourier transform give:

$$\mathbf{d}(\mathbf{q},\omega)=\alpha(\omega)\mathbf{E}_0(\mathbf{q},\omega)-4\pi\alpha(\omega)\int\frac{d^3l d^3R_{ab}}{l^2-\frac{\omega^2}{c^2}}g(\mathbf{q}-\mathbf{l})\left[\mathbf{l}\left[\mathbf{l},\mathbf{d}(\mathbf{q},\omega)\right]\right]. \quad (1.6)$$

where $\mathbf{q}$ is the Fourier transform variable.

Applying inverse Fourier transform to (1.6) and rearranging lead us to the expression:

$$A_{ij}d_j(\mathbf{R}_a,\omega)=\alpha(\omega)E_i^0(\mathbf{R}_a,\omega), \quad (1.7)$$

where

$$A_{ij}(\mathbf{R}_a,\omega)=2\pi\delta(\mathbf{R}_a)\delta_{ij}-4\pi\alpha(\omega)\iint\frac{d^3l d^3q}{l^2-\frac{\omega^2}{c^2}}\left(l^2\delta_{ij}-l_il_j\right)g(\mathbf{q}-\mathbf{l})\exp(i\mathbf{q}\mathbf{R}_a). \quad (1.8)$$

Multiplying (1.8) by A_{ki}^{-1} and rearranging give:

$$d_k=\alpha(\omega)A_{ki}^{-1}E_i^0(\mathbf{R}_a,\omega). \quad (1.9)$$

Tensor A_{ki}^{-1} defines relation between the dipole moment of the sphere in the chain and the external field acting on this sphere.

Substituting (1.9) into (1.2) leads to the expression for the current density, generated by all the spherical particles at some certain point in the space, characterized by $\mathbf{r}'$:

$$\mathbf{j}(\mathbf{r}',\omega) = -i\omega\sum_b \alpha(\omega)\hat{A}^{-1}\mathbf{E}^0(\mathbf{R}_b,\omega)\delta(\mathbf{r}'-\mathbf{R}_b) \quad (1.10)$$

Now let us define analytical expression for A_{ij}. Applying Fourier transform to (1.8) for $A_{ij}(\mathbf{q},\omega)$ we obtain:

$$A_{ij}(\mathbf{q},\omega) = \delta_{ij} - 4\pi\alpha(\omega)\int\frac{d^3l}{l^2-\frac{\omega^2}{c^2}}g(\mathbf{q}-\mathbf{l})\left(l^2\delta_{ij}-l_il_j\right) \quad (1.11)$$

where $g(\mathbf{q}-\mathbf{l})$ is the Fourier transform of the probability density $g(\mathbf{R}_{ab})$. Probability density at Oyz plane is defined by two-dimensional Gaussian distribution:

$$\eta(y,z) = \frac{1}{2\pi\Delta^2}\exp\left(-\frac{R_\perp^2}{2\Delta^2}\right),$$

with parameter Δ, representing particle coordinate deviation along y and z axes. Since there is no correlation between distant spheres, therefore $f(x_{ba}) = 0$, where $x_{ba} >> n^{-1}$ ($n = \frac{N}{L}$ is linear probability density along the x axis, L is the chain length).

Consequently $g(\mathbf{R}_{ab})$ can be written as:

$$g(\mathbf{R}_{ab}) = \frac{1}{2\pi L\Delta^2}\exp\left(-\frac{\mathbf{R}_\perp^2}{2\Delta^2}\right)\left(1-f(x)\right) \quad (1.12)$$

where $\mathbf{R}_\perp$ is located at Oyz plane.

Applying the Fourier transform to (1.12) gives:

$$g(\mathbf{q}-\mathbf{l}) = \frac{1}{(2\pi)^4 L\Delta^2}\int d^3R\exp\left(-\frac{R_\perp^2}{2\Delta^2}\right)\left(1-f(x)\right)\exp\left(-i(\mathbf{q}-\mathbf{l})\mathbf{R}\right) \quad (1.13)$$

Substituting (1.13) into (1.11) leads to:

$$A_{ij}(\mathbf{q},\omega) = \delta_{ij} - \frac{4\pi\alpha(\omega)}{(2\pi)^4 L\Delta^2}\iint\frac{d^3l d^3R}{l^2-\frac{\omega^2}{c^2}}\left(l^2\delta_{ij}-l_il_j\right)\left(1-f(x)\right)\exp\left(-i(\mathbf{q}-\mathbf{l})\mathbf{R}-\frac{R_\perp^2}{2\Delta^2}\right). \quad (1.14)$$

Integration over $\mathbf{R}$ and l_y gives:

$$A_{ij}(\omega) = \left(1+\frac{\Theta(\omega)}{2}\right)\left[\delta_{ij}-e_ie_j\right]+\left(1-\Theta(\omega)\right)e_ie_j; \quad (1.15)$$

where

$$\Theta(\omega) = \frac{\alpha(\omega)}{L}\iint dl_x dl_z f(l_x)\left[1-\Phi\left(\sqrt{\frac{l_x^2+l_z^2}{2}}\Delta\right)\right]\frac{l_x^2}{\sqrt{l_x^2+l_z^2}}\exp\left(\frac{l_x^2\Delta^2}{2}\right)$$

where Φ is Errors function.

Taking into consideration, that:

$$\left[(\delta_{ki}-e_ke_i)B+e_ke_iC\right]\left[(\delta_{ij}-e_ie_j)\frac{1}{B}+e_ie_j\frac{1}{C}\right] = \delta_{kj},$$

helps us to obtain $A_{ki}^{-1}(\omega)$:

$$A_{ki}^{-1}(\omega)=\frac{\delta_{ki}-e_k e_i}{\left(1+\frac{\Theta(\omega)}{2}\right)}+\frac{e_k e_i}{\left(1-\Theta(\omega)\right)}. \quad (1.16)$$

Substituting (1.16) into (1.9) gives the expression for the dipole moment of the spherical particle:

$$\begin{aligned} d_k(\mathbf{k},\omega) &= \alpha(\omega)\left[\frac{\delta_{ki}-e_k e_i}{\left(1+\frac{\Theta(\omega)}{2}\right)}+\frac{e_k e_i}{\left(1-\Theta(\omega)\right)}\right]E_i^0(\mathbf{k},\omega) \\ &= \frac{\alpha(\omega)}{\left(1+\frac{\Theta(\omega)}{2}\right)}\delta_{ki}E_i^0(\mathbf{k},\omega)+\frac{3\alpha(\omega)\Theta(\omega)}{\left(1+\frac{\Theta(\omega)}{2}\right)\left(1-\Theta(\omega)\right)}e_k e_i E_i^0(\mathbf{k},\omega) \end{aligned} \quad ; \quad (1.17)$$

where polarizability of the particle is defined by:

$$\alpha(\omega)=R^3\frac{\varepsilon(\omega)-1}{\varepsilon(\omega)+2}, \quad (1.18)$$

where R is the diameter of the spherical particle and $\varepsilon(\omega)$ is dielectric conductivity function.
Distribution of the emitted radiation with respect to an angle and a frequency can be written as:

$$d^2E(\mathbf{k},\omega)=(2\pi)^6\frac{\omega^2 d\omega d\Omega}{c^3}\left|\left[\mathbf{n},\mathbf{j}(\mathbf{k},\omega)\right]\right|^2, \quad (1.19)$$

where the expression for the current density, generated in the chain, can be derived from (1.17):

$$\begin{aligned} j_k(\mathbf{k},\omega)=-\frac{i\omega}{(2\pi)^3}\alpha(\omega)&\left[\sum_{b=1}^{N-1}\frac{1}{1+\frac{\Theta(\omega)}{2}}\delta_{ki}E_i^0(\mathbf{R}_b,\omega)\right. \\ &\left.+\frac{3}{2}\sum_{b=1}^{N-1}\frac{\Theta(\omega)}{\left(1+\frac{\Theta(\omega)}{2}\right)\left(1-\Theta(\omega)\right)}e_k e_i E_i^0(\mathbf{R}_b,\omega)\right]\exp(-i\mathbf{k}\mathbf{R}_b) \end{aligned} \quad ; \quad (1.20)$$

where N is number of the spherical particles in the chain and

$$\Theta(\omega)=\frac{\alpha(\omega)}{L}\iint dl_x dl_z f(l_x)\left[1-\Phi\left(\sqrt{\frac{l_x^2+l_z^2}{2}}\Delta\right)\right]\frac{l_x^2}{\sqrt{l_x^2+l_z^2}}\exp\left(\frac{l_x^2\Delta^2}{2}\right). \quad (1.21)$$

The value of $\theta(\omega)$ can have an imaginary part, so the denominators in (1.19) and (1.20) are not zero. However, they can be very small, that implies the peaks of radiation intensity, defined by the condition:

$$\begin{cases} 1+\frac{\theta(\omega)}{2}=0; \\ 1-\theta(\omega)=0. \end{cases} \quad (1.22)$$

In (1.19) and (1.20), according to the local field approach, particles of the chain are considered to interact with each other. We would like to emphasize that taking into account the interaction between the grating elements (spheres, in our case) means, inter alia, taking into consideration the so-called shadowing effect (see the report of Prof. X. Artru at RREPS-2009 and [13]).

3. Local field effects in Smith-Purcell radiation from the periodic chain of spheres.
Since (1.19) and (1.20) are obtained we shall consider the case when the chain of spheres is periodic and, as a first approximation, electron self-field only acts on each particle in the chain. Due to this fact

shadowing effect is not taken into consideration. Other parameters remaining the same: we consider dielectric or metal spherical particles of different sizes, characterized by arbitrary dielectric function; electron energy losses are assumed to be negligible to its kinetic energy: $E_{los} << E_{kin}$; wavelength of emitted radiation must be much greater than the size of the particles: $\lambda_{rad} >> R$.

Assuming that an electron travels with constant velocity v at a height $z = h$ above the chain of spheres, aligned along x axis, electron charge density can be written as:

$$\rho_0(\mathbf{r},t) = e\delta(x - vt)\delta(y)\delta(z - h) \tag{2.1}$$

Current density generated in the material of the spheres at the moment t and at the certain point in the space, characterized by $\mathbf{r}$, can be calculated as:

$$\mathbf{j}_0(\mathbf{r},t) = e\mathbf{v}\delta(x - vt)\delta(y)\delta(z - h) \tag{2.2}$$

Applying the Fourier transform to (2.2) gives:

$$\mathbf{j}_0(\mathbf{q},\omega) = \frac{e\mathbf{v}}{(2\pi)^4}\int e^{i\omega t - i\mathbf{q}\mathbf{r}}\delta(x - vt)\delta(y)\delta(z - h)dtd^3r \tag{2.3}$$

As it was proposed, to a first approximation the electron self-field acts only, therefore:

$$\mathbf{E}_{loc}(\mathbf{r},\omega) \approx \mathbf{E}_0(\mathbf{r},\omega) \tag{2.4}$$

Using microscopic Maxwell's equations electron self-field can be obtained as:

$$\mathbf{E}_0(\mathbf{q},\omega) = -\frac{4\pi i}{\omega}\left[\frac{\left[\mathbf{q}\left[\mathbf{q}\mathbf{j}_0(\mathbf{q},\omega)\right]\right]}{q^2 - \frac{\omega^2}{c^2}} + \mathbf{j}_0(\mathbf{q},\omega)\right]. \tag{2.5}$$

Substituting (2.3) into (2.5) and applying Fourier transform to modified (2.5) lead us to the expression for the electron self-field acting at the spherical particle located at $\mathbf{R}_b$:

$$\begin{aligned}\mathbf{E}_0(\mathbf{R}_b,\omega) = &-\frac{ie}{4\pi^3\omega}\int d^3q\frac{\left[\mathbf{q}\left[\mathbf{q}\mathbf{v}\int e^{i\omega t + i\mathbf{q}(\mathbf{R}_b - \mathbf{r})}\delta(x - vt)\delta(y)\delta(z - h)dtd^3r\right]\right]}{q^2 - \frac{\omega^2}{c^2}} \\ &-\frac{ie\mathbf{v}}{4\pi^3\omega}\int d^3q\int e^{i\omega t + i\mathbf{q}(\mathbf{R}_b - \mathbf{r})}\delta(x - vt)\delta(y)\delta(z - h)dtd^3r.\end{aligned} \tag{2.6}$$

Distribution of Smith-Purcell radiation with respect to an angle and a frequency can be written as:

$$d^2E(\mathbf{n},\omega) = \frac{\omega^2 d\omega d\Omega}{c^3}\left|\int d^3r'\exp\left(-i\mathbf{n}\mathbf{r}'\frac{\omega}{c}\right)\left[\mathbf{n},\mathbf{j}(\mathbf{r}',\omega)\right]\right|^2, \tag{2.7}$$

where

$$\mathbf{j}(\mathbf{r}',\omega) = -i\omega\sum_b \mathbf{d}(\mathbf{R}_b,\omega)\delta(\mathbf{r}' - \mathbf{R}_b) = -i\omega\sum_b \alpha(\omega)\mathbf{E}_0(\mathbf{R}_b,\omega)\delta(\mathbf{r}' - \mathbf{R}_b). \tag{2.8}$$

Substituting (2.8) into (2.7) gives:

$$\frac{d^2E(\mathbf{n},\omega)}{d\omega d\Omega} = \frac{\omega^4|\alpha|^2}{c^3}\left\|\left[\mathbf{n},\sum_b \mathbf{E}_0(\mathbf{R}_b,\omega)\exp(-i\mathbf{k}\mathbf{R}_b)\right]\right\|^2. \tag{2.9}$$

Substituting (2.6) into (2.9) and integrating over the $\mathbf{r},\mathbf{q},t$; bearing in mind that $y_b = z_b = 0$, lead to Smith-Purcell radiation distribution:

$$\begin{aligned}\frac{d^2E(\mathbf{n},\omega)}{d\omega d\Omega} = &\frac{e^2}{c}\frac{k^6R^6}{\pi^2\beta^4}\left(\frac{\varepsilon(\omega) - 1}{\varepsilon(\omega) + 2}\right)^2\left|\sum_{b=0}^{N-1}\exp\left(i\left(\beta^{-1} - n_x\right)kdNb\right)\right|^2 \\ &\times\left[\left(-\frac{n_y}{\gamma}K_1\left(\frac{kh}{\beta\gamma}\right)\right)^2 + \left(-\frac{n_z}{\gamma^2}K_0\left(\frac{kh}{\beta\gamma}\right) + \frac{n_x}{\gamma}K_1\left(\frac{hk}{\beta\gamma}\right)\right)^2 + \left(\frac{n_y}{\gamma^2}K_0\left(\frac{kh}{\beta\gamma}\right)\right)^2\right]\end{aligned} \tag{2.10}$$

Taking into account, that:

$$\left|\sum_{b=0}^{N-1} \exp\left(i\left(\beta^{-1} - n_x\right)kdNb\right)\right|^2 = \frac{\sin^2\left(\left(\beta^{-1} - n_x\right)\frac{kdN}{2}\right)}{\sin^2\left(\left(\beta^{-1} - n_x\right)\frac{kd}{2}\right)}, \qquad (2.11)$$

helps to transform (2.10) to:

$$\frac{d^2E(\mathbf{n},\omega)}{d\omega d\Omega} = \frac{e^2}{c}\frac{k^6R^6}{\pi^2\beta^4}\left(\frac{\varepsilon(\omega)-1}{\varepsilon(\omega)+2}\right)^2 \frac{\sin^2\left(\left(\beta^{-1} - n_x\right)\frac{kdN}{2}\right)}{\sin^2\left(\left(\beta^{-1} - n_x\right)\frac{kd}{2}\right)} \times\left[\left(\frac{n_y}{\gamma}K_1\left(\frac{kh}{\beta\gamma}\right)\right)^2 + \left(\frac{n_x}{\gamma}K_1\left(\frac{hk}{\beta\gamma}\right) - \frac{n_z}{\gamma^2}K_0\left(\frac{kh}{\beta\gamma}\right)\right)^2 + \left(\frac{n_y}{\gamma^2}K_0\left(\frac{kh}{\beta\gamma}\right)\right)^2\right], \qquad (2.12)$$

where R is the spherical particle diameter, $\varepsilon(\omega)$ is the dielectric conductivity function of the sphere's material, h is the impact parameter or the distance between the chain and the charged particle, d is the spacing between the spherical particles, K_1, K_0 are Bessel's functions of first and second order accordingly.

Expressions for ultrarelativistic and nonrelativistic cases can be performed as well. In nonrelativistic case, when $kh \ll \beta \ll 1$, (2.12) can be transformed as:

$$\frac{d^2E(\mathbf{n},\omega)}{d\omega d\Omega} = \frac{e^2}{c}\frac{k^4R^6}{\pi^2h^2\beta^2}\left(\frac{\varepsilon(\omega)-1}{\varepsilon(\omega)+2}\right)^2 \frac{\sin^2\left(\left(\beta^{-1} - n_x\right)\frac{kdN}{2}\right)}{\sin^2\left(\left(\beta^{-1} - n_x\right)\frac{kd}{2}\right)}\left(1-n_z^2\right), \qquad (2.13)$$

where n_x, n_z are x and z components of the unit vector between $\mathbf{v}$ and $\mathbf{k}$.

In ultrarelativistic case, when $\gamma \gg 1$, (2.12) can be written as:

$$\frac{d^2E(\mathbf{n},\omega)}{d\omega d\Omega} = \frac{e^2}{c}\frac{k^4R^6}{\pi^2h^2}\left(\frac{\varepsilon(\omega)-1}{\varepsilon(\omega)+2}\right)^2 \frac{\sin^2\left(\left(1 - n_x\right)\frac{kdN}{2}\right)}{\sin^2\left(\left(1 - n_x\right)\frac{kd}{2}\right)}\left(1-n_z^2\right)\left[\frac{kh}{\gamma}K_1\left(\frac{kh}{\gamma}\right)\right] \qquad (2.14)$$

where K_1 is Bessel's function of the first order.

Detailed analysis of the results will be performed in the Discussion section.

4. Discussion

Expressions (1.19) and (1.20) for the distribution of diffraction radiation and the current density in the chain of spheres were obtained. They are valid for the particle chains, consisting of different numbers of spherical particles. The possible sharp enhancement of the radiation intensity can be observed at the frequencies determined in (1.22).

Expression for the distribution of Smith-Purcell radiation (2.12) defined in rather general form, what allows us to obtain expressions for ultrarelativistic and nonrelativistic cases. The radiation distribution is proportional to the square of the sphere's polarizability, where polarizability itself can be written as [14],[15]:

$$\alpha(\omega) = R^3\frac{\varepsilon(\omega)-1}{\varepsilon(\omega)+2}. \qquad (3.1)$$

Expression for the distribution of Smith-Purcell radiation is valid for the spherical particles of the diameter R, made from materials, characterized by the dielectric conductivity function $\varepsilon(\omega)$. The formula (3.1) is obtained in the frames of Reyleigh theory and describes polarizability of small particles very well.

Taking into account the condition for the local field effects:

$$\frac{\sin^2\left(\frac{p\varphi N}{2}\right)}{\sin^2\left(\frac{p\varphi}{2}\right)} \xrightarrow{N>>1} 2\pi N\sum_m \delta(p\varphi - 2\pi m), \tag{3.2}$$

where $p = \beta^{-1} - n_x$ and $\varphi = kd$, leads us to the definition of the Smith-Purcell relation:

$$\frac{d}{\lambda}(\beta^{-1} - n_x) = m, \quad m = 1,2... \tag{3.3}$$

where n_x is the x component of the unit vector between $\mathbf{v}$ and $\mathbf{k}$, m is the order of emitted radiation.

Smith-Purcell radiation from the chain of spheres was analyzed graphically. We are considering two geometries. First can be described as:

$$n_x = \sin\theta\cos\phi, n_y = \sin\theta\sin\phi, n_z = \cos\theta;$$

the second one as:

$$n_x = \cos\theta', n_y = \sin\theta'\cos\phi', n_z = \sin\theta'\sin\phi'.$$

In Figure 1(a, b) the angular distribution of Smith-Purcell radiation from the chain of polyethylene spheres for two different wavelengths is performed. Two narrow peaks on the sides of the broad one correspond to Smith-Purcell radiation. The peak at $\varphi = 0$ relates to the forward radiation.

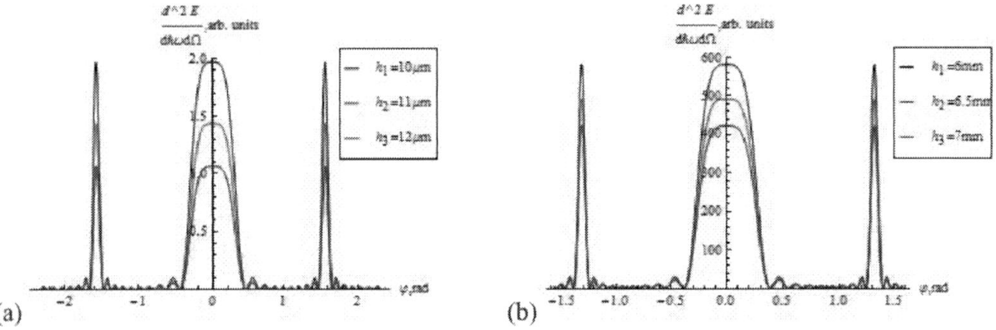

Figure 1. Angular distribution of Smith-Purcell radiation for (a) $\lambda = 700nm$ and (b) $\lambda = 3mm$ in ultrarelativistic case. Distribution of radiation is presented in first geometry for three different values of the impact parameter and the rest parameters chosen as (a): $\gamma = 100$, $d = 700nm$, $R = 100nm$, $N = 10$, $\varepsilon_{HDPE} = 2,37$, $m = 1$, $\theta = \frac{\pi}{2}$, $\phi \in [-\pi, \pi]$; (b): $\gamma = 100$, $d = 4mm$, $R = 0,5mm$, $N = 10$, $\varepsilon_{HDPE} = 2,37$, $m = 1$, $\theta = \frac{\pi}{2}$, $\phi \in [-\pi, \pi]$.

Analysis of the radiation distribution lets us assume that Smith-Purcell radiation distributed over the cone and the cone's axis coincides with the axis x of both geometries. Moreover second geometry provides us with information how exactly the Smith-Purcell radiation is distributed over the cone, what is shown in Figure 2(a, b).

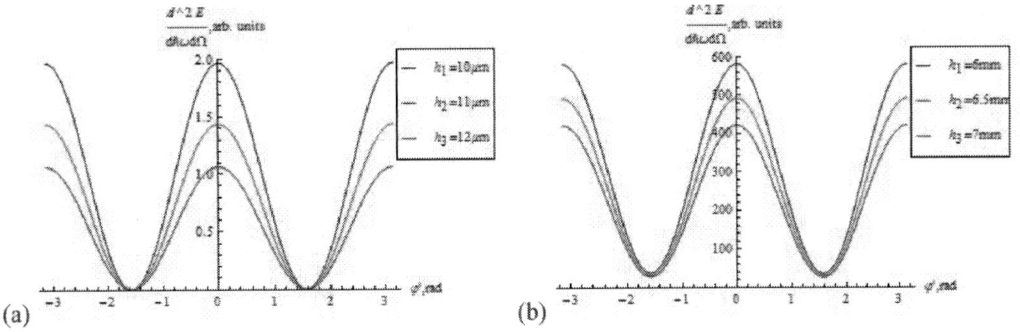

Figure 2. Angular distribution of Smith-Purcell radiation over the cone in second geometry for (a) $\lambda = 700nm$ and (b) $\lambda = 3mm$. These distributions correspond to the peaks of Smith-Purcell radiation in the first geometry (a) $\varphi = \pi / 2$ and (b) $\varphi = \arccos(0.25)$.

From (3.3) it is possible to derive the region of validity for (2.12). Smith-Purcell radiation for a nonrelativistic charged particle ($\beta << 1$) is possible when:

$$\lambda \sim \frac{d}{\beta m} >> d, \qquad (3.4)$$

and for an ultrarelativistic charge($\gamma >> 1$) when:

$$\frac{\gamma^{-2}}{2}\frac{d}{m} << \lambda << 2\frac{d}{m}. \qquad (3.5)$$

Acknowledgements

This work was supported by the Federal Agency for Science and Innovation (grant № 02.120.21.4158). One of the authors (A.T.) is grateful to CRDF for the financial support under the BRHE programme (grant №.Y4-P-11-05).

References

[1] Frank I M 1942 Doppler effect in refracting material Izv. Akad. Nauk Ser. Fiz. **6** 3

[2] Smith S J and Purcell E M 1953 Visible light from localized surface charges moving across a grating *Phys. Rev.* **92** 1069

[3] Garcia de Abajo F G 2000 Smith–Purcell radiation emission in aligned nanoparticles *Phys. Rev. E* **61** 5743

[4] Yamaguti S, et al. 2002 Photonic crystals versus diffraction gratings in Smith-Purcell radiation *Phys. Rev. B* **66** 195202

[5] Ochiai T and Ohtaka K 2005 Electron energy loss and Smith-Purcell radiation in two and three-dimensional photonic crystals *OPTICS EXPRESS* 7698 **13** №19

[6] Ryazanov M I, Strikhanov M N and Tishchenko A A 2008 Local field effect in diffraction radiation from a periodical system of dielectric spheres *Nucl. Instr. and Meth. B* **266** 3811

[7] Ochiai T and Ohtaka K 2004 Relativistic electron energy loss and induced radiation emission in two dimensional metallic photonic crystals. II. Photonic band effects *Phys Rev B* **69** 125107

[8] Ochiai T and Ohtaka K 2004 Relativistic electron energy loss and induced radiation emission in two dimensional metallic photonic crystals. I. Formalism and surface plasmon polaration *Phys Rev B* **69** 125106

[9] Ochiai T and Ohtaka K 2006 Theory of unconventional Smith-Purcell radiation in finite-size photonic crystals *OPTICS EXPRESS* 7397 **14** №16

[10] Lekomtsev K V and Tishchenko A A 2009 Local field effects in Smith–Purcell radiation from a chain of spherical particles *Proc. of Scientific Session MEPHI – 2009 V.* **1** P. 161

[11] Horiuchi N 2006 Exotic radiation from a photonic crystal excited by an ultrarelativistic electron beam *Phys. Rev. E* **74** 056601

[12] Ryazanov M I and Tishchenko A A 2006 Clausius-Mossotti-type relation for planar monolayers *JETP* **103** 539

[13] Artru X and Ray C 2008 Interference and shadow effects in the production of light by charged particles in optical fibers *NIMB* **266** 3725-3732

[14] Ryazanov M I 1984 *Electrodynamics of condensed matter* (Moscow, M: Nauka)

[15] Landau L D and Lifshiz E M 1984 *Electrodynamics of Continuous Media* (second ed., Pergamon, Oxford)

Investigation of the surface current excitation by a relativistic electron electromagnetic field

G Naumenko[1], A Potylitsyn[2], Yu Popov[2], L Sukhikh[2] and M Shevelev[1]
[1]Nuclear Physics Institute of Tomsk Polytechnic University, Lenina str. 2a Tomsk, 634050, Russia

[2]Tomsk Polytechnic University, Lenina str. 2, Tomsk, 634050, Russia

E-mail: naumenko@npi.tpu.ru

Abstract. Surface current method and pseudo-photon ones are widely used in the problems of diffraction and transition radiation of relativistic electron in conductive targets. The simple analysis disclosed the contradiction between these methods in respect to the surface current excitation on target surfaces. This contradiction was resolved experimentally by the measurement of a surface current on the upstream and downstream target surfaces in diffraction radiation geometry. The experimental test showed, that no surface current is induced on the target downstream surface under the influence of a relativistic electron electromagnetic field in contrast to the upstream surface. This is important for the understanding of a forward transition and diffraction radiation nature and electromagnetic field evolution in interaction processes.

1. About problem

In principle the radiation emitted by relativistic electrons passing through or near material targets can be calculated using the macroscopic Maxwell equations. However so far such calculations for thick conductive targets, when the thickness is much larger than a skin-layer, are absent. Therefore phenomenological concepts, like "surface current viewpoint" and "pseudo-photons", are widely used for calculation of transition radiation (TR) and diffraction radiation (DR) from conductive targets. Furthermore these concepts are useful for the intuitive understanding of the main features of these phenomena. Let us consider in more details both these concepts in respect to the transition and diffraction radiation of relativistic electrons passing through or near conductive targets.

1.1. Surface current viewpoint

In this approach TR and DR are considered as a radiation of a current, induced on a conductive target surface by the relativistic particle electromagnetic field. This technique for DR from inclined conducting strip is described in [9]. In [8] the similar approach was used both for backward and forward TR and is irrespectively to the thickness of the target surface.

For diffraction radiation this viewpoint was more pronounced in the article of B.M. Bolotovskiy [2], where he places primary emphasis upon the radiation formation length effect, when a charged particle moves over the conductive screen in parallel to the axis z with a velocity v (figure 1). It seems the formation length notion was introduced by M. Ter-Mikaelian in [1] as a result of interference between a charged particle electromagnetic field and a radiation field emitted by this particle.

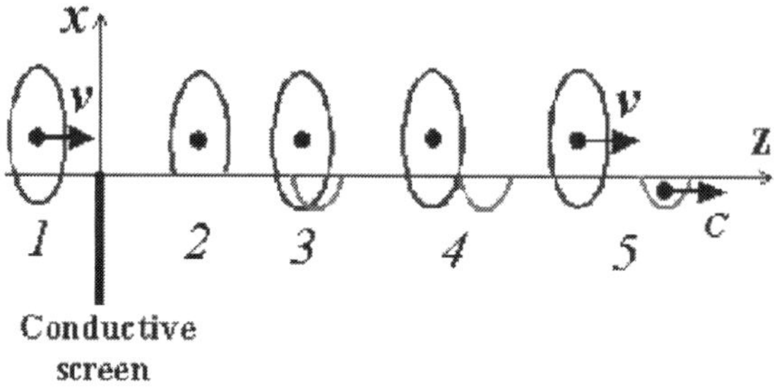

Figure 1. Presentation of the radiation formation length effect by B.M. Bolotovskil in [11].

According to [2] The ellipse 1 represents the field of the particle. It hits the half-plane screen and induces current in screen, which in turn emit Diffraction Radiation. "It results the peculiar features of the radiation field. The radiation field is such that close to the screen it kills part of the particle field". At positions 3, 4, 5 we see the separation of the radiation field and the Coulomb field at the distance ~γ2λ, due to their different velocities. It is important, that this point of view assumes the obligatory surface electric current on the target downstream surface irrespectively to the thickness of the conductive target. The association between forward diffraction radiation (FDR) and surface current is evident from the exact solution of Maxwell equations for diffraction radiation field from ideally conductive infinite thin semi-surface in [3], where the diffraction radiation field is expressed directly as a function of surface current:

$$A(R)=\frac{2\pi e^{i\omega R}}{R}J(k_0,q_0), \quad (1)$$

where $k_0=-\omega\cdot\sin\psi\cdot\cos\varphi$, $q_0=-\omega\cdot\cos\psi$, φ and ψ are the angles of observation direction in spherical system and J is the surface electric current density. However, this is not obvious for a thick target, if the target thickness is much larger than a skin-layer.

1.2. Pseudo-photon method

On the other hand, the pseudo-photon method suggested by Fermi [4] and extended by Williams [5] is of considerable current use as the approach for electromagnetic processes theoretical investigations (see for instance [1] and [10]). According to this approach the charged particle field may be replaced by the field of photons, which in this case are named pseudo-photons (In [10] is used the term "virtual quanta". One should distinguish this term from ones in quantum theory.). This approach provides a good accuracy for ultra-relativistic particles when the particle velocity becomes close to the light velocity (v→c) (see [1]). In this case the longitudinal component of the particle electric field is negligible and the particle electromagnetic field being a transversal ones has the same properties as a real photon field.

Let us remind you the properties an interaction of real photons with a thick conductive mirror of a high reflectivity in geometry, shown in figure 1. It is clear that the real photons are reflected almost fully, they don't penetrate through the target and they don't induce a surface current on the target downstream surface. If the particle electromagnetic field and the real photons have the same properties, we may expect the same character of interactions of the particle electromagnetic field with a thick conductive mirror, namely, we may expect, in contrast to the surface current viewpoint, the absence of a surface current on the target downstream surface.

1.3. Deduction
Both these concepts are in principal contradiction in respect to the surface current induction. The resolution of this contradiction may clarify the understanding of the nature of the forward transition radiation (FTR), FDR and other important phenomena. One of the ways of this problem resolution is an experimental test of the surface current on the downstream target surface.
It is not necessary to make an absolute surface current measurement, because we are sure, that backward transition and backward diffraction radiation (BDR) is emitted by the surface electric current on the upstream target surface. We may make the relative measurements and compare a surface current on the both surfaces.
To exclude problems, which may be brought up due to the direct contact of relativistic electrons with a target material, the experiment in DR geometry is more preferable.

2. Experiment

2.1. Experimental setup
The experiment was carried out on the extracted electron beam of Tomsk Nuclear Physics Institute microtron with parameters presented in Table 1.

Table 1. Electron beam parameters.

Electron energy	6.1 MeV ($\gamma=12$)	Bunch period	380 psec
Train duration	$\tau \approx 4\ \mu$sec	Bunch population	N_e=6·10^8
Bunches in a train	n_b≈1.6·10^4	Bunch length	σ≈1.3~1.6mm

Beam parameters of microtron allow using coherent properties of a radiation to investigate DR characteristics and surface current induction in millimeter wavelength region. This possibility increases the surface electric current and diffraction radiation intensity by the 8 orders (proportional to the bunch population) and makes these values achievable for measurement using existing sensors.
For the surface electric current measurement we use the well-known technique, which is applied for the surface current measurement in strip-line beam position monitors [6]. Figure 2 shows the scheme and photography of the surface current sensor being built-in in diffraction radiation target.

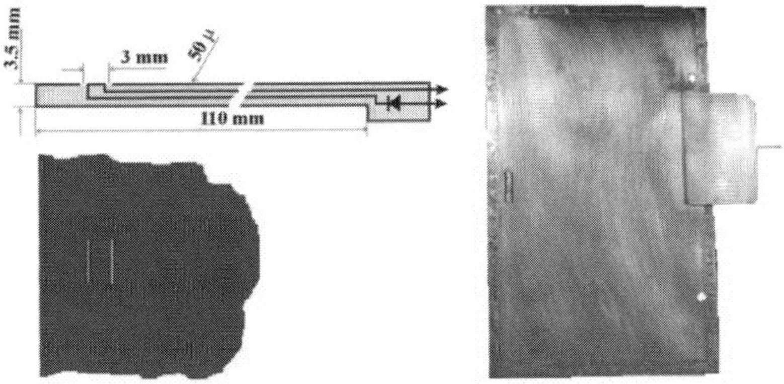

Figure 2. Surface electric current sensor built-in in diffraction radiation target (scheme and photography).

This sensor registered a surface current component perpendicular to the slits.
To have a similarity of BDR and surface current dependence on impact-parameter the sensor strip length L was chosen 3 mm (the quarter of the average wavelength of a registered BDR). The slit width

was 0.4 mm. According to [6] the strip-line pickup usually operate in spectral interval between the 3 dB points of the first lobe $4/3 \cdot L < \lambda < 8L$ with maximum of the sensitivity in the point $\lambda = 4L$, i.e. for our sensor $4mm < \lambda < 24mm$ with maximum in the point $\lambda = 12L$.

The DR target with built-in surface current sensor was tested on the real photon beam from the pulse emitter of 6.7 mm wavelength radiation. The test shows that in case of real photons a surface current on the downstream target surface is absent within the experimental error (≈3% of the pulse amplitude from sensor with upstream surface sensor position). The sensitivity of sensor on wavelength 6.7 mm was ≈50 mV/(mA/cm) (experimentally estimated lower limit).

Since BDR is generated by the electric current on the upstream target surface, we may check the correlation between the surface current and BDR intensity by simultaneous measurement of the surface current and BDR. Therefore the scheme of experiment (figure 3) provides the measurement of BDR and surface current simultaneously.

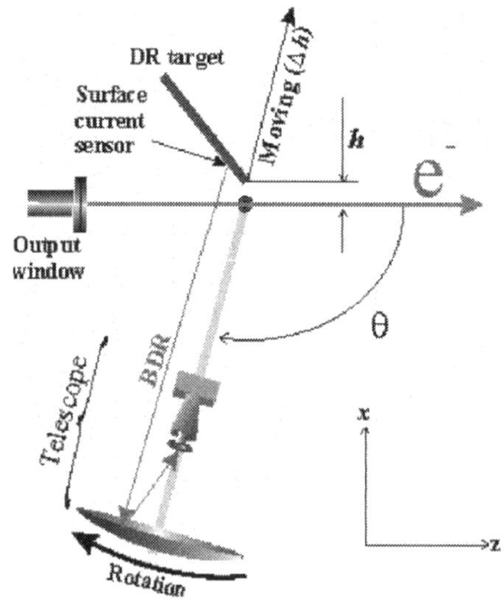

Figure 3. Scheme of experiment for study of the surface current induction under the influence of a relativistic electron electromagnetic field.

For measurement of BDR angular distribution (see figure 4) we had used the parabolic telescope described in [7], which provides the measurement a BDR in far field zone mode.

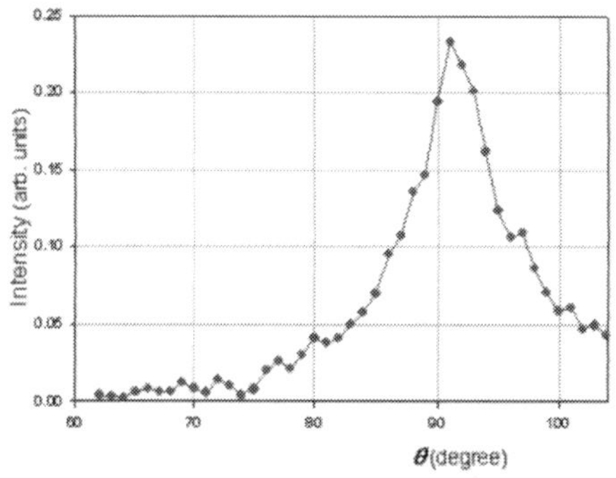

Figure 4. Angular distribution of BDR.

The parabolic telescope and the target with the built-in surface electric current sensor may be moved in horizontal plane to change the impact-parameter h. Dependences of BDR intensity on impact-parameter were measured in the maximum of BDR angular distribution. The background from surface current sensor was measured with a strip of sensor stopped up by a conductive foil using conductive glue.
The Faraday cup signal allows measuring a zero point of impact parameter using the electron beam intensity suppression when electrons cross the target.

2.2. Results
The measurements were done for upstream and downstream orientations of surface current sensor attended by comparison of results for both cases in the same units. The measured simultaneously a surface current and BDR intensity as a function of impact-parameter for both orientations of sensor (upstream and downstream) after background subtraction are presented in figures 5 and 6. The solid line in figure 5 is the fit by equation for DR intensity $W = A \cdot \exp(-4\pi h / \gamma\lambda)$ with A=0.58, γ=12 and λ=15.3 mm. Maximal surface electric current density was 0.15 mA/cm (estimated lower limit).

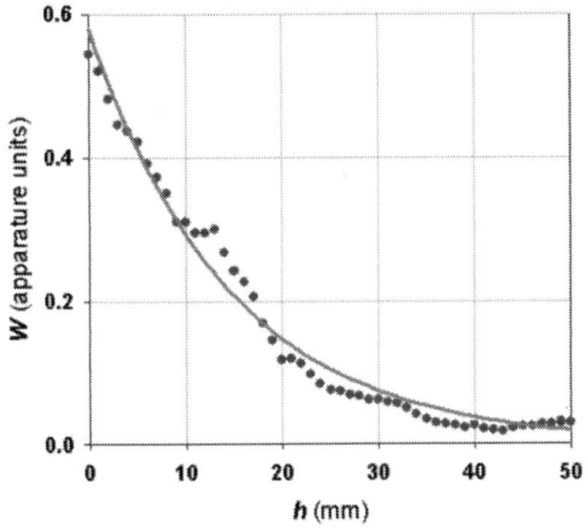

Figure 5. Measured dependences of BDR on impact-parameter. Solid line is the fit $W = A \cdot \exp(-4\pi h / \gamma\lambda)$ with A=0.58, γ=12 and λ=15.3 mm.

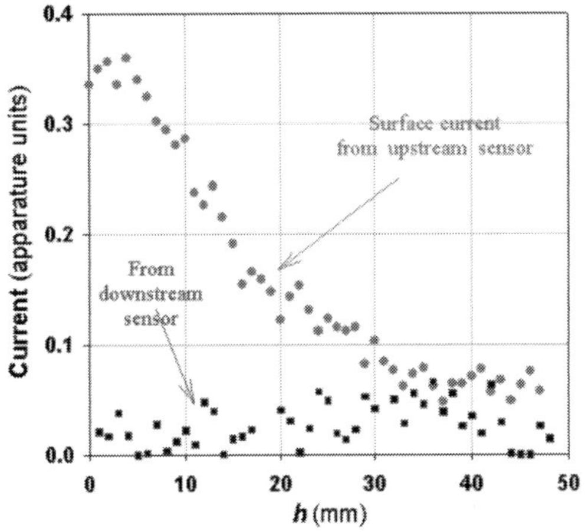

Figure 5. Measured dependences on impact-parameter of the current for upstream and downstream orientation of the current sensor.

The results presented in figures 5 and 6 show a good correlation between surface current on the upstream surface and BDR in dependence on impact-parameter. Within the experimental error the surface current on the downstream surface is absent.
One may argue that in used geometry of experiment the upstream and downstream surfaces are in asymmetrical conditions in respect to the surface current induction due to the inclination of the target. In our opinion it is not matter, but for a confirmation of this contention we had measured the similar dependence for the symmetrical geometry, when the target is perpendicular to the electron beam direction. In this case we cannot measure BDR, because BDR cone coincides with electron beam. Only surface current on upstream and downstream surfaces was measured. Results of this measurement shown in figure 7 confirm a contention that no surface current on the downstream surface is induced.

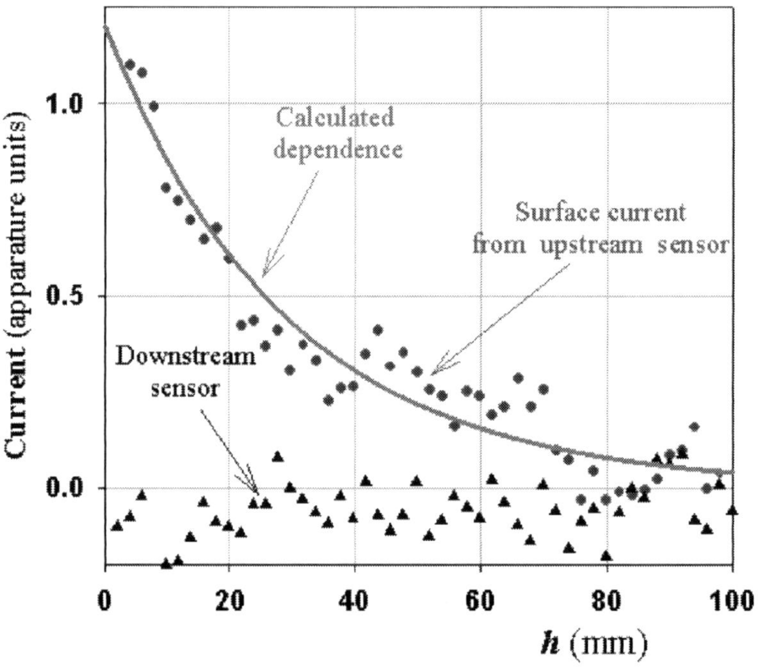

Figure 7. Measured dependences on impact-parameter of the surface current for upstream and downstream orientation of the current sensor for target perpendicular to the electron beam. The solid lines is the calculated dependence using [3] for γ=12 and λ=16.5 mm.

Fortunately this geometry is convenient for calculation of the measured dependence using the expression for a surface current from [3]. We are interested by the *x*-component (in notation of figure 3) of a surface current when a target is perpendicular to the electron beam in DR geometry. In this case the expression for the *x*-component of a surface current from [3] may be simplified to the formula (1).

$$J_{kq,x} = \frac{\sqrt{\lambda}\left(i\xi + \gamma'\sqrt{4\pi^2 - \lambda^2 q^2}\right)}{\sqrt{\sqrt{4\pi^2 - \lambda^2 q^2} - k\lambda \cdot (i\xi + k\gamma'\lambda)}} \cdot bq\lambda\,, \tag{1}$$

where $bq\lambda = -i\dfrac{e\sqrt{\gamma'\lambda}}{8\pi^3\gamma}\left(\gamma'\sqrt{4\pi^2 - \lambda^2 q^2} - i\xi\right)\cdot \exp\left(-\dfrac{a}{\gamma'\lambda}\xi\right)$,

$\xi = \sqrt{\lambda^2 q^2\left(\gamma^2 - 1\right) + 4\pi^2}$, $\gamma' = \sqrt{\gamma^2 - 1}$, k and q are the Fourier variables corresponding to a spatial variables x and y, e is the electron charge, a is an impact-parameter. Here the light velocity equal to 1 is assumed.

To turn to the spatial presentation we may make the backward Fourier transformation:

$$J_x(x,y) = \int_{-\infty}^{\infty}\int_{-\infty}^{\infty} \frac{1}{(2\pi)^2} e^{-i(q\cdot y + k\cdot x)} J_{kq,x}\, dq\, dk \tag{2}$$

The final calculation of expression (2) can be performed numerically.

In figure 8 is shown the spatial distribution of the x-component surface current density $\left|J_x(x,y)\right|$ calculated using (2) for the experimental conditions: γ=12, λ=16 mm and impact-parameter a=0.2 $\gamma\cdot\lambda$.

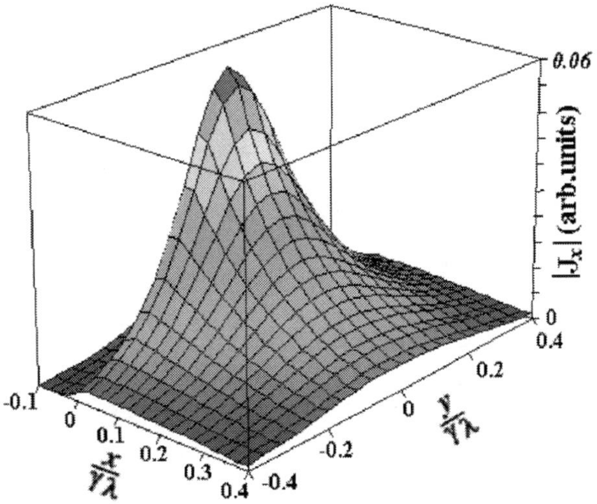

Figure 8. Calculated spatial distribution of the x-component surface current density for γ=12, λ=16 mm and a=0.2 $\gamma\cdot\lambda$.

It is seen from the figure 8 that a target polarization stretches to the dimensions comparable with $\gamma\lambda$. Using the equation (2) we had calculated the dependence of the surface current density on the impact-parameter with γ=12 and λ=16.5 mm in the point which corresponds to the sensor position on the target (the solid line in figure 7). Calculated values were normalized by the experimental data in centre of dependence. The chosen wavelength provides the minimal discrepancy between the calculated and experimental data. Let's notice, that this wavelength is within the interval of the sensor sensitivity.

3. Discussion

In traditional interpretation a forward transition and diffraction radiation from perfectly conducting targets are considered as radiation emitted by surface currents (for instance in [3]). These surface currents are induced by the electromagnetic field of the relativistic electron. However, as it is mentioned above, for $\gamma >> 1$ the electron field properties become close to those of an electromagnetic field in free space, and pseudo-photons are reflected by the mirror almost like real photons and no surface current is induced on a forward surface of a target.

The present experiment has confirmed the second viewpoint based on pseudo-photon method. In this case we need to find another nature of forward transition and diffraction radiation. One may attribute forward transition radiation (FTR) to the current of the travelling electron after traversal of the screen. Indeed this radiation is practically the same as that of a suddenly accelerated electron (see for instance [11]). At the same time this current re-creates the Coulomb field. We can then say that FTR is generated during the transit of electron from the unstable "naked" state to the stable "dressed" state (here the term "transition radiation" seems particularly appropriate). We can find a similar interpretation of this phenomenon in [1]. As for forward DR, it can be viewed as a consequence of the transit from a "half-naked" state to the "dressed" state.

Acknowledgment

This work was partly supported by the warrant-order 1.226.08 of the Ministry of Education and Science of the Russian Federation and by the Federal agency for science and innovation, contract 02.740.11.0245

References

[1] Ter-Mikaelian N L 1972 *High-Energy Electromagnetic Processes in Condensed Media* (Wiley, New York)
[2] BolotovskiI B N 1982 *Preprints of Lebedev Institute of Physics, Russian Academy of Sciences* **140** 95
[3] Kazantsev A P and Surdutovich G I 1963 *Dokl. Akad. Nauk, USSR* **147** 74
[4] Fermi E 1924 *Zh. Phyz* **29** 315
[5] Wiliams E and Danske K 1935 *Vidensk. Selsk.* **13** 4
[6] Sargsyan V 2004 Comparison of Stripline and Cavity Beam Position Monitors, *TESLA Report* 03
[7] Kalinin B N, Naumenko G A, Potylitsyn A P et al 2006 *JETP Letters* **84** 3 110
[8] Reiche S and J. B. Rosenzweig J B 2001 *Transition Radiation for Uneven, Limited Surfaces.* (Particle Accelerator Conference, Chicago)
[9] Brownell J H and Walsh J 1998 *Phys. Rev. E* **57** 1 1075
[10] Jackson J D *Classical Electrodynamics, 3rd ed.* (J. Willey&Sons, New-York, 1998)
[11] Shul'ga N F and Syshchenko V V 2000 *Journal Physics of Atomic Nuclei* **63** 11 2018

Investigation of coherent Čerenkov radiation generated by 6.1 MeV electron beam passing near the dielectric target

A. P. Potylitsyn, Yu. A. Popov, L. G. Sukhikh
Tomsk Polytechnic University, Lenin avenue, 30, Tomsk, 634050, Russia
E-mail: Sukhikh@interact.phtd.tpu.ru

G. A. Naumenko, M. V. Shevelev
Nuclear Physics Institute at Tomsk Polytechnic University, Lenin avenue 2, Tomsk, 634050, Russia

Abstract. Coherent Čerenkov radiation generated by the 6.1 MeV bunched electron beam travelling near dielectric target (teflon) have been investigated theoretically and experimentally. It has been shown that the longitudinal bunch form-factor plays important role in the characteristics of coherent Čerenkov radiation. We have also compared the characteristics of coherent Čerenkov radiation and the ones of coherent diffraction radiation from a metal target in the similar conditions and have shown that coherent Čerenkov radiation is more intensive. These facts make coherent Čerenkov radiation a promising mechanism for a new noninvasive beam diagnostic technique.

1. Introduction

Čerenkov radiation (ČR) that appeares while a charged partilce travels in a media with a velocity that exceeds the speed of light in this media is developed both theoretically and experimentally and is widely used in particle detectors for nuclear physics. The radiation cone in transparent medium is defined by the following condition:

$$\cos\theta = \frac{1}{\beta n} \tag{1}$$

where θ is the radiation angle, β is the electron velocity in the speed of light units, $n = \sqrt{\varepsilon}$ is the refractive index.

ČR may also appear while a charged particle moves in the vacuum in a vicinity of the medium due to the fact that this interaction is caused by the charged particle's electromagnetic field that have the transverse dimensions of about $\gamma\lambda$ (γ is the particle Lorentz-factor, λ is the radiation wavelength). The latter is used in the schemes of Free Electron Laser (FEL) based on relativistic beam ($\gamma \simeq 2$) moving in a cylindrical waveguide [1].

In such well-known radiation mechanism there are still a couple of open questions while dealing with ČR caused by the moderately relativistic electons traveling near the dielectric

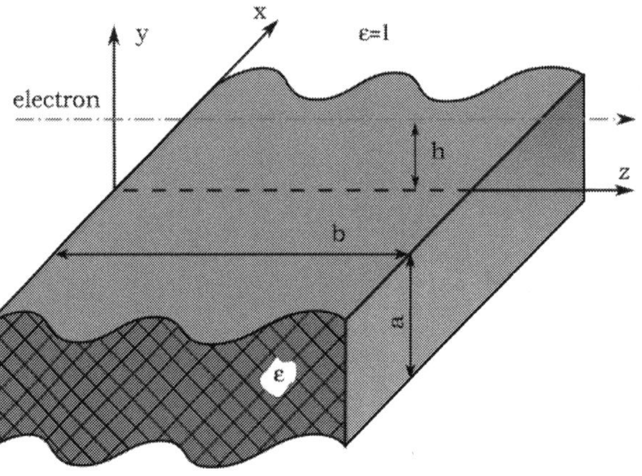

Figure 1. The scheme of theoretical simulations and some defenitions.

medium. The problem is a coherency of ČR. In reference [2] it was claimed that the longitudinal beam size does not influence on the radiation intensity that makes it impossible to define the longitudinal beam size. This point of view contradicts to the data presented in reference [3], where authors measured properties of coherent ČR (CČR) and [4] where authors used CČR for the measurement of electron bunch longitudinal size.

The main goal of this paper is to show theoretically and experimentally that the longitudinal beam size influences drastically on CČR spectrum and intensity. We have also carried out a comparison at the similar conditions of the characteristics of CČR and coherent diffraction radiation (CDR) that is nowadays used in beam diagnostics.

2. Theoretical background

In our theoretical estimations we will follow a recent paper [5] in which rather simple and physically clear method for the solution of a problem of ČR and DR that may appear simultaneously is developed. We will apply this method to CČR problem to show that the longitudinal bunch size plays an important role. One should specify that in this theoretical estimation we are interested in the radiation inside a target bulk to simplify the problem.

According to the mentioned method [5] the magnetic field of polarization radiation $\mathbf{H}^{\mathbf{pol}}(\mathbf{r},\omega)$ that appears while a charged particle travels in the vacuum near a target with arbitrary shape and characteristics may be written in a general as:

$$\mathbf{H}^{\mathbf{pol}}(\mathbf{r},\omega) = \mathrm{curl}\frac{1}{c}\int\limits_{V_T} \mathbf{j}^{(\mathbf{0})}_{\mathbf{pol}}(\mathbf{r}',\omega)\frac{e^{i\sqrt{\varepsilon(\omega)}\frac{\omega}{c}|\mathbf{r}-\mathbf{r}'|}}{|\mathbf{r}-\mathbf{r}'|}d^3r'. \tag{2}$$

We should mention that this formula is the exact solution of Maxwell equations without any assumptions or simplifications. Here c is the speed of light; $\mathbf{j}^{(\mathbf{0})}_{\mathbf{pol}}(\mathbf{r}',\omega) = \sigma(\omega)\mathbf{E_e}(\mathbf{r}',\omega)$ is the polarization current density, $\sigma(\omega) = \frac{(\varepsilon(\omega)-1)\omega}{4\pi i}$ is the conductivity of the target material ($\varepsilon(\omega)$ is the permittivity); $\mathbf{E_e}(\mathbf{r}',\omega)$ is the external field, in our case electron one; $\frac{e^{i\sqrt{\varepsilon(\omega)}\omega|\mathbf{r}-\mathbf{r}'|/c}}{|\mathbf{r}-\mathbf{r}'|}$ is the Green function with $\mathbf{r}'$ – the radiation point and $\mathbf{r}$ – the detection point. The integration is performed over the whole target volume V_T.

In the case when the electron (or a bunch of electrons) travels near the target with impact-parameter h (see Fig. 1) and in one direction the target may be assumed to be an infinite one (see

Fig. 1) and the detector is situated in the far-field zone the previous formula may be simplified:

$$
\mathbf{H}^{\mathbf{pol}}(\mathbf{r},\omega) = \frac{2\pi i}{c}\frac{e^{\sqrt{\varepsilon(\omega)}r\frac{\omega}{c}}}{r}\mathbf{k}\times
\int\limits_{0}^{b} dz' \int\limits_{-a}^{0} dy' \mathbf{j}^{(\mathbf{0})}_{\mathbf{pol}}(k_x, y', z', \omega) e^{-i(k_y y' + k_z z')}. \tag{3}
$$

Here $\mathbf{k}$ is the radiation wave-vector, $\mathbf{j}^{(\mathbf{0})}_{\mathbf{pol}}(k_x, y', z', \omega)$ is the Fourier transform of the polarization current density, a and b are the target sizes.

From now let us assume that we have a longitudinal distribution of N_e electrons ($N_e \gg 1$) that move in one direction with the uniform velocity. In this case the bunch current density that is needed to find the electrical bunch field may be written as:

$$
\mathbf{j}(\mathbf{r},t) = e\mathbf{v}\sum_n \delta(x)\,\delta(y-h)\,\delta(z-z_n-vt). \tag{4}
$$

Here e is the electron charge, $\mathbf{r_n} = \{0, h, z_n\}$ is the position of n-th electron in the bunch, $\mathbf{v} = \{0, 0, v\}$ is the electrons velocity vector, h is the impact-parameter.

Full Fourier-transform of the bunch current density may be written as:

$$
\mathbf{j}(\mathbf{k},\omega) = \frac{e}{(2\pi)^3}\frac{\mathbf{v}}{v}e^{-ik_y h}\delta\left(\frac{\omega}{v} - k_z\right)\sum_n e^{-ik_z z_n} \tag{5}
$$

Full Fourier-transform of the bunch electrical field $\mathbf{E_e}(\mathbf{k},\omega)$ that is convinient to use in order to find partial Fourier-transform $\mathbf{E_e}(k_x, y, z, \omega)$ that is needed for radiation field determination may be written as [6]:

$$
\mathbf{E_e}(\mathbf{k},\omega) = \frac{2ei}{(2\pi)^2\omega}\frac{\mathbf{v}\omega^2/c^2 - \mathbf{k}(\mathbf{k},\mathbf{v})}{k^2 - \omega^2/c^2}e^{-ik_y h}\sum_n e^{-ik_z z_n} \tag{6}
$$

Partial Fourier-transform $\mathbf{E_e}(k_x, y, z, \omega)$ may be found as:

$$
\mathbf{E_e}(k_x, y, z, \omega) = -\frac{ie}{2\pi v}\frac{\exp\left[-(h-y)\sqrt{k_x^2 + \frac{\omega^2}{v^2}\gamma^{-2}}\right]}{\sqrt{k_x^2 + \frac{\omega^2}{v^2}\gamma^{-2}}}
\left\{k_x, -i\sqrt{k_x^2 + \frac{\omega^2}{v^2}\gamma^{-2}}, \frac{\omega}{v}\gamma^{-2}\right\}\sum_n e^{i\frac{\omega}{v}(z-z_n)} \tag{7}
$$

Here we assumed that all electrons moves outside of the target, that made us to choose $h > 0$. The radiation field in the medium according to Eq. (3) taking into account Eq. (7) and $\mathbf{k} = w/c\sqrt{\varepsilon}\mathbf{e}$ may be written as:

$$
\mathbf{H}^{\mathbf{pol}} = \frac{e\beta\gamma}{4\pi c}\sqrt{\varepsilon}(\varepsilon-1)\frac{e^{\sqrt{\varepsilon}r\omega/c}}{r}\mathbf{h}\frac{e^{ib\frac{\omega}{v}(1-\sqrt{\varepsilon}\beta e_z)} - 1}{1-\beta\sqrt{\varepsilon}e_z}\times
\frac{(e^{a\frac{\omega}{v\gamma}(-i\beta\gamma\sqrt{\varepsilon}e_y + \sqrt{1+\varepsilon(\beta\gamma e_x)^2})} - 1)e^{-h\frac{\omega}{v\gamma}\sqrt{1+\varepsilon(\beta\gamma e_x)^2}}}{\sqrt{1+\varepsilon(\beta\gamma e_x)^2}(-i\beta\gamma\sqrt{\varepsilon}e_y + \sqrt{1+\varepsilon(\beta\gamma e_x)^2})}
\sum_n e^{-i\frac{\omega}{v}z_n}, \tag{8}
$$

where:

$$\mathbf{h} = \{\gamma^{-1}e_y + ie_z\sqrt{1+\varepsilon(\beta\gamma e_x)^2}, e_x(\beta\gamma\sqrt{\varepsilon}e_z - \gamma^{-1}),$$
$$-e_x(i\sqrt{1+\varepsilon(\beta\gamma e_x)^2} + \beta\gamma\sqrt{\varepsilon}e_y)\}. \tag{9}$$

and

$$\mathbf{e} = \{\sin\theta\sin\phi, \sin\theta\cos\phi, \cos\theta\} \tag{10}$$

Thus, the spectral-angular density of radiation may be written as:

$$\frac{d^2W}{\hbar d\omega d\Omega} = \frac{cr^2}{\hbar}\frac{1}{\sqrt{\varepsilon}}|\mathbf{H}^{\mathbf{pol}}|^2 \tag{11}$$

From Eqs. (8) and (11) one may see that only one term in spectral-angular density of radiation that is ČR term depends on the target longitudinal size b. If the ČR condition is fulfilled and b is finite the pole may be eliminated as follows [5]:

$$\left|\frac{e^{ib\frac{\omega}{v}(1-\sqrt{\varepsilon}\beta e_z)} - 1}{1-\beta\sqrt{\varepsilon}e_z}\right|^2 \rightarrow \left|b\frac{\omega}{\beta c}\right|^2. \tag{12}$$

From this equation one may clearly see that the dependence of spectral-angular density of ČR from target longitudinal size is quadratic for the Čerenkov angle and fixed radiation wavelength.

The sum in Eq. (8) may be treated in the following way:

$$\left|\sum_{n=1}^{N_e} e^{-i\frac{\omega}{v}z_n}\right|^2 = \begin{cases} N, & n = m; \\ \sum_{n=1}^{N_e-1} e^{-i\frac{\omega}{v}z_n} \sum_{m=1}^{N_e} e^{i\frac{\omega}{v}z_m}, & n \neq m. \end{cases} \tag{13}$$

In the case when $n = m$ we have simple incoherent radiation. In the case when $n \neq m$ we have coherent radiation and may treat the sums in the following way:

$$\sum_{n=1}^{N_e-1} e^{-i\frac{\omega}{v}z_n} = \sum_{n=1}^{N_e-1}\int_{-\infty}^{\infty}\delta(z-z_n)e^{-i\frac{\omega}{v}z}dz =$$
$$(N_e-1)\int_{-\infty}^{\infty}\rho(z)e^{-i\frac{\omega}{v}z}dz \tag{14}$$

Where, in the case of a Gaussian bunch with rms σ_z, one may obtain:

$$\rho(z) = \frac{1}{N_e-1}\sum_{n=1}^{N_e-1}\delta(z-z_n) \simeq \frac{1}{\sqrt{2\pi}\sigma_z}\exp\left[-\frac{z^2}{2\sigma_z^2}\right] \tag{15}$$

The total spectral-angular density of polarization radiation from the bunch of N_e electrons may be written as:

$$\frac{d^2W}{\hbar d\omega d\Omega} = \frac{cr^2}{\hbar}\frac{1}{\sqrt{\varepsilon}}|\mathbf{H}^{\mathbf{pol}}|^2 N_e\left(1+(N_e-1)|f_z(\sigma_z)|^2\right), \tag{16}$$

where $|f_z(\sigma_z)|^2 = \left|\int_{-\infty}^{\infty}\rho(z)e^{-i\frac{\omega}{v}z}dz\right|^2$ is the form-factor of the electron bunch in z direction. For the Gaussian bunch with rms σ_z the form-factor looks like following:

$$|f_z(\sigma_z)|^2 = \exp\left[-\frac{\omega^2\sigma_z^2}{\beta^2c^2}\right] \tag{17}$$

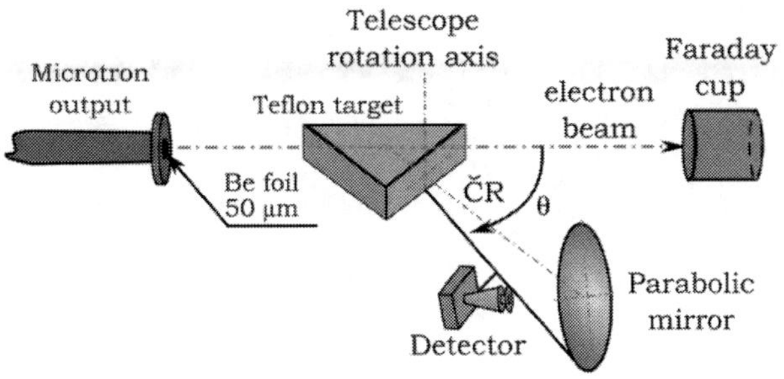

Figure 2. Experimental scheme and some definitions.

From Eq. (16) one may clearly see that for a case when the bunch travels near the target the longitudinal form-factor drastically changes the intensity of polarization radiation. One may also see that longitudinal form-factor does not depend on ČR condition that contradicts the predictions made in reference [2]. This simple estimation proves the fact that the coherency of any kind of polarization radiation strongly depends on the longitudinal bunch profile.

3. Experiment

We have carried out experimental investigations in Tomsk Polytechnic University at Nuclear Physics Institute microtron. The experimental scheme is shown in Fig. 2. The electron beam extracted into air through $50\,\mu$m Be foil was used. The train of bunches with electron energy 6.1 Mev ($\gamma \approx 12$), consisting of $n_b = 10526$ bunches (the maximal bunch population is about $N_e = 10^8$ electrons) with $\tau = 4\,\mu$s duration travels near the teflon target. The average beam current was about 30 mA. The transverse sizes of electron beam in the extraction point are about $\sigma_x \times \sigma_y = 4 \times 4\,\mathrm{mm}^2$. The longitudinal distribution of electrons in the bunch is a Gaussian with rms $\sigma_z = 1.1$ mm [9] (there is a misprint in paper [9] and "$\sigma_z = 6.8$ mm" should be read as "$6\sigma_z = 6.8$ mm").

The detecting system consisted of so-called "telescope", which represented a paraboloidal mirror (diameter 170 mm, focal distance 151 mm) in focus of which the detector was set up. Such telescope allows to measure the angular radiation characteristics equal to wave-zone ones [10]. The radiation from each train has been detected using DP-21M detector. The last is based on a wide-band antenna, high-frequency low barrier diode and preamplifier. The average sensitivity of the detector in the radiation wavelength region $11 \div 17$ mm is approximately equal to 0.3 Volt/mWatt [11]. The measured region was limited by coherent threshold in the smaller wavelengths and by the beyond-cutoff waveguide (diameter 15 mm) that passes wavelengths lower 25 mm used to decrease accelerator RF background. Incoherent radiation can not be measured by the detector. The measured radiation yield was averaged over 20 trains. The statistic error was less than 10% during the experiments. The wire scanner was used to find the electron beam axis which was used as a starting point to determine the impact-parameter that was equal to $h = 25$ mm. ČR in diffraction geometry should be polarized only in horizontal plane, i.e. the plane of electron velocity vector and radiation wave-vector. During the experiment a grid polarizer was used. The Faraday cup was used to measure beam current.

Telflon (Polytetrafluoroethylene – PTFE) target with the length equal to 247 mm and the height equal to 74 mm was used during the experiment. The target was a prism based on right-angled isosceles triangle in order to decrease ČR refraction losses. Teflon refractive index in abovementioned wavelength region is equal to $n = 1.45$ [7] and $n = 1.439$ in region less than 5 mm according to [8].

At the first we have found the CČR peak position while scanning over θ angle. Fig. 3

shows the result of θ scan. The measured horizontal polarization component of radiation is shown in Fig. 3 by the red circles. The theoretical simulation result obtained using Eq. (16) taken into account Snell's law for $n = 1.45$ is shown by the solid line. Fig. 3 also shows the vertical polarization component of CČR by the blue squares. From Fig. 3 one may clearly see that the vertical polarization component is suppressed as expected . The measured peak is situated at the angle of $\theta = 46.5° \pm 0.05°$ while theoretical value is $\theta = 46.7°$ that confirms the nature of measured radiation. The measured peak is wider than theoretical: measured FWHM $= 4.3° \pm 0.14°$ and simulated FWHM $= 2.72°$. That may be explained by the angular divergence of the electron beam that is due to beam scattering in the beryllium foil and in the air and by the finite angular acceptance of the telescope. The beam divergence was not taken into account during theoretical simulations. The comparison of the peak power of measured and simulated CČR was not performed because of simplification made during theoretical estimation such as infinite target size in x direction (see Fig. 1).

There are main point of interest to check for CČR as it was mentioned before. That is the coherency of radiation and its nature. In order to find whether the radiation is coherent or not the beam current dependence was measured. The latter is shown in Fig. 4. Measured radiation is shown by the red cirles and the solid line shows the fit by the function $y = a + bx + cx^2$. One may see from Fig. 4 that the fit function agrees very well with measured data and, therefore, radiation is coherent.

One may argue that such coherency is caused by the transverse beam size, not longitudinal one, just as it was considered in reference [2]. In the case when electron beam travels in a vicinity of the target the transverse form-factor is not so critical as longitudinal one if $\sigma_y \lesssim h \ll \gamma\lambda$. These conditions were fulfilled for our measurements ($\sigma_y \simeq 4\,\text{mm}$, $h = 25\,\text{mm}$, $\gamma\lambda \simeq 150\,\text{mm}$). If longitudinal form-factor plays some role and our estimations are true the radiation coherent threshold should be at the wavelength of about 9 mm. Fig. 5 shows by the red histogram bars the spectrum of CČR measured at the angle of radiation maximum. For spectrum measurements the low-pass filters, described in [9] were used. The filters were set up in order instead of beyond-cutoff waveguide at the detector horn. The solid line shows the radiation form-factor from

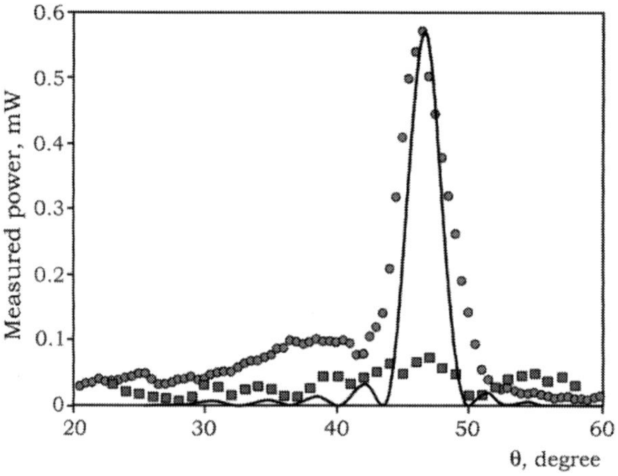

Figure 3. The angular dependence of CČR: the red circles – horizontal polarization component, the blue squares – vertical polarization component, the solid line – theoretical simulation. Target No.1

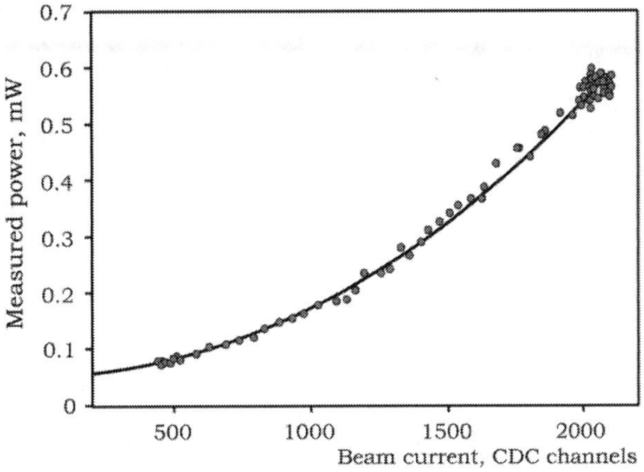

Figure 4. The current dependence of CČR: the red circles – experimental data, the solid line – quadratic fit $y = a + bx + cx^2$.

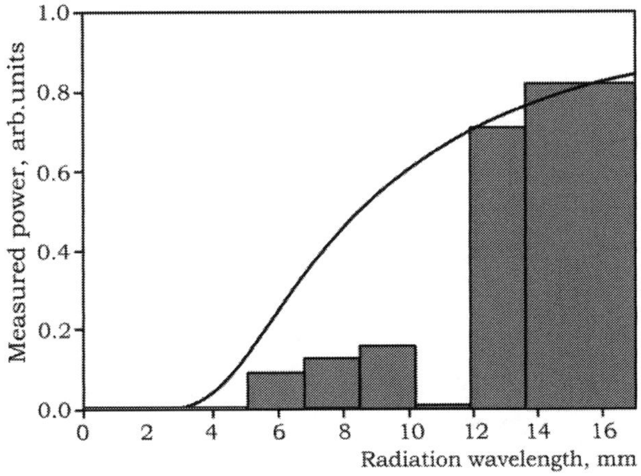

Figure 5. The spectrum of CČR measured at radiation peak angle $\theta = 46.5°$. The red histogram bars – experimental data, the solid line – form-factor from Eq. (17) ($\sigma_z = 1.13$ mm, $\sigma_{tr} = 2$ mm). Target No.1

Eq. (17) for $\sigma_z = 1.1$ mm. One may see from Fig. 5 that there is no radiation with wavelengths less than 5 mm that corresponds to the form-factor dependence. The lack of radiation in the region from $\lambda = 5$ to $\lambda = 11$ mm comparing with form-factor dependence is no actually clear. It may be explained by the detector sensitivity that is unknown in the waveband $6 \div 11$ mm. In spite of some disagreement, one may clearly see that the coherency of ČR is caused by the longitudinal bunch size, not transverse one.

In order to compare CČR and CDR from a metal target which is used widely for beam diagnostics we have measured both in the corresponding conditions. The CDR target was a 20 μm thick copper on a woven-glase reinforced substrate of a square form with 170 mm side. The CDR target was inclined at an angle of 45° with respect to the beam direction. The impact-parameter was equal to 25 mm, too. The measured distributions are shown in the same

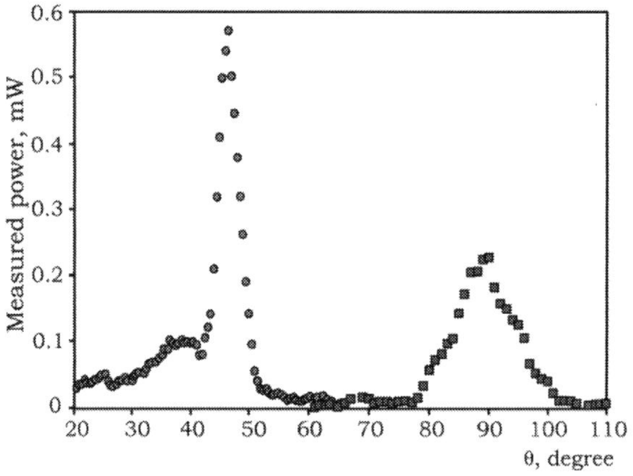

Figure 6. The comparison of experimentally measured CČR (the red circles) and CDR (the blue squares) in corresponding conditions.

Fig. 6 for convinience. One may clearly see in Fig. 6 that the measured power of CČR is larger than CDR one in a factor of 2.5. The CČR angular distribution is narrower than CDR one: $FWHM_{C\check{C}R} = 4.3 \pm 0.14°$ and $FWHM_{CDR} = 12 \pm 0.3°$. It may be shown that the angular width of CČR peak almost doesn't depend of the particle Lorentz-factor in contrast of CDR one. Due to this fact a power of a part of CČR cone (as in our case) may be higher than one from a whole cone of CDR for low values of Lorentz-factor. The theoretical FWHM of CDR is equal to 8.3°. One should mention that for both CČR and CDR measured FWHM is larger than simulated one in a factor of 1.5.

4. Conclusion

In conclusion we would like to point that the coherency of polarization radiation and its dependence on the longitudinal bunch profile seems to be a basic feature that does not depend on the radiation mechanism (transition radiation, diffraction radiation, Čerenkov radiation, etc.). From this point of view the predictions made in reference [2] concerning that coherent Čerenkov radiation does not depend on longitudinal bunch size seems to be incorrect as it was shown theoretically. We demonstrated experimentally for 6.1 MeV bunched electron beam the square dependence of coherent Čerenkov radiation intensity on a bunch population.

The "longitudinal coherency" and high linear polarization may be used for new diagnostic methods based on the Čerenkov radiation beam. Development of the similar noninvasive methods seems to be promising due to the fact that the coherent Čerenkov radiation distribution is narrower and more intensive than the distribution of the coherent diffraction radiation from the metal target which is used nowadays for the same purpose.

The diagnostic scheme with coherent Čerenkov radiation that goes at an angle to beam direction seems to be more promising that the scheme proposed in [4], where coherent Čerenkov radiation goes in forward direction and one needs additional mirrors to extract it. This additional optics adds some radiation (e.g., transition one) that may interfere with coherent Čerenkov radiation and increase a background.

Acknowledgments

This work was partly supported by Federal Agency for Science and Innovations of Russian Federation, Grant No. 02.740.11.0245 and by Federal Agency for Education of Russian

Federation, Grant No. П1143.

Reference

[1] J.E. Walsh, T.C. Marshall, S.P. Shlesinger, Phys. Fluid. **20**(4), 709 (1977).
[2] J. Zheng, C. X. Yu, Z. J. Zheng and K. A. Tanaka, Phys. Plasmas **12**, 093105 (2005).
[3] T. Takahashi, Y. Shibata, K. Ishi, M. Ikezawa, M. Oyamada, Y. Kondo, Phys. Rev. E **62**(6), 8606 (2000).
[4] S.V. Shchelkunov, T. C. Marshall, J. L. Hirshfield, and M. A. LaPointe, Phys. Rev. ST Accel. Beams **8**, 062801 (2005)
[5] D.V. Karlovets, A.P. Potylitsyn, Pis'ma v ZhETP **90**(5), 368 (2009) (D.V. Karlovets, A.P. Potylitsyn, JETP Lett. in press)
[6] D.V. Karlovets, A.P. Potylitsyn, ZhETP **134**(5), 887 (2008) (D.V. Karlovets, A.P. Potylitsyn, JETP **107**(5), 755 (2008))
[7] James W. Lamb, International Journal of Infrared and Millimeter Waves **17**(12), 1997 (1996)
[8] Mohammed N. Afsar, Hua Chi, and Igor I. Tkachov, in *Proceedings of Millimeter and Submillimeter Waves II*, 1995, edited by Mohammed N. Afsar (SPIE) **2558**, 73 (1995)
[9] B.N. Kalinin, D.V. Karlovets, A.S. Kostousov, G.A. Naumenko, A.P. Potylitsyn, G.A. Saruev and L.G. Sukhikh, Nucl. Instrum. Methods Phys. Res., Sect. B **252**, 62 (2006).
[10] B. N. Kalinin, G. A. Naumenko, A. P. Potylitsyn, G. A. Saruev, L. G. Sukhikh and V. A. Cha, Pis'ma v Zh. Eksp. Teor. Fiz. **84**(3), 110 (2006) [JETP Lett. **84**(3), 110 (2006)].
[11] V.A. Cha, B.N. Kalinin, E.A. Monastyrev, G.A. Naumenko, A.P. Potylitsyn, G.A. Saruev and L.G. Sukhikh, in *Proceedings of International Conference on Charged and Neutral Particles Channeling Phenomena II*, 2006, edited by Sultan B. Dabagov (SPIE), **6634**, 663416 (2007).

Coherent bremsstrahlung in a bent crystal vs. experiments on radiation at volume reflection

M V Bondarenco

Kharkiv Institute of Physics & Technology, 1 Academichna St., Kharkiv 61108, Ukraine

Abstract. Coherent bremsstrahlung in a bent silicon crystal in the orientation (111) is calculated. The radiation spectrum is averaged over the angular distribution of particles in the incident beam. Comparison of the dipole radiation theory predictions with the available experimental data is made.

1. Introduction

Recent experiments [1, 2] on radiation from relativistic electrons and positrons in thin bent crystals were searching for spectrum features inherent to radiation at volume reflection. It was suggested that the spectrum from the side of high frequencies may be treated as a coherent bremsstrahlung in a bent crystal (CBBC), which is simpler than radiation from vicinity of the volume reflection point, and a formula for the CBBC spectrum from a single electron was derived in [3]. However, the found spectrum profile appears to be sensitive to the angle of the particle entrance to the crystal, on the scale of the crystal bending angles, which necessitates taking into account angular divergence of the beam in experiments [1, 2]. In addition, [3] explicitly dealt with the simplest case of (110) orientation, involving only a single-well inter-planar continuous potential, whereas in experiments [1, 2] crystal orientation with the active plane (111) was employed. In the present note it is desired to adapt the theory of [3] to the experimental conditions, and to draw a comparison with the data.

2. CBBC from a single particle in a crystal with orientation (111)

2.1. Generic formula for CBBC [3]

The spectral density of dipole radiation (emitted in the forward cone $\sim 1/\gamma$) on a bent crystal with inter-planar continuous potential of arbitrary profile results from Eq. 40 of [3] by an obvious substitution $\frac{F^2_{\max 0}}{n^{3+2\epsilon}} \to n\left(\frac{\pi^2 V_n}{d}\right)^2$:

$$\frac{dE_{\mathrm{CBBC}}}{d\omega} = \frac{\pi e^2 R}{m^2 d}\frac{E'^2}{E^2}\sum_{n=1}^{\infty} nV_n^2 \Big\{ \Theta\left(nq_- - q_{\min}\right) D\left(\frac{q_{\min}}{nq_-}, \frac{\omega}{E}\right) + \Theta(nq_- + q_{\min})\Theta(nq_+ - q_{\min}) D\left(\frac{q_{\min}}{nq_+}, \frac{\omega}{E}\right) + \Theta(-nq_- - q_{\min})\left[D\left(\frac{q_{\min}}{nq_+}, \frac{\omega}{E}\right) - D\left(\frac{q_{\min}}{n|q_-|}, \frac{\omega}{E}\right)\right]\Big\}. \tag{1}$$

Here d is the distance between active crystalline planes, L and R are the crystal thickness and the active plane bending radius, m and e – the electron mass and charge, E and ω – the initial electron's and the photon's energies

$$q_{\min} = q_{\min}(\omega) = \frac{\omega m^2}{2EE'}, \qquad E' = E - \omega, \qquad q_{\pm} = \frac{2\pi}{d}\left(\frac{L}{2R} \pm |\theta_0|\right), \tag{2}$$

$\Theta(v)$ – the Heavyside unit-step function, and

$$D\left(v, \frac{\omega}{E}\right) = (1-v)\left(\frac{2 - v + 2v^2}{3} + \frac{\omega^2}{2EE'}\right). \qquad (0 \le v \le 1) \tag{3}$$

Finally, V_n are the Fourier series coefficients for the inter-planar continuous potential in a straight crystal, i. e.

$$V(x) = \sum_{n=1}^{\infty} V_n \cos\frac{2\pi n x}{d} \quad \Leftrightarrow \quad V_n = \frac{2}{d}\int_{-d/2}^{d/2} dx V(x) \cos\frac{2\pi n x}{d}, \tag{4}$$

(presuming the origin of x-coordinate to be chosen in a point of symmetry of the potential).

2.2. Silicon inter-planar continuous potential at orientation (111)

Silicon has a lattice of diamond type, with the lattice constant $5.431\mathring{A}$, thus the period in direction $\langle 111 \rangle$ equals $d = \frac{5.431\mathring{A}}{\sqrt{3}} \approx 3.1355\mathring{A}$. However, for this orientation there are two atomic planes falling within a period. Packing densities of those planes are equal, hence the potential derivative discontinuities, as well potentials themselves, are equal at each plane, and the potential function is symmetrical with respect to center of any of the wells. The well width ratio is $\frac{d_L}{d_S} = 3$, and the sum of the widths $d_L + d_S = d$. As for the well depths and shape, we will infer them from [4], presuming the well shapes to be parabolic (neglecting here the thermal smear-out of the potential) – see Fig. 1.

$$V_L \approx 26.5\text{eV}, \qquad V_S \approx 7.5\text{eV}. \tag{5}$$

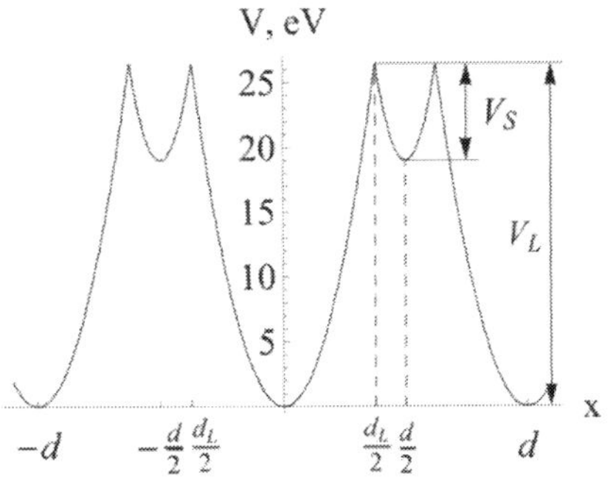

Figure 1. Continuous potential of silicon crystal in planar orientation (111).

2.3. Potential decomposition to Fourier series

Having specified the periodic potential, we can evaluate its Fourier series coefficients. Calculation of the integral in (4) with our potential function yields

$$\begin{aligned} V_n &= (-1)^n (d_L + d_S)\left(\frac{2}{\pi n}\right)^2 \left\{\left(\frac{V_L}{d_L} + \frac{V_S}{d_S}\right)\cos\frac{\pi n d_S}{d_L + d_S} + \frac{d_L + d_S}{\pi n d_S}\left(\frac{V_L d_S}{d_L^2} - \frac{V_S}{d_S}\right)\sin\frac{\pi n d_S}{d_L + d_S}\right\} \\ &= (-1)^n \left(\frac{4}{\pi n}\right)^2 \left\{\left(\frac{V_L}{3} + V_S\right)\cos\frac{\pi n}{4} + \frac{4}{\pi n}\left(\frac{V_L}{9} - V_S\right)\sin\frac{\pi n}{4}\right\}. \end{aligned} \tag{6}$$

With numerical values (5), one estimates

$$V_1 \approx -12.1\,\text{eV},$$

$$2\left(\frac{V_2}{V_1}\right)^2 \approx 0.02, \quad 3\left(\frac{V_3}{V_1}\right)^3 \approx 0.11, \quad 4\left(\frac{V_4}{V_1}\right)^2 \approx 0.075, \quad 5\left(\frac{V_5}{V_1}\right)^2 \approx 0.02, \quad \sum_{n=2}^{\infty} n\left(\frac{V_n}{V_1}\right)^2 \approx 0.25.$$

Therefore, at typical radiation frequencies the first Fourier harmonic of the potential dominates, but within its range the 3rd and the 4th harmonics in sum contribute $\sim 20\%$. This is about as much as contributes the 2nd harmonics in case of orientation (110).

In investigations of particle dynamics and radiation in the crystal the continuous potential properties enter through two basic parameters: the critical radius R_c and the non-dipole radiation angle θ_V. For perturbative treatment of the particle passage, the potential shape-independent definitions of R_c, θ_V are

$$R_c = \frac{E}{|V_1|\pi^2/d} \approx \frac{E}{4\text{GeV/cm}} \tag{7}$$

(concordant with $4V_L/d_L \approx 4.5\text{GeV/cm}$, $4V_S/d_S \approx 3.8\text{GeV/cm}$), and

$$\tilde{\theta}_V = \frac{2\sqrt{2}}{m}|V_1| \approx 0.67 \cdot 10^{-4}\text{rad} \tag{8}$$

(cf. $\theta_V = \frac{V_L}{m} \approx 0.5 \cdot 10^{-4}\text{rad}$).

3. Beam and bent crystal parameters in IHEP and CERN SPS experiments

The parameters of experiments [1, 2] are listed in Table 1. As one realizes immediately, the beam spread in all cases is significant compared to the crystal half-bending angle entering (2), so it is necessary to average intensity (1) with distribution $f(\theta_0) = \frac{1}{\sqrt{2\pi}\sigma} e^{-\theta_0^2/2\sigma^2}$.

Table 1. Bent crystal and the beam parameters

Experiment	Energy, E	Beam divergence	Critical radius, R_c	Critical angle, $\theta_c = \sqrt{\frac{d}{2R_c}}$	Crystal thickness, L	Bending radius, R	Bending angle, $\frac{L}{R}$
IHEP e^+	10 GeV	$\Gamma = 0.1$mrad	2.5cm	80μrad	0.65mm	1.3m	0.5mrad
CERN e^+	180 GeV	$\sigma = 25\mu$rad	0.5m	20μrad	0.84mm	12m	70μrad
CERN e^-	180 GeV	$\sigma = 25\mu$rad	0.5m	20μrad	0.90mm	8m	110μrad

To judge about the reliability of CBBC theory, it is expedient to form dimensionless ratios quantifying various effects ignored by Eq. (1) (see [3]). Those are gathered in Table 2.

Parameter R/R_c is made large by setup, to sharpen manifestation of the volume reflection effect, and thus, the deflection is highly non-perturbative. Nonetheless, a lot of radiation is generated away from the volume reflection area, given the significant crystal thickness, and may be viewed as emitted by a perturbatively deflected particle (smallness of numbers in the third

Table 2. Robustness of the dipole CBBC theory

Experiment	Steering strength, R/R_c	Coherence intervals per crystal thickness, $\frac{L}{\sqrt{2Rd}}$	Trajectory wiggling on the lattice scale, $\frac{4\sqrt{2}R^2 d}{\pi^3 L^2 R_c}$	Non-dipole degree of radiation, $\frac{R}{L}\tilde{\theta}_V$	Multiple scattering effect on radiation, $\frac{L}{R\tilde{\theta}_V}\sqrt{\frac{\sqrt{2Rd}}{0.13\text{mm}}}$
IHEP e^+	50	23	0.01	0.13	3.5
CERN e^+	27	10	0.03	0.95	0.86
CERN e^-	18	13	0.01	0.6	1.25

column of Table 2). Non-dipole and multiple scattering spoiling effects are rather moderate, except that the multiple scattering is too large in the IHEP experiment.

In any case, it is interesting to confront the dipole CBBC theory with the data from both experiments. In advance, let us assess main parameters of the radiation spectrum [3], which are collected in Table 3. One notices that the turnover frequency, related with the volume reflection, is substantially lower than the spectrum extent, thus the domain of CBBC must be formidable. Also, the BH background is seen to be subdominant. Let us now average intensity (1) over the charged particle incidence angles and look over the correspondence with the data (Figs. 2-4).

Table 3. Radiation spectrum dimensions

Experiment	End-of-spectrum at normal incidence, $\omega_{\text{end}} = \frac{2\pi\gamma^2 L}{Rd}$	Smear of the end-point, $\frac{4\pi\gamma^2}{d}\sigma$	Turnover frequency, $\omega_{\text{v.r.}} = 4\gamma^2\sqrt{\frac{2}{R_c d}}$	CBBC spectrum height at turnover $\frac{dE_{\text{CBBC}}}{d\omega}(\omega_{\text{v.r.}})$	Bethe-Heitlter contri-bution, $\frac{4}{3}\frac{L}{9.4\text{cm}}$	Coherent radiation intensity at $\omega \to 0$, $\frac{2e^2}{3\pi}\gamma^2\theta^2_{\text{v.r.}}$
IHEP e^+	0.77GeV	0.3GeV	0.15GeV	0.07	0.009	0.01
CERN e^+	35GeV	25GeV	12GeV	0.37	0.03 [2]	0.19
CERN e^-	56GeV	25GeV	12GeV	0.35	0.03 [2]	0.075

The agreement with CERN data is encouraging in the high-frequency part, and a turnover occurs approximately in the place predicted by the theory. Still, to some degree, there can be non-dipole effects. In the case of IHEP, we encounter a big normalization disagreement, yet there is no signature of a turnover in the expected place. These data also disagree with the calculation of [5]. One may attribute the discrepancy to a destruction of the radiation coherence by multiple scattering, but quantitative theory of such influence is lacking.

4. Conclusions

The dipole CBBC theory appears to be a satisfactory first approximation for radiation from relativistic charged particle passing through bent crystals, and owing to its simplicity it may be used as a guide at the initial stage of experiment planning. Admittedly, it does not take into account many effects, which need more dedicated theoretical investigations.

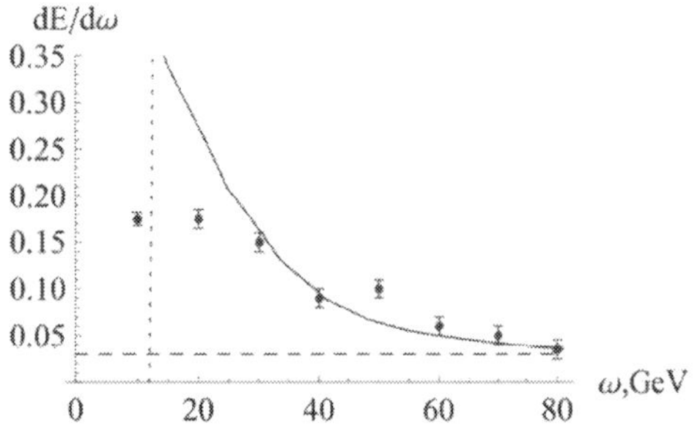

Figure 2. Points: CERN e^+ data [2]. Solid curve: θ_0-averaged CBBC plus the BH background (dashed line), for $R = 12$m, $L = 0.84$mm. Dotted line – turnover frequency.

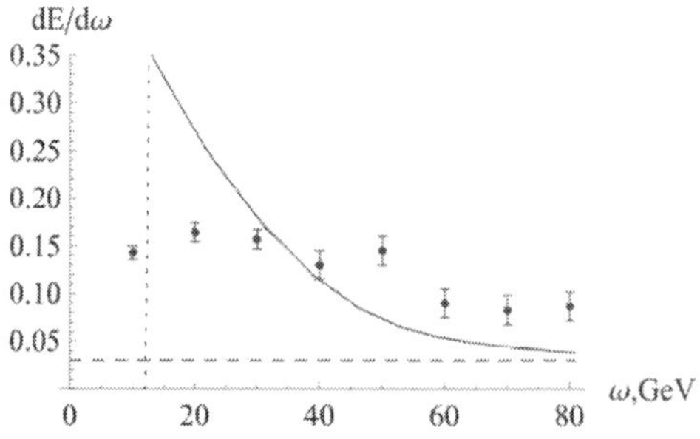

Figure 3. Same as Fig. 2, but for CERN e^- experiment [2], with $R = 8$m, $L = 0.9$mm.

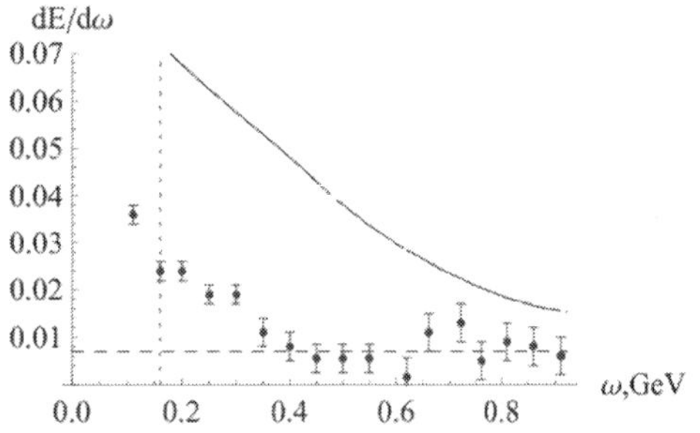

Figure 4. Same as Fig. 2, but for IHEP e^+ experiment [1], with $R = 1.3$m, $L = 0.65$mm.

To verify the expectation that the high-energy part of CERN experiments is actually CBBC, it would be desirable to perform additional measurements, in the same energy range, but with a better collimated beam, and with a variation of the target orientation. That might check the existence of a sharp spectrum end and its dependence on the beam and crystal parameters.

References

[1] Afonin A G et al. 2008 *JETP Lett.* **88** 414.
[2] Scandale W et al. 2009 *Phys. Rev.* A **79** 012903.
[3] Bondarenco M V 2009 Coherent bremsstrahlung in a bent crystal *Preprint* physics.acc-ph/0908.4503
[4] Biryukov V M, Chesnokov Yu A, Kotov V I 1994 *Phys. Usp.* **37** 937.
[5] Chesnokov Yu A, Kotov V I, Maisheev V A, and Yazynin I A 2008 *JINST* **3** P02005.

Contribution of incoherent effects to the orientation dependence of bremsstrahlung from rapid electrons in crystal

N F Shul'ga[1], V V Syshchenko[2] and A I Tarnovsky[2]

[1] Akhiezer Institute for Theoretical Physics of the NSC "KIPT", Akademicheskaya Street, 1, Kharkov 61108, Ukraine

[2] Belgorod State University, Pobedy Street, 85, Belgorod 308015, Russian Federation

E-mail: shulga@kipt.kharkov.ua, syshch@bsu.edu.ru, syshch@yandex.ru

Abstract. The bremsstrahlung cross section for relativistic electrons in a crystal is split into the sum of coherent and incoherent parts (the last is due to a thermal motion of atoms in the crystal). Although the spectrum of incoherent radiation in crystal is similar to one in amorphous medium, the incoherent radiation intensity could demonstrate substantial dependence on the crystal orientation due to the electrons' flux redistribution in the crystal. In the present paper we apply our method of the incoherent bremsstrahlung simulation developed earlier to interpretation of some recent experimental results obtained at the Mainz Microtron MAMI.

1. Introduction

It is well known (see, e.g. [1, 2, 3]) that high energy electron beam incident on an oriented single crystal produces the coherent radiation that is due to the spatial periodicity of the lattice atoms, and the incoherent one, that is due to the thermal spread of atoms from their positions of equilibrium in the lattice. For the first look, the incoherent part of radiation is similar to the last in amorphous medium (with Bethe-Heitler spectrum), and do not depend on the crystal orientation in relation to the particles beam.

However, in [4, 5] it was paid attention to the fact that some features of the particle's dynamics in the crystal (channeling effect etc.) could lead to various substantial orientation effects in the hard range of the spectrum, where (for $\varepsilon \sim 1$ GeV electrons) the incoherent part is predominant. The semi-numerical approach developed in [4, 5] was used for interpretation of early experimental data [6].

The ideass of [5] had been referred by the authors of recent experiments [7] to interpret some of their results. In our article we present the results of simulation of the incoherent radiation under the conditions of the experiment [7]. A good agreement with the experimental data confirms the interpretation given in [7].

For the reader convenience, in the next section we outline some theoretical ideas of our approach.

2. Bremsstrahlung in dipole approximation

Radiation of relativistic electron in matter develops in a large spatial region along the particle's momentum. This region is known as the coherence length (or formation length) [1, 2] $l_{\mathrm{coh}} \sim 2\varepsilon\varepsilon'/m^2c^3\omega$, where ε is the energy of the initial electron, ω is the radiated photon frequency, $\varepsilon' = \varepsilon - \hbar\omega$, m is the electron mass, c is the speed of light. In the large range of radiation frequencies the coherence length could exceed the interatomic distances in crystal:

$$l_{\mathrm{coh}} \gg a. \tag{1}$$

In this case the effective constant of interaction of the electron with the lattice atoms may be large in comparison with the unit, so we could use the semiclassical description of the radiation process. In the dipole approximation the spectral density of bremsstrahlung is described by the formula [2]

$$\frac{dE}{d\omega} = \frac{e^2\omega}{2\pi c^4}\int_{\delta}^{\infty}\frac{dq}{q^2}\left[1 + \frac{(\hbar\omega)^2}{2\varepsilon\varepsilon'} - 2\frac{\delta}{q}\left(1 - \frac{\delta}{q}\right)\right]\left|\mathbf{W}_q\right|^2, \tag{2}$$

where $\delta = m^2c^3\omega/2\varepsilon\varepsilon' \sim l_{\mathrm{coh}}^{-1}$, and the value

$$\mathbf{W}_q = \int_{-\infty}^{\infty}\dot{\mathbf{v}}_{\perp}(t)e^{icqt}dt \tag{3}$$

is the Fourier component of the electron's acceleration in the direction orthogonal to its initial velocity.

The main contribution to the integral in (2) is made by the small values of q, $q \sim \delta$. On the other hand, the characteristic distances on which the electron's acceleration in the field of the atom in (3) would substantially distinct from zero are equal by the order of magnitude to the atomic radius R. The corresponding time intervals in which the integrand in (3) is distinct from zero are $\Delta t \sim R/c$. But under the condition (1) in the frequency range of our interest $\delta \ll R^{-1}$. So we could present $\mathbf{W}_q$ in the form [8]

$$\mathbf{W}_q = c\sum_n \boldsymbol{\vartheta}_n e^{icqt_n}, \tag{4}$$

where $\boldsymbol{\vartheta}_n$ is the two-dimensional electron scattering angle under collision with the n-th atom, t_n is the time moment of the collision.

Consider now the radiation of the electron incident onto the crystal under small angle ψ to one of its crystallographic axes. It is known [1, 2] that averaging of the equation for $\left|\mathbf{W}_q\right|^2$ over the thermal vibrations of atoms in the lattice leads to the split of this value (and so the radiation intensity) into the sum of two terms describing coherent and incoherent effects in radiation:

$$\left\langle\left|\mathbf{W}_q\right|^2\right\rangle = c^2\sum_{n,m} e^{icq(t_n - t_m)}\left\langle\boldsymbol{\vartheta}(\boldsymbol{\rho}_n + \mathbf{u}_n)\right\rangle\left\langle\boldsymbol{\vartheta}(\boldsymbol{\rho}_m + \mathbf{u}_m)\right\rangle \tag{5}$$

$$+c^2\sum_n\left\{\left\langle(\boldsymbol{\vartheta}(\boldsymbol{\rho}_n + \mathbf{u}_n))^2\right\rangle - (\langle\boldsymbol{\vartheta}(\boldsymbol{\rho}_n + \mathbf{u}_n)\rangle)^2\right\}, \tag{6}$$

where $\boldsymbol{\rho}_n = \boldsymbol{\rho}(t_n) - \boldsymbol{\rho}_n^0$ is the impact parameter of the collision with the n-th atom in its equilibrium position $\boldsymbol{\rho}_n^0$, $\boldsymbol{\rho}(t)$ is the trajectory of the electron in the plane orthogonal to the crystallographic axis (which could be obtained by numerical integration of the equation of motion), and $\mathbf{u}_n$ is the thermal shift of the n-th atom from the position of equilibrium. In the range of radiation frequencies for which

$$l_{\mathrm{coh}} \ll a/\psi, \tag{7}$$

where a is the distance between two parallel atomic strings the closest to each other, the incoherent term (6) makes the main contribution into the bremsstrahlung intensity (2).

The radiation by the uniform beam of particles is characterized by the radiation efficiency, that is the radiation intensity (2) integrated over impact parameters of the particles' incidence onto the crystal in the limits of one elementary cell. So, the efficiency is the classical analog of the quantum cross section. In the further consideration we shall compare the radiation efficiency in the crystal to the Bethe-Heitler efficiency of bremsstrahlung in amorphous medium.

For further computational details see [4, 5].

3. Origin of the orientation dependence of the incoherent bremsstrahlung

When charged particles are incident onto the crystal under small angle θ to one of the atomic planes densely packed with atoms, the channeling phenomenon could takes the place (see, e.g., [2, 3]). Under planar channeling the electron moves in the potential well formed by the attractive continuum potential of the atomic plane (see figure 1, left panel). The largest incidence angle, for which the capture into the channel is possible, is called as the critical channeling angle θ_c [2, 3].

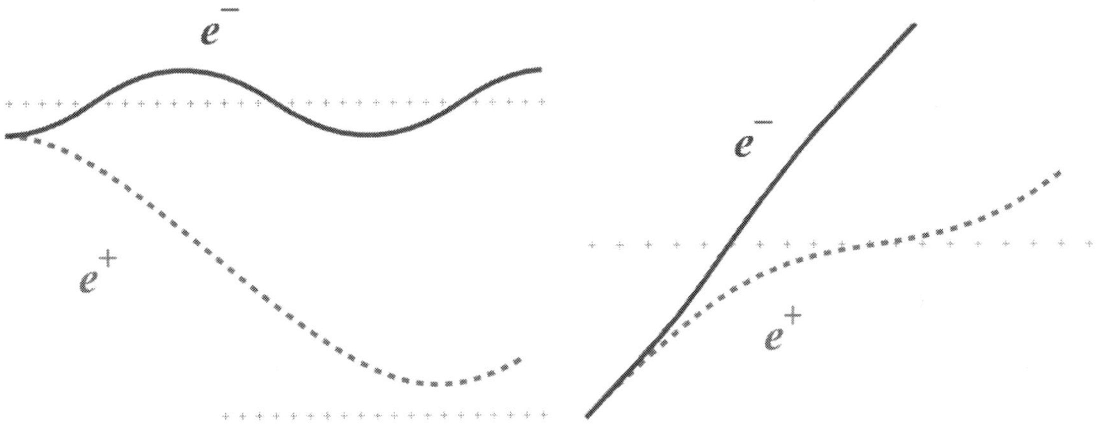

Figure 1. Typical trajectories of the electrons (——) and positrons (- - - -) under planar channeling (left) and above-barrier motion (right). Pluses mark the positions of atomic strings (perpendicular to the plane of the figure) forming the atomic planes of the crystal.

Under $\theta \ll \theta_c$ the most part of the incident electrons would move in the planar channeling regime. These electrons will collide with atoms at small impact parameters more frequently then in amorphous medium, that leads to the increase of the incoherent bremsstrahlung efficiency (see figure 2). For $\theta \sim \theta_c$ the above-barrier motion in the continuum potential takes the place for the most part of the particles (figure 1, right panel). Above-barrier electrons rapidly cross the atomic plane, with reduced number of close collisions with atoms comparing to the case of amorphous medium. This leads to the decrease of the incoherent bremsstrahlung efficiency (figure 2). For the positron beam the situation is opposite.

4. Results and discussion

In one of the experiments of [7] the integral intensity of radiation (with the photon energy $\hbar\omega \geq 15$ MeV, that eliminates the contribution of the channeling radiation) had been measured for different orientations of the silicon crystals in relation to 855 MeV electron beam. Note that under scanning of the goniometer angle ϕ both the angle of incidence to the [100] axis and the

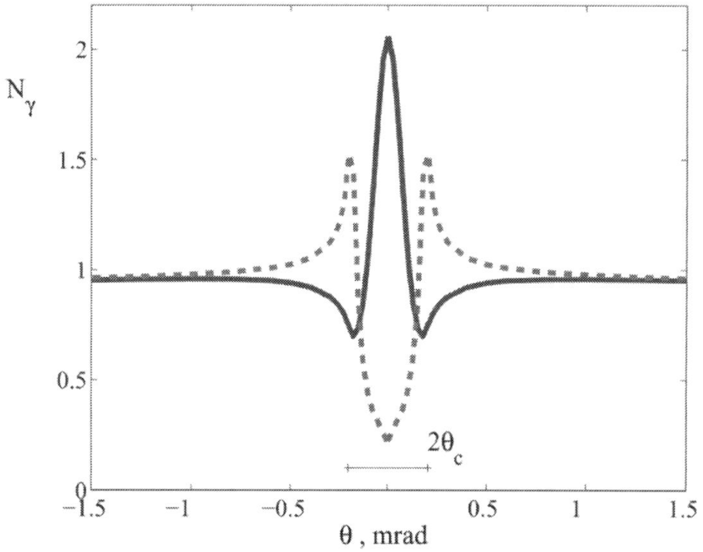

Figure 2. Incoherent bremsstrahlung efficiency (in ratio to the Bethe-Heitler efficiency in amorphous medium) from 1 GeV electrons (——) and positrons (- - - -) vs incidence angle θ to $(0\bar{1}1)$ plane of Si crystal, as a result of simulation.

angle to $(0\bar{1}1)$ plane (and other planes of the crystal) were changing. The zero angle between the plane $(0\bar{1}1)$ and the axis of the electron beam is achieved near $\phi = 26$ mrad.

The experimental data ([7], Fig. 14(b)) and the results of our simulation are presented on the figure 3.

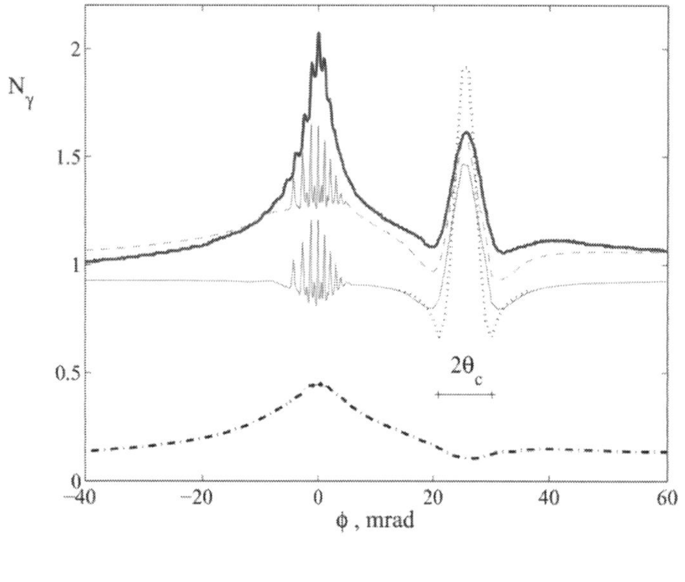

Figure 3. Coherent (— · —) and incoherent (——) contributions to the relative efficiency of radiation (- - - -) according to simulation, in comparison to the experimental results [7] (thick curve).

The right peak, and the gaps surrounding it, are caused by the contribution of incoherent radiation, as it was illustrated on the figure 2. Dotted curve presents the result of simulation without account of the incoherent multiple scattering of the electron on the thermal vibrations of the lattice atoms; taking that into account leads to some softening of the curve.

The left peak is caused by the effect of many crystallographic planes with common [100] axis. The fine structure of this peak is also determined by the orientation dependence of the *incoherent* radiation, as described above (see figure 4). The crystallographic planes responsible to the origin of specific local maxima in the incoherent radiation intensity are named on the figure.

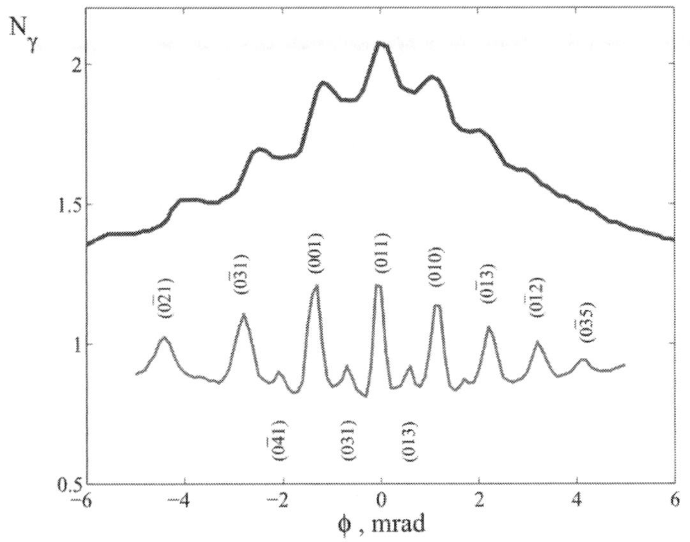

Figure 4. Region near $\phi = 0$ of the previous figure. Only experimental data (upper curve) and incoherent contribution (lower curve) are shown.

The coherent contribution was calculated using standard Born theory of the coherent bremsstrahlung [2]. We normalized the experimental curve to the right edge of the calculated one, far from the conditions of planar channeling, where Born theory of the coherent bremsstrahlung is surely valid. For better agreement with the experiment in the range near the left peak, it seems the most suitable to carry out the simulation of the coherent bremsstrahlung based on the same semiclassical approach, as used for the study of the incoherent one [8].

Acknowledgments
This work is supported in part by the internal grant of Belgorod State University.

References

[1] Ter-Mikaelyan M L 1972 *High-Energy Electromagnetic Processes in Condensed Media* (New York: Wiley-Interscience)
[2] Akhiezer A I and Shul'ga N F 1996 *High-Energy Electrodynamics in Matter* (Amsterdam: Gordon and Breach)
[3] Uggerhøj U I 2005 *Rev. Mod. Phys.* **77** 1131
[4] Shul'ga N F and Syshchenko V V 2005 *Nucl. Instr. and Meth.* B **227** 125
[5] Shul'ga N F, Syshchenko V V and Tarnovsky A I 2008 *Nucl. Instr. and Meth.* B **266** 3863
[6] Sanin V M, Khvastunov V M, Boldyshev V F and Shul'ga N F 1992 *Nucl. Instr. and Meth.* B **67** 251
[7] Backe H, Kunz P, Lauth W and Rueda A 2008 *Nucl. Instr. and Meth.* B **266** 3835
[8] Shul'ga N F, Truten' V I and Syshchenko V V 1996 *Nucl. Instr. and Meth.* B **119** 55

Simulations of electron and positron planar channeling for BTF and SPARC beams

A A Babaev[1] **and S B Dabagov**[2,3]

[1] Nuclear Physics Institute, Tomsk Polytechnic University, Lenin Ave. 2a, Tomsk 634050, Russia

[2] INFN Laboratori Nazionali di Frascati, Via E. Fermi 40, Frascati 00044, Italy

[3] RAS PN Lebedev Physical Institute, Leninsky pr. 53, Moscow 119991, Russia

E-mail: krass58ad@mail.ru

Abstract. Presently new experiments on channeling radiation by both 150 MeV electrons and 400 MeV positrons are planned at the LNF facilities. These experiments require the preliminary theoretical treatment. The channeling radiation spectrum is formed by transitions between bound energy levels of transverse motion of a channeled particle. The intensity of channeling radiation lines depends on populations of bound energy levels. These populations change during the projectile motion in a crystal that, in turn, influences the channeling radiation intensities. In this manuscript we present theoretical model and computer simulations to investigate the bound energy spectra of planar-channeled electrons and positrons as well as to obtain the populations of bound states. Solving the kinetic equations we explore the dynamics of bound state populations.

1. Introduction

When the charged particles move in a crystal under small angles to the crystallographic planes or axes, the channeling motion may take place [1]. The motion of relativistic light projectiles along axes (in the case of axial channeling) or along planes (in the case of planar channeling) is mainly described within classical mechanics approach. However, motion in transverse to these axes or planes direction requires quantum-mechanical treatment for both low and moderate projectile energies. Hence, interaction of electrons and positrons with a crystal potential under the channeling conditions forms the spectrum of bound energy states for transverse motion [2]. The transitions between these energy levels at channeling result in channeling radiation (CR) [3]. The intensities of CR lines depend on populations of energy levels at the time of radiation.

Hence, to estimate the line intensity, one should know the dynamics of populations for bound states. The population features can be described by "the kinetic equations" [2]. In this work we present the solutions of kinetic equations for planar-channeled 150 MeV electrons and 400 MeV positrons for planned experiments at LNF BTF and SPARC facilities [4].

The DAΦNE Beam Test Facility (BTF) is an electron/positron transfer line, by which the beam accelerated from the Linac, is transported to an experimental hall of DAΦNE where the beam tests as well as the experiments can be performed. The facility can provide electron/positron beam in a wide range of intensity: from single particle per bunch up to 10^{10} particles per pulse. The BTF is operating since 2002: during these years, tens of high energy physics experiments from all over Europe have been hosted. The main applications of the facility are: high energy detector calibration, low energy calorimetry, low energy electromagnetic interaction studies, detector efficiency and aging measurements, tests of beam diagnostic devices. The BTF facility provides electrons and positrons with energy ranging from 20 to 800 MeV (750 MeV for positrons) [5]. The DAΦNE BTF is the unique European Facility that at present time is able to deliver positron beams in the moderate energy range which is of special interest for channeling experiments. In fact, according to well accepted channeling radiation theories a strong photon peak should appear at X-ray and γ frequencies; experimental studies on positron channeling in various crystalline structures have been planning within the CUP project [6].

Our simulations have been also performed for low-emittance electron beams of SPARC (FEL project of LNF INFN), another very modern free electron laser facility, representing a special interest due to advanced parameters of the beam, to be utilized as a powerful radiation source. The SPARC project foresees the realization of a free electron laser operating at 500 nm driven by a high brightness photo-injector at beam energy of 150-200 MeV. Presently various coherent radiation applications, to be realized within PlasmonX and Thomson projects, are under construction at LNF; electron channeling in crystals as a powerful coherent radiation source is one of the planning experiment [7].

2. Bound energy states and wave functions

The crystal potential forms as a sum of separate atomic potentials, and when a particle moves along a crystal plane under small glancing angle, the potential can be replaced by the averaged planar potential - "continuous planar potential" [1-3,8]. It is notable that the potential averaging can be analytically performed keeping rather good agreement with experimentally observed data. There are many various approximations for continuous potentials; in this work we use the Moliere approximation for screened atomic potential [8].

Let choose the OX axis of rectangular coordinate frame (Cartesian coordinate system) perpendicular to the crystallographic plane, and the OY, OZ axes to be coincided with the crystallographic axes in this crystallographic plane (figure 1). The origin of OX axis is fixed at the channel center. In this case the averaged planar crystal potential may be represented by Fourier expansion:

$$V_{\rm pl}(\tilde{x},y,z)=V_{\rm pl,0}(\tilde{x})+\sum_{k_y,k_z\neq 0}V_{{\rm pl},k_y,k_z}(\tilde{x})\exp(ik_y y+ik_z z)\,, \tag{1}$$

where $V_{\rm pl,0}$ is the continuous planar potential, $V_{{\rm pl},k_y,k_z}(\tilde{x})\exp(ik_y y+{\rm i}k_z z)$ is the harmonic of space-periodical potential, k_y and k_z are the reciprocal lattice vectors corresponding OY and OZ directions in the plane, $\tilde{x}$ is the distance from the plane. Thus, the full crystal potential is the sum of planar potentials (1). In our calculations we take into account the crystal potentials of only two neighboring to the projectile planes.

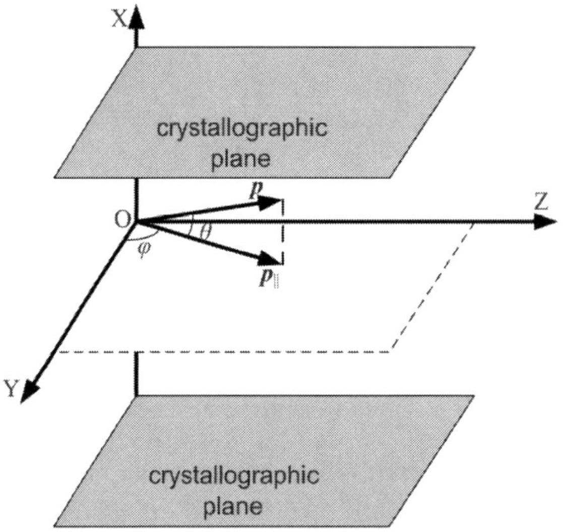

Figure 1. The coordinate frame: OX axis is perpendicular to the crystal plane, OY and OZ axes coincide with the axes in crystallographic planes.

The total energy E of the projectile can be presented as a sum of longitudinal energy $E_{\|}$ and transverse energy ε. The longitudinal energy is $E_{\|}=\left[\left(m_0c^2\right)^2+\left(p_{\|}c\right)^2\right]^{1/2}-m_0c^2$, where $p_{\|}=p\cos\theta$ is the longitudinal momentum, $p=\left(\gamma^2-1\right)^{1/2}m_0c$ is the full particle momentum, $\gamma=E_{\text{kin}}/m_0c^2+1$, E_{kin} is the total kinetic energy of particle, θ is the angle between momentum and crystallographic planes (figure 1), m_0c^2 is the rest energy of projectile (electron or positron). In the first approximation, one can consider that the particle moves in a channel formed by the continuous potential only. The transverse energy has discrete values and is defined by Schrodinger-type equation [2]:

$$\left(-\frac{1}{2E_{\|}}\frac{\mathrm{d}^2}{\mathrm{d}x^2}+U_0(x)\right)\psi(x)=\varepsilon\psi(x), \tag{2}$$

where $U_0(x)=eV_0(x)=e\left(V_{\text{pl},0}(a_x/2-x)+V_{\text{pl},0}(a_x/2+x)\right)$ is the potential energy of projectile in the continuous crystal potential, e is the charge of projectile, a_x is the distance between planes.

Due to the transverse periodicity of the continuous potential V_0, both wave functions and potential energy U_0 can be presented by Fourier expansions [2]:

$$\psi_i(x)=\frac{1}{\sqrt{a_x}}\sum_n C_n^i\exp\left[\mathrm{i}\left(\kappa-ng_0\right)x\right], \tag{3}$$

$$U_0(x)=\sum_m U_{0,m}\exp\left(\mathrm{i}mg_0x\right). \tag{4}$$

In expressions (3), (4) $g_0=2\pi/a_x$ is the reciprocal lattice vector, which corresponds to OX direction, κ is the quasimomentum defined as:

$$\kappa = \frac{p_\perp}{\hbar} - N g_0 .$$

The integer number N is defined by the following requirement $-\pi / a_x \le \kappa \le \pi / a_x$; $p_\perp = p \sin\theta$ is the initial transverse momentum of channeled particle. When the expansions (3) and (4) are substituted into equation (2), the coefficients C_n^i as well as the corresponding energies ε_i can be defined [2, 9]. The energy level is characterized by number i and quasimomentum κ. Thus, the bound energy levels form the zone structure: i-th zone contains the energy levels, which differ by quasimomentum κ. Due to this fact, each zone is characterized by specific width.

For the purposes of future LNF BTF and SPARC experiments, we calculate the bound energy zone spectra for (220) planar channeled 150 MeV electrons and 400 MeV positrons (figure 2, only sub-barrier zones are shown). It should be underlined that when the projectile energy exceeds 100 MeV, the bound energy spectrum is characterized by large number (>10) of very narrow zones (to compare, see, for example, the results of calculations with relatively small energies in [2]). Actually, the width of zones seems "visible" for top sub-barrier zones only (the width becomes comparable to the difference between neighbor zones).

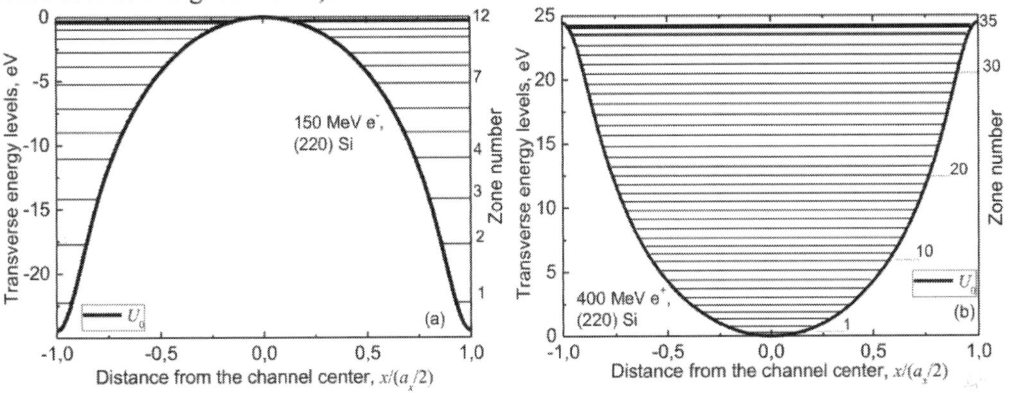

Figure 2. The potential energy profiles (thick curve) and bound energy spectrum (horizontal lines) for 150 MeV electrons (a) and 400 MeV positrons (b) under (220) planar channeling in Si crystal.

3. Populations of bound states

The initial distribution of channeled particles over transverse energy states is formed when particles penetrate into a crystal. Let us describe the transverse motion of a free particle by the plane wave

$$\psi_0 = \frac{1}{\sqrt{a_x}} \exp\left(\frac{\mathrm{i}}{\hbar} p_\perp x \right).$$

The probability $P_{i,0}$ of particle capture into the i-th zone defines by the expression:

$$P_{i,0} = \left| \int_{-a_x/2}^{a_x/2} \psi_0(x) \psi_i(x) \mathrm{d}x \right|^2 . \quad (5)$$

It is notable that only elastic captures have been considered, i.e. the transverse momentum of a free particle corresponds to its quasimomentum in the bound state. The inelastic captures were considered in [10]. Here, it should be mentioned, then the initial populations of even states equal zero (ground state corresponds to number 1), when only elastic captures are considered. In general, taking into account the inelastic captures changes slightly the populations [10]. This means, populations of even states does not equal zero when inelastic capture is included in consideration, although it remains relatively small in comparison to the populations of sub-barrier odd states.

In figure 3 the initial populations of (220) Si planar channeled 150 MeV electrons and 400 MeV positrons are presented. As known, channeling of charged particles in crystals takes place when the particle glancing angle with respect to the crystal plane less than the critical angle of channeling, $\theta \le \theta_L$, where θ_L is the critical angle (Lindhard angle) [2]. For 150 MeV electrons $\theta_L = 0.033^o$ and for 400 MeV positrons $\theta_L = 0.02^o$. In figure 3 the initial populations are presented at different values of the ratio θ / θ_L, i.e. at different values of initial transverse momentum $p_\perp$. The results of calculations demonstrate that populations of a state change strongly at different angles θ. Hence, the angular divergence of a projectile beam may influence the populations of bound energy zones and CR intensity. The populations of bound states decrease when the angle θ approaches the angle θ_L.

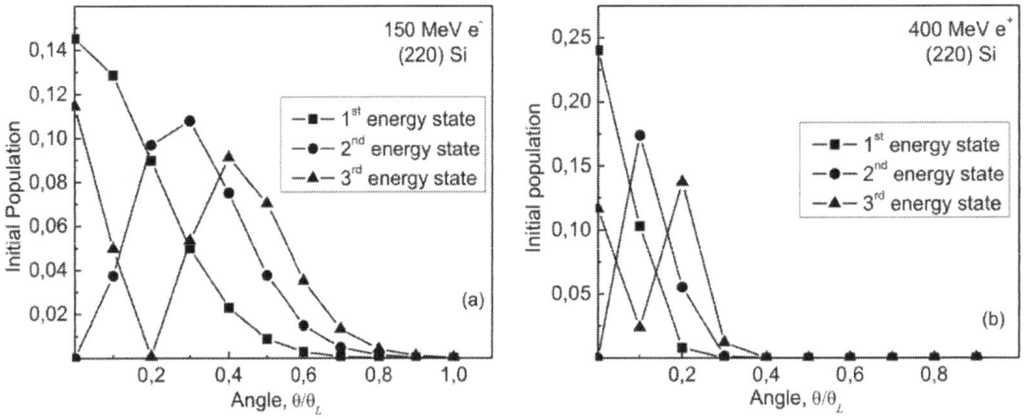

Figure 3. The initial populations of lowest bound energy states for (220) Si planar channeled 150 MeV electrons (a) 400 MeV positrons (b) at different values of ratio θ / θ_L.

At motion in a crystal, particle can change its transverse energy under the influence of periodic part of crystal potential, which can be determined as a sum of periodic parts of the planar potentials (1). To describe the evolution of bound state populations P_i one can use a system of the kinetic equations written in the form [2,9,10]

$$\frac{\mathrm{d}P_i}{\mathrm{d}t} = \sum_j w_{ij} P_j - \Gamma_i P_i \,, \tag{6}$$

where w_{ij} is the transition probability from j-th to i-th state during the small time dt, $\Gamma_i = \sum_j w_{ji}$ is the full probability to leave the i-th state during the time dt. Now our aim is to calculate the quantities w_{ij}.

The periodic part of crystal potential is approximated by summing the periodic parts of two neighboring to the projectile planes. At the time moment t of particle penetration into crystal, this particle has managed by the periodic potential [9]

$$V(x,t) = \sum_{k_y, k_z \neq 0} V_{k_y, k_z}(x) \exp\left[\mathrm{i}\left(k_y \cos\varphi + k_z \sin\varphi\right) v_{\|} t\right],$$

where φ is the angle between longitudinal momentum $p_{\|}$ (corresponding longitudinal velocity $v_{\|}$) and OY axes (figure 1). The full probability of $j \rightarrow i$ transition during time of motion t is defined by

$$W_{ij} = \frac{1}{\hbar^2}\left|\int_0^t \left(\int_{-a_x/2}^{a_x/2} \psi_j^*(x)V(x,t)\psi_i(x)\mathrm{d}x \right) \exp\left[\frac{\mathrm{i}}{\hbar}\left(\varepsilon_i - \varepsilon_j\right)t\right]\mathrm{d}t\right|^2,$$

from which the transition probability can be reduced as follows

$$w_{ij} = \frac{\mathrm{d}W_{ij}}{\mathrm{d}t}. \quad (7)$$

To solve the kinetic equations (6), the expansions (3) and (4) are cut off on the number $N_{\max}$ and $-N_{\max} \le n \le N_{\max}$ in expression (3), $-N_{\max} \le m \le N_{\max}$ in expression (4). Hence, the equation (2) defines $2N_{\max}+1$ lowest bound energy levels ε_i [9]. In our simulations we used $N_{\max}$=50. The system (6) includes only N_{eq} equations to define the populations P_i when $\varepsilon_i \le 2U_{0,\max}$, where $U_{0,\max}$ is the top of the potential barrier. It was suggested, as in [11], if the projectile transits to the level with energy $\varepsilon_i > 2U_{0,\max}$, it will be not more included in simulations (as dechanneled forever); i.e. actually, we solve the system of kinetic equations in a form

$$\frac{\mathrm{d}P_i}{\mathrm{d}t} = \sum_{j=1}^{N_{eq}} w_{ij}P_j - \sum_{j=1}^{2N_{\max}+1} w_{ji}P_i, \qquad i = 1 \ldots N_{eq} \quad (8)$$

The system (8) was solved numerically for planar channeled 150 MeV electrons and 400 MeV positrons in 50 μm Si (220) crystal. In our simulations we suppose that projectiles move parallel to (220) planes (θ=0°, φ=10°) and we did not take into account the angular divergence of projectile beam as well as inelastic processes, i.e. the quasimomentum of the initial state corresponds to one of the final state (in other words, we neglect the momentum transferred to the crystal). In figure 4 the evolution of populations for specified states are presented. Our simulations demonstrate that populations of top sub-barrier states decrease significantly due to the continuous dechanneling processes. The effective populations of lowest states, characterizing by short living time of the projectile in these states, decrease less intensive and even can slightly grow due to strong transition mixture between energy levels.

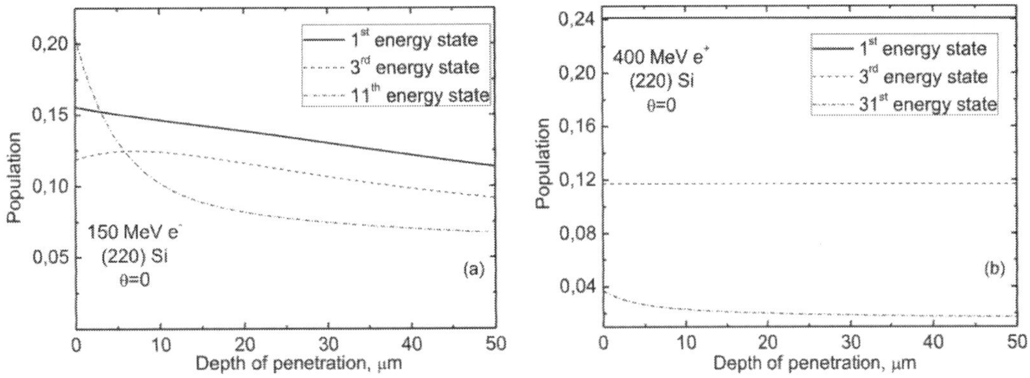

Figure 4. The evolution of bound state populations at (220) planar channeling of 150 MeV electrons (a) and 400 MeV positrons (b) over 50 μm Si crystal.

To estimate the influence of both angular divergence of projectile beam and inelastic transitions between energy levels (i.e. transitions with quasimomentum change) at channeling on the population dynamics, we simulate planar channeling of 4 MeV electrons in 5 μm Si (220) crystal (figure 5). We have compared our results with analogous results of simulations presented in [11]. Actually, we deal with a number of discreet energy levels, whereas authors [11] deal with the finite broad zones. Our results demonstrate greater populations of bound states than the results for corresponding zones. The difference is about 10%. Moreover, the 2nd state remains unpopulated during channeling, but the results of the above mentioned paper show the population of 2nd zone equals zero at initial time moment and then it grows due to transitions from other zones. However, it is notable that inelastic transitions for 150 and 400 MeV projectiles play the smaller role than for 4 MeV projectiles, because in general the zones for low-energy projectiles are wider than ones for relativistic projectiles. Hence, we may expect the good precision of our approximations for 150 MeV electrons and 400 MeV positrons even having neglected the inelastic transitions. Detailed simulations will be published elsewhere.

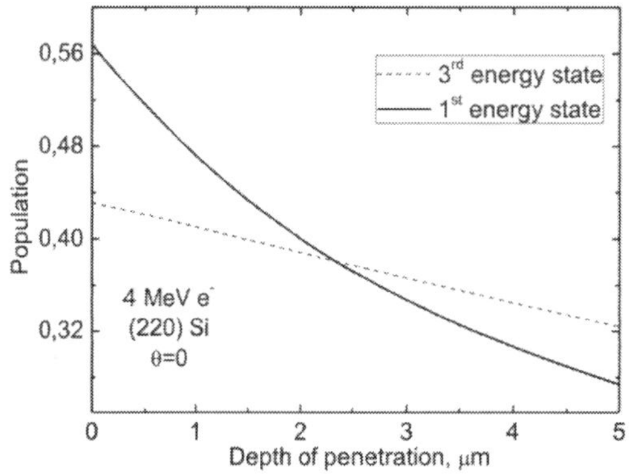

Figure 5. The evolution of bound state populations during (220) planar channeling of 4 MeV electrons over 5 μm Si crystal.

4. Conclusion

Nowadays theories of both charged particles channeling and channeling radiation are fairly well developed. Researches in this area are largely of applied nature, with the exception for channeling of ultrarelativistic particles. The development of numerical methods for analyzing the channeling in various crystals has allowed the optimal parameters of studied samples to be determined. The use of spectroscopy of channeling radiation as a sensitive method for solids studying is still has a practical interest. However, the examination of channeling of moderate-energy particles taking into account main fundamental scattering processes has to be continued due to the lack of correlated theoretical and experimental results [12].

In this work the simulations of electron and positron planar channeling in condition of planned LNF BTF/SPARC experiments were analyzed. We have both calculated the bound energy spectrum of projectiles and described the dynamics of the bound states population during motion of projectiles in a crystal. The method to calculate the probabilities of projectile transitions between bound states of transverse motion based on consideration of crystal potential as a sum of periodic terms instead of usually used approach [2] has been developed.

Our results demonstrate rather good agreement with the results on population dynamics of bound states previously published by other authors. Nevertheless, to obtain a better accuracy, in the future we plan to take into account the angular projectile beam divergence as well as the finite width of bound energy zones.

References

[1] Lindhard J 1965 *Kgl. Dan. Viden. Sels. Mat.-Fys. Medd.* **34** 14

[2] Bazylev V A and Zhevago N K 1987 *Radiation from fast particles in a matter and external fields* Moskow Nauka (in Russian)

[3] Kumakhov M A 1986 *Radiation from channeled particles in crystals* Moskow Energoatomizdat

(in Russian)
[4] Dabagov S B, Ferrario M, Palumbo L and Serafini L 2007 *Int J. Mod. Phys A* **22** 23 4280
[5] Mazzitelli G and Valente P 2003 *Nucl. Instr. Meth. A* **513** 3 516
[6] Quitnieri L, Buonomo B, Dabagov S B, Mazzitelli G and Valente P 2007 *Proc. of SPIE* **6634** 66340U
[7] Alesini D et al 2003 *Nucl. Instr. Meth. A* **507** 1-2 345
[8] Gemmel Donald S 1974 *Reviews of Moderm Physics* **44** 1 129
[9] Babaev A and Dabagov S B 2009 Dynamics of bound state populations for channeled electrons/positrons *Preprint* LNF -09-07(IR)
[10] Bogdanov O V, Korotchenko K B and Pivovarov Yu L 2007 *JETP Letters* **85** 11 555
[11] Tulupov A V 1984 *JETP Letters* **59** 4 797
[12] Dabagov S B and Zhevago N K 2008 *Rivista del Nuovo Cimento* **31** 9 491

Basic Channeling with Mathematica©: a New Computer Code

O V Bogdanov, E I Fiks, K B Korotchenko, Yu L Pivovarov, T A Tukhfatullin

Tomsk Polytechnic University, Tomsk, 634050 Russia

korotchenko@tpu.ru

Abstract. The Mathematica© software is a powerful tool to solve numerous problems in physics, essentially when the analytical solution is not possible. The channeling and related phenomena is just that case. Here we suggest to introduce a "standard" in calculations of channeling and related phenomena using Mathematica© software. The new computer code "Basic Channeling with Mathematica©" BCM–1.0, at the first stage, will include a set of packages to make symbolic calculation of the planar channeling phenomena.

1. Introduction

The analysis of channelling and related phenomena (e.g. channelling radiation, diffracted channelling radiation, etc.) is connected with cumbersome analytical and numerical calculations. The new computer code BCM 1.0 is created to avoid these difficulties and introduce "the standard" in these calculations. The "standard" problems in planar channeling are:

- calculation of planar periodic potential function $U(\boldsymbol{r}_{\perp})$ for chosen set of crystallographic planes of desired crystal;
- computation of classical trajectories of channeled charged particles in defined periodic potential function $U(\boldsymbol{r}_{\perp})$;
- computation of eigenvectors (i.e. wave functions) and eigenvalues (i.e. energy levels) of channeled electrons (positrons) transverse motion in defined periodic potential function $U(\boldsymbol{r}_{\perp})$;
- calculation of transverse energy levels initial populations;
- calculation of transitions matrix elements of channeled electrons (positrons).

Our solution experience of these problems using Mathematica© software has allowed us to attack the problem of creation the library of packages for the radiation problems at the channeling solution – "Basic Channeling with Mathematica© "(BCM).

The first version of the new computer code – BCM 1.0 includes the following packages:

- "Potential.m" – computation of continuous potential function of the crystallographic planes $U(y)$.
- "Trajectory.m" – computation of classical channeled charged particles trajectories in planar periodic potential function $U(y)$.
- "Eigen.m" – solution of eigenvalue problem to determine transverse energy band structure and eigenfunctions from the Schrödinger equation with relativistic mass of channeled electron (positron) in periodic potential function $U(y)$.

- "Population.m" – computation of initial populations of transverse energy bands using "Eigen.m".
- "Matrix.m" – computation of transitions matrix elements (in dipole approximation) using "Eigen.m".

2. Planar channeling potential function

Good description for planar potential function $U(y)$ of real crystal one can obtain from formula given by Doyle and Turner [1] for single atom potential function by fitting the electron scattering form factor with experimental parameters a_j and b_j

$$U_{At}(\vec{r}) = -(2\hbar^2/\sqrt{\pi}m_e)\sum_j a_j c_j^{-3/2} \exp(-r^2/c_j) \ . \tag{1}$$

Here, $4\pi^2 c_j = b_j + 8\pi^2\rho_1^2$ and ρ_1 is the amplitude of thermal vibrations of atom. Integration of the formula (9) along channeling plane (*XZ*–plane) leads to separate plane potential function

$$U_{Pl}(y) = -A\sum_j a_j c_j^{-1/2} \exp(-y^2/c_j) \ , \tag{2}$$

where $A = 16\sqrt{\pi}r_o 27.2116 d/a$, d is the interplanar spacing, a is a lattice constant, r_o is the Bohr radius.

In the case of complex crystal, the periodic potential function $U(y)$ is obtained by summation of potential functions $U_{Pl}(y)$ over the single planes for all kinds of atoms in crystal. For example, total planar potential function for (111) LiF is

$$U(y) = \sum_m (U_F(y+md) + U_{Li}(y+md-d/2)) \, , \tag{3}$$

where m is the minimal number of single planes for the necessary precision of the total planar potential function $U(y)$ computation.

Next we may calculate Fourier components of the potential function $U(y)$

$$U_m = \int_{-d/2}^{d/2} U(y)\exp(-im|\mathbf{g}|y)dy \, , \tag{4}$$

3. Classical trajectory of planar channelled charged particles in crystals

The motion equation for the classical relativistic charged particle in the field with potential function $U(y)$ for the transverse direction (in the approximation of $v_\perp \equiv \dot{y} << v_\parallel ; \ v_\perp/c << 1/\gamma$) has the form (see in Fig.1)

$$\gamma m\ddot{y} = -\frac{\partial U(y)}{\partial y}, \quad \gamma = \frac{1}{\sqrt{1 - v_\parallel^2/c^2}} \tag{5}$$

where $v_\parallel$ is the average electron's velocity along the channel and $v_\perp$ is the velocity of transverse motion. The initial conditions are: the point of incidence into a crystal $y(0) \equiv y_0$ and transverse momentum $p_\perp(0) = p_\parallel \sin\theta_0 \cong p_\parallel\theta_0$, where $p_\parallel = \gamma m v_\parallel$ and θ_0 is the electron incidence angle with respect to the channeling plane. They determine the integral of motion equation, the so–called transverse energy:

$$\varepsilon_\perp = U(y) + \frac{p_\perp^2}{2\gamma m} = U(y_0) + \frac{p_\parallel^2\theta_0^2}{2\gamma m} \, . \tag{6}$$

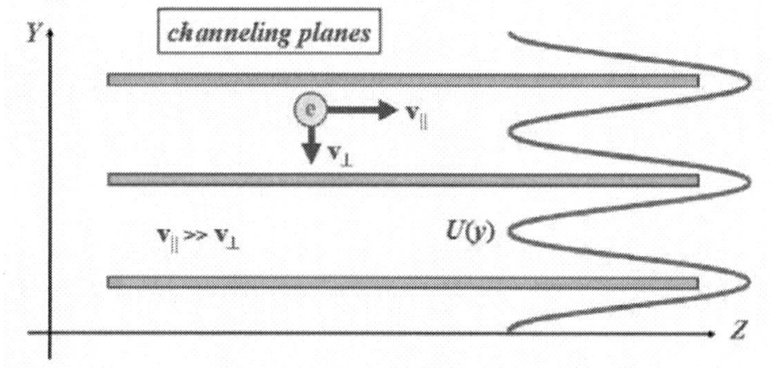

Figure 1. Motion model of planar channelled classical relativistic charged particle in a crystal.

4. Eigensystem problem: a theoretical background

The wave function $\Psi(\boldsymbol{r},t)$ of the planar channeled electron with an energy E > 1 MeV may be written as [2]

$$\Psi(\boldsymbol{r},t) = \exp(-\mathrm{i}(p_{||}z - Et)/\hbar)\phi(y) \ . \tag{7}$$

Here, Y–axis is perpendicular to the channeling plane, $p_{||}$ is the longitudinal momentum (along the channeling plane and Z–axis – see Fig.1). The wave function (7) is the solution of the Klein–Gordon equation with potential function $U(y)$, which transforms into Schrödinger equation with relativistic mass γm_e when $2E_{||} >> (E_{\perp} - U(y))$ [2]

$$2m_e\gamma(E_{\perp} - U(y))\phi(y) = -\hbar^2\partial_y^2\phi(y) \,. \tag{8}$$

Here, γ is relativistic factor, $E_{||} = \sqrt{(cp_{||})^2 + m_e^2c^4}$ and $E_{\perp}$ is the energy of electron transverse motion. In the periodic potential function $U(y)$ wave function $\phi(y)$ and energy of transverse motion of electron $E_{\perp}$ are continuous functions of Y–component electron wave vector k_y, which lies in the first Brillouin zone (e.g., see[3]). Therefore, $\phi(y) \to \phi^i(y,k_y)$, $E_{\perp} \to E_{\perp}^i(k_y)$.

The standard method of solving the Schrödinger equation (5) in accordance with the Bloch form of wave function of electron $\phi^i(y,k_y)$ in the periodic potential function $U(y)$ consists in use of the Fourier expansion of wave function $\phi^i(y,k_y)$

$$\phi^i(y,k_y) = \sum_m C_m^i(k_y)\exp\{\mathrm{i}(k_y + m\,|\,\boldsymbol{g}\,|)y\}/d \,, \tag{9}$$

and potential function $U(y)$

$$U(y) = \sum_m U_m \exp\{\mathrm{i}m\,|\,\boldsymbol{g}\,|\,y\} \,, \tag{10}$$

Here, $\boldsymbol{g}$ is the reciprocal lattice vector of channeling planes, U_m are the Fourier components of potential function $U(y)$. Then we substitute equations (9) and (10) into (8) and obtain eigensystem problem [1], [4–5]:

$$\sum_m A_{mn}C_m^i(k_y) = E_{\perp}^i(k_y)C_n^i(k_y), \quad A_{mn} = U_{m-n} + \delta(m,n)(\hbar^2(m\,|\,\boldsymbol{g}\,| + k_y)^2 / 2m_e\gamma) \,, \tag{11}$$

where $\delta(m,n)$ is the Kronecker delta.

5. Calculation of planar channeling potential function

With new computer code BCM 1.0 calculation of planar channeling potential function $U(y)$ and its Fourier components U_m starts with work directory setting and package "Potential.m" attaching (see Fig.2).

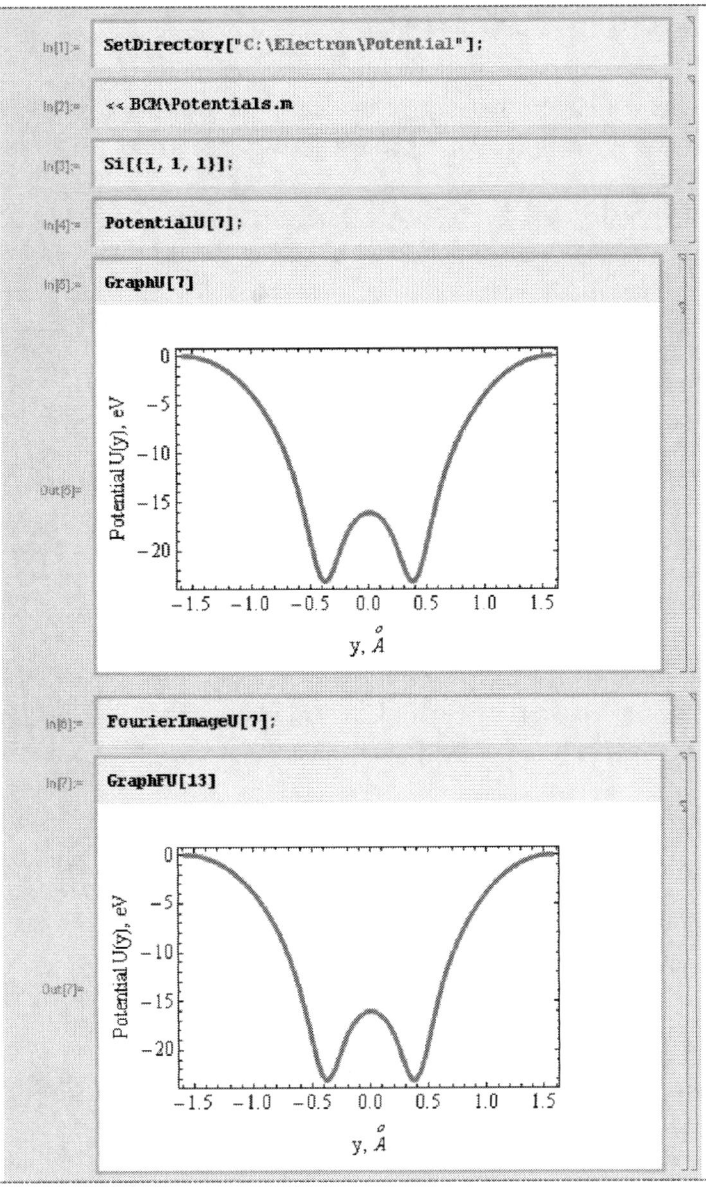

Figure 2. Computation example with package "Potential.m".

The package **"Potential.m"** consist of following subroutines:

- **ElementName[{*Plane*}]** – to choice a crystal (now only Si, LiF and Ge are available) and plane (in form of Miller indices). For example: Si[{1,1,1}] – Fig2.

- **PotentialU[*Np*]** – computation of the periodic planar potential function $U(y)$ (in eV of Å) for the selected crystallographic plane (i.e. (111) Si for above mentioned example). Here, $Np = 2m+1$ (m from Eq. (11)) is the minimal number of single planes for the necessary precision of the total planar potential function $U(y)$ computation. For example: PotentialU[7] – Fig2.
- **GraphU[*Np*]** – to display the plot of function $U(y)$. For example: GraphU[7] – Fig2.
- **FourierImageU[*Np*]** – computation of the Fourier components $U(m)$ of the planar potential function $U(y)$ using formula (5). For example: FourierImageU[7] – Fig2.
- **GraphFU[*NN*]** – to display the plot of function $U(y)$ calculated by his Fourier components $U(m)$. Here, *NN* is number of Fourier components for the plot of the planar potential function $U(y)$. For example: GraphFU[13] – Fig2.

The functions PotentialU[*Np*] and FourierImageU[*Np*] save the results of calculation potential function $U(y)$ and its Fourier components $U(m)$ in special folder with name forming by crystallographic plane. For the selected plane, i.e. (111) Si (for above mentioned example) "Data-Si-111" into directory specified by you – "C:\Electron\Potential" (see in Fig2).

6. Calculation of classical traectory

To solve the differential equation (5) with the help of the software suite Mathematica© we have developed sufficiently simple and high–performance algorithm [7–8].

For solution differential equation (5) using BCM 1.0 the package **"Trajectory.m"** is used. This package consists of subroutines:

- **ElementParticle[{*Plane*}, *m*, *q*]}]** – to choice the particle parameters: mass *m* (in MeV) and charge *q* (in chadge of electron); and the crystal (now only Si, LiF and Ge are available) and plane (in form of Miller indices). For example: ElementParticle[Si[{1,0,0}, 0.511, 1].
- **Period[*Xo*, *Θo*, *Eo*]** –computation the period (in sec.) of transverse motion of charged particle with initial parameters: coordinate *Xo* (in *d*, i.e. in the interplanar spacing), initial angle with respect to the channeling planes *Θo* (in Lindhard critical angle θ_L) and energy *Eo* (in MeV). For example: **Period** [0.25, 0.2, 1000] equal 8.30866×10^{-15} sec. (for above mentioned example).
- **FourierImagePath[*Xo*, *Θo*, *Eo*]** – computation of the Fourier components of $y = y(t)$ is classical trajectory of planar channelled charged particle in crystal with initial parameters *Xo*, *Θo*, *Eo* .
- **GraphPath[*Xo*, *Θo*, *Eo*]** – to display the trajectory plot $y = y(t)$ of channelled particle with initial parameters: *Xo*, *Θo* and *Eo*.

This package uses the period planar potential function $U(y)$ calculated by "Potentials.m".

7. Solution of eigensystem problem

The experience of solving the problem (11) with the help of the software suite Mathematica© allowed us develop sufficiently simple and high–performance algorithm [7–8].

For solution of eigensystem problem (11) in BCM 1.0 there is the package **"Eigen.m":**

- **EigenSystem["*Name*", {*Plane*}, *NN*, *Np*]** – computation the Fourier components $C_m^i(k_y)$ of the wave functions $\phi^i(y,k_y)$ for i^{th} quantum state and eigenvalues, i.e. energies $E_\perp^i(k_y)$ (in eV) of channeled electrons occupying these i^{th} quantum state. Here, *NN* is the number of Fourier components $C_m^i(k_y)$ and *Np* is the minimal number of single planes for the necessary precision of the total planar potential function $U(y)$ computation.

For example: EigenSystem["Si", {111}, 20, 7] – Fig.3.

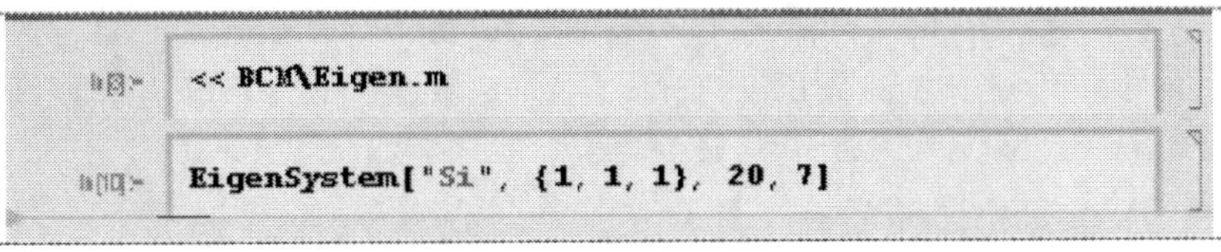

Figure 3. Computation example with package "Eigen.m".

This package uses the Fourier component $U(m)$ of planar potential function $U(y)$ calculated earlier with "Potentials.m". This package saves the results of calculation: Fourier components $C^i_m(k_y)$, energies $E^i_\perp(k_y)$ and matrix A_{mn} from Eq.(11) in special folder with name "alm" into directory specified by you. For example in Fig2. – "C:\Electron\Potential\Data-Si-111\".

8. Calculation of dipole transition matrix elements: package "Matrix.m"

The planar channeling of relativistic particles is accompanied by spontaneous radiation due to transitions between the energy levels $E^i_\perp(k_y)$ and $E^f_\perp(k_y)$ of the transverse motion. For calculation of channeling radiation probability in dipole approximation, the transition matrix elements are required

$$y_{if}(k^i_y,k^f_y)=\langle\phi^f(y,k^f_y)|y|\phi^i(y,k^i_y)\rangle. \tag{12}$$

In packages of BCM-01, by performing calculations of initial population and transition matrix elements, we divide the first Brillouin zone into $N=10$ equal parts.

Then we denote $y_{if}(k^i_y,k^f_y)\to$ yif[*i*, *ni*, *f*, *nf*] where *i* and *f* are the quantum state numbers of electron transverse motion at planar channeling, *ni* and *nf* are the numbers of nth part of the ith and f^{th} quantum states.

By performing calculation using the package **"Matrix.m":**

- **MatrixElement[*Name, gamma*]** – to calculate matrix elements yif[*i*, *ni*, *f*, *nf*] in dipole approximation for electrons with $\gamma=$ *gamma* channeled along the crystallographic planes *Name*, and for transition between the quantum states $i_{ni}\to f_{nf}$.

For example: MatrixElement[(111)LiF, 20] – calculation of dipole transition matrix element at planar channeling electron with γ=20 along (111)LiF. Then, yif[0, 5, 3, 10] is the matrix element for transition $3_{10}\to 0_5$ at planar channeling electron with γ=20 for (111)LiF.

9. Calculation of transverse quantum states initial populations: package "Population.m"

For calculation of channeling radiation from relativistic electron beam entering a crystal, it is necessary to know initial populations of transverse energy levels. It is the probability that a free electron entering at a small angle with respect to the crystallographic planes may be captured to one of the possible transverse energy levels

$$P_i(k^i_y,k_o)=\left|\langle\phi^o(y,k_o)\|\phi^i(y,k^i_y)\rangle\right|^2/d, \tag{13}$$

where $\phi^o(y,k_o)=\exp(-ik_oy)$ is the wave function for transverse motion of incident free electron with wave vector $\boldsymbol{k}=(k_o,k_\parallel)$, . The wave vector $\boldsymbol{k}$ components of free electron are defined as: $k_\parallel=p_\parallel/\hbar$, $k_o=p_\perp(0)/\hbar\cong p_\parallel\theta_o/\hbar$.

In the package "Population.m" we denote $P_i(k_y^i, k_o) \rightarrow$ pP[*i*, *ni*, *nk*], where *i* is the quantum state numbers of electron transverse motion at planar channeling, *ni* is the number of nth part of the ith quantum state and *nk* is the number of nth part of the θ_L – Lindhard critical channeling angle.

We divide the Lindhard critical angle θ_L into $N_L = 15$ equal parts. The package calculate initial populations till *nk* = 30, that is till initial angle $\theta_0 = 2\theta_L$.

By performing calculation using the package **"Population.m":**

- **InitialPopulation[*Name*, *gamma*]** – to calculate initial populations pP[*i*, *ni*, *nk*] of quantum states (with numbers i_{ni}) by free electrons with γ = *gamma* entering a crystal at angle $\theta_o = \theta_L nk/30$ with respect to the channeling planes *Name*.

For example: InitialPopulation[(111)LiF, 20] – calculation of initial population at planar channeling of electrons with γ=20 along (111)LiF. Then, pP[3, 10, 5] is initial population of quantum state 3_{10} by free electrons entering a crystal at angle $\theta_o = 5\theta_L/30$ with respect to the channeling planes (111)LiF.

10. Comments for users

The packages "Matrix.m" and "Population.m" use the matrix A_{mn} calculated earlier with package "Eigen.m".

Before using the packages "Potentials.m", "Trajectory.m", "Eigen.m", "Matrix.m" and "Population.m" one must put these packages in special folder of Mathematica© which is reserved for external packages. For example, put all packages in folder "BCM" and put this folder in directory "C:\Program Files\Wolfram Research\Mathematica\6.0\AddOns\Packages". Then you can use computer code BCM 1.0 as in Figs.2,3.

11. Conclusions

The 1st version of new computer code BCM–1.0 is presented. To start calculations, one needs the Mathematica© 6.0 version, necessary packages from [9] and standard personal computer (e.g. Intel(R) Core(TM)2 CPU 2 GHz, 2 Gb of RAM).

The beta–version of BCM–1.0 is available for testing [9].

The future developments will include:

- Calculation of planar CR spectra (classical).
- Calculation of planar CR spectra (quantum).
- Calculation planar DCR and PXRC.
- Calculation of axial CR spectra (classical).
- Calculation of axial CR spectra (quantum).
- Calculation of axial DCR.

Then we plan to include inelastic processes, e.g. multiple scattering, energy loss, dechanneling etc.

The work is partially supported by Russian Science and Innovations Federal Agency under contract No 02.740.11.0238

12. Acknowledgements

The authors are grateful to Prof. Yu.P.Kunashenko for valuable recommendations and testing of beta–version.

References

[1] Doyle P A and Turner P S 1967 *Acta Cryst.* A24, 390

[2] Bazylev V A, Zhevago N K 1987 *Radiation from fast particles in a matter and external fields* (Moscow: Nauka)

[3] Baryshevsky V G, Feranchuk I D, Ulyanenkov A P 2005 *Parametric X–ray Radiation in Crystals* (Berlin: Springer–Verlag)

[4] Ashcroft N W, Mermin N D 1979 *Solid State Physics* (New York: Holt, Rinehart and Winston)

[5] Buxtont B F and Loveluck J E 1977 *J. Phys. C: Solid State Phys.* **10** 3941

[6] Klein R K, Kephart J O, Pantell R H, Park H, Berman B L, Swent R L, Datz S and Fearick R W 1985 *Phys.Rev.*B **31** 68

[7] Bogdanov O V, Korotchenko K B, Pivovarov Yu L 2007 *JETP Letters* **85** No. 11 555

[8] Korotchenko K B, Pivovarov Yu L Tukhfatullin T A 2008 *Nucl. Inst. & Meth. in Phys. Res.* **B266** 3753

[9] BCM–1.0 betta–version: http://portal.tpu.ru/

Mechanisms of high energy charged particles beams deflection by a bent crystal

N F Shul'ga[1], V I Truten'[1] and I V Kirillin[2]

[1]Akhiezer Institute for Theoretical Physics of NSC KIPT, Kharkov, Ukraine

[2]Karazin Kharkov National University, Kharkov, Ukraine

E-mail: shulga@kipt.kharkov.ua

Annotation. From the single point of view we have considered various mechanisms of high energy charged particles beams bending by a bent crystal. These mechanisms are based on multiple scattering of fast particles on bent atomic strings (stochastic mechanism), planar channelling and reflection of particles from crystal atomic planes (the effect of volume reflection). The results of computer simulation which allow to make a comparison between these beams bending mechanisms are presented. The consideration is performed for both positively and negatively charged particles. It is shown that stochastic mechanism is the most effective for negatively charged particles beam bending. The results of simulation under conditions that have been used in last CERN experiments on observation of stochastic mechanism of positively and negatively charged particles beams bending are presented.

1. Introduction

When fast charged particles are moving in a crystal under a small angle to one of its axes or planes correlations between consecutive collisions of particles with lattice atoms appear. As a result of these correlations the particle motion in a crystal is basically defined by continuous potential of atomic strings or atomic crystal planes near to which particle is moving [1-3]. Strong intracrystalline fields allow to make on small distances an essential impact on beam particles characteristics, leading to considerable effects in scattering and radiation. Special interest represents high energy particles passage through the bent crystal because in this case an effective beam bending by a small sized crystal is possible. There are several mechanisms of high energy charged particles bending by a bent crystal, connected with finite motion (channeling) and infinite (above barrier) motion in relation to the bent atomic strings or crystal atomic planes [4-6]. Method of computer simulation of particles motion in bent crystal plays an important role at the research of these beams bending mechanisms (see for example [7-10] and references therein).

In the present work the method allowing to consider various mechanisms of high energy charged particles beams bending by a bent crystal from a single point of view is suggested. This method is based on the development of the simulation method of particles motion in the field of the bent crystal atomic strings offered in [10,11]. We have employed this developed method to a case of particles motion near crystal atomic planes. Thus crystal atomic planes are considered as the planes formed by crystal atomic strings, located periodically in a plane. On the basis of this method it is possible to consider from one point of view not only mechanisms of beams bending by a bent crystal, connected with particles plane channeling, volume reflection and their multiple scattering on crystal atomic strings, but also to investigate intermediate cases.

2. The motion of high energy charged particles in bent crystal

Fast charged particle motion in a crystal near one of crystal axes is basically defined by continuous potential of crystal atomic string located parallel to this axis. In such field the particle can commit both finite (channeling), and infinite (above barrier motion in a perpendicular plane) in relation to atomic strings. Character of particle motion in a periodic field of atomic strings is rather complicated and it essentially depends on conditions of particle entrance into a crystal. Thus depending on entry conditions the particle can execute both regular (plane channeling in a periodic field of atomic strings) and chaotic (the phenomenon of dynamic chaos [12]) motion relatively to the atomic strings. Due to incoherent effects in scattering on thermal fluctuations of lattice atoms and on crystal electronic subsystem mutual transitions between various regimes of movement are possible. Such transitions are possible also due to curvature of crystal atomic strings. Therefore the important role at the research of fast particles motion in a bent crystal plays a method of numerical simulation.

The simulation in this article was carried out on the base of classical physics because the inequality $N_c Ze^2 / \hbar c >> 1$ in considerable case is fulfilled [3]. Here $N_c \propto R / \psi a$ is a number of string atoms, with which particle effectively interacts passing the string, Z is the atomic nucleus charge, a is the lattice constant, R is the atomic potential screening radius and ψ is an angle between particle momentum and atomic string axes. The simulation consisted in numerical solving of two-dimensional motion equation

$$\dot{\mathbf{v}}_\perp(t) = -\frac{c^2}{\varepsilon}\frac{\partial}{\partial \rho}U(\rho), \qquad (1)$$

which describes particle motion in periodic crystal atomic strings potential energy in orthogonal to the strings axis plane. In equation (1) m is a particle mass, $\mathbf{v}_\perp(t)$ is its orthogonal velocity as a function of time, $U(\rho)$ is continuous potential energy of atomic strings [1]. At equation (1) numerical solution the particle trajectory within elementary cell was partitioned in many path sections and in each of them the crystal was considered as straight. In the end of each path section the change in particle velocity generated by crystal bending and incoherent scattering on thermal fluctuations of lattice atoms and crystal electronic subsystem was taken into account.

3. Simulation of fast charged particles beams passage through a bent crystals at various entry conditions

The character of fast charged particles motion in a crystal essentially depends on conditions of particle entrance in a crystal. Thus depending on an incident beam orientation with respect to the crystal axes and planes various mechanisms of the charged particles beams bending by a bent crystal can be realized.

To show this we will present results of simulation (see figures 1-3) of passages of positively (b) and negative (c) charged particles with 400 GeV energy through the bent silicon crystal with thickness of 1 cm at various entry conditions of an input of particles in a crystal with respect to the $<111>$ axis. Simulation was performed for a 1000 particles beam which divergence was $\psi_c / 2$. The crystal bend was performed by z axis (which is parallel to the $<111>$ axis) bending in the direction of the x axis (which is orthogonal to the $(1\bar{1}0)$ planes). The bend radius was 40 m. Broad spots in figures 1a, 2a, and 3a correspond to various entry conditions of the beam input in the crystal. The size of these spots corresponds to divergence of incident beam particles. The centers of these spots correspond to incidence of the beam along the $<111>$ axis (figure 1a), crystal disorientation on angles

$\theta_x = 0, \theta_y = 10\psi_c$ (figure 2a) and disorientation on $\theta_x = 4\psi_c, \theta_y = 10\psi_c$ (figure 3a). The point on crossing of lines in the right part of the mentioned figures corresponds to the final direction of the <111> crystal axis. Points are the results of simulation.

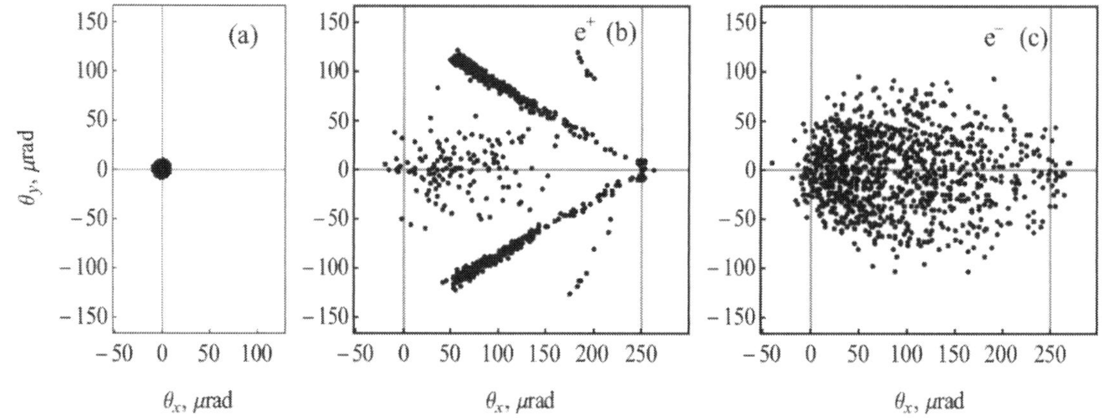

Figure 1. Angular distribution of 400 GeV charged particles before (a) and after (b, c) passing 1 cm Si bent crystal for the case of stochastic mechanism of beam deflection by a bent crystal.

The results of simulation that are presented in figure 1 illustrate the stochastic mechanism of the beam deflection by a bent crystal, connected with multiple scattering of particles on the bent atomic strings. Due to this mechanism practically all beam particles deflect towards the bend of the crystal axis on the angles considerably exceeding the critical axial channeling angle. Thus for positively charged particles the considerable part of beam particles due to multiple scattering on the bent atomic strings is captured in plane channels and then follows the bend of these planes. For negatively charged particles such mechanism of particle capture in plane channels in the considered case is not realized. At the same time, as figure 1c shows, the multiple scattering on the bent atomic strings leads to considerable beam deflection comparing to its initial position.

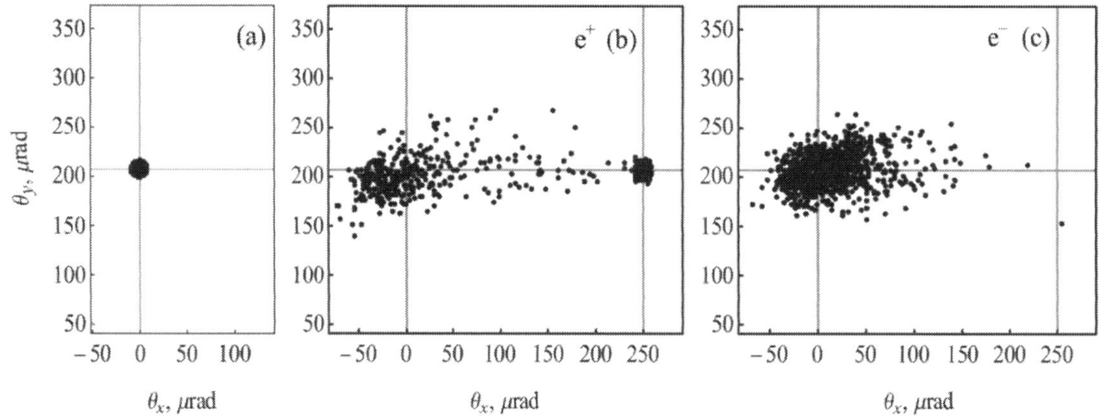

Figure 2. Angular distribution of 400 GeV charged particles before (a) and after (b, c) passing 1 cm Si bent crystal in planar channeling conditions.

The results of simulation with initial conditions that correspond to possibility of particles capture in plane channels at the beam input in the crystal are presented in figure 2. Thus the considerable part of positively charged particles captured in the channel follows the bend of crystal planes and bends on the crystal bend angle. The small exit of such particles from the bent plane channel is caused by incoherent scattering on an electronic subsystem of a crystal and on atomic strings, forming crystal planes. For negatively charged particles there is no beam deflection in the considered case. Primarily it is connected with intensive multiple scattering of such particles on thermal fluctuations of lattice atoms that leads to significant instabilities in particles motion in plane channels.

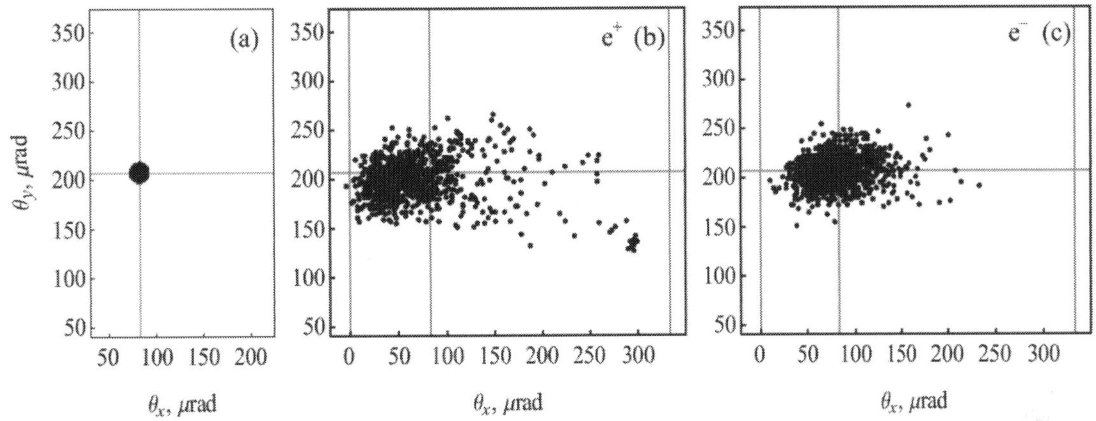

Figure 3. Angular distribution of 400 GeV charged particles before (a) and after (b, c) passing 1 cm Si bent crystal for the case of volume reflection mechanism of beam deflection by the bent crystal.

The results of simulation presented in figure 3 correspond to above barrier particles motion at the beam input in the crystal concerning $(1\bar{1}0)$ planes under the angle exceeding the critical angle of plane channeling θ_c ($\theta_x \approx 8\theta_c$). For positively charged particles the effect of their reflection from the bent crystal planes (see figure 3b) in this case takes place. Thus deflection of the major part of beam particles occurs on angles on the order of θ_c in the direction opposite to the crystal planes bent direction. The similar effect takes place and for negatively charged particles. However, it is expressed much more weakly (see figure 3c), than for positively charged particles.

The possibility of stochastic mechanism of high energy positively and negatively charged particles beam deflection by a bent crystal was recently confirmed experimentally at CERN accelerator [13,14]. The results of computer simulation under these experimental conditions are represented in figure 4. These results of computer simulation are in good agreement with experimental data.

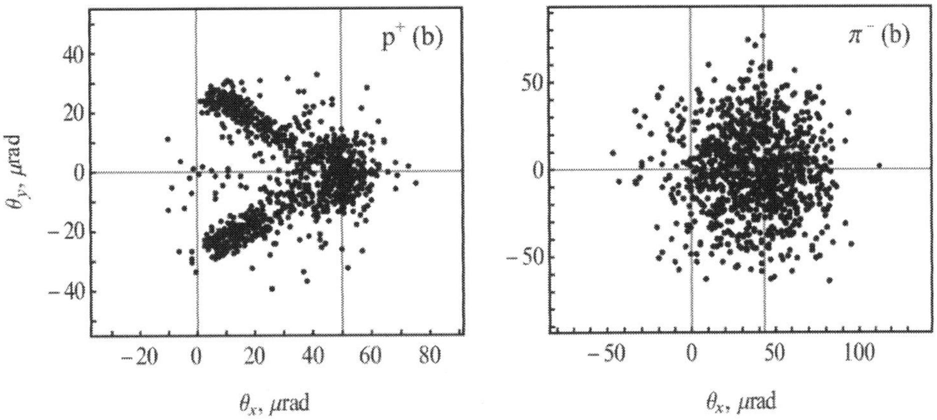

Figure 4. Angular distribution of 400 GeV protons after passing 2 mm Si bent crystal with 40 m curvature radius (a) and 150 GeV π^--mesons after passing 1.172 mm Si bent crystal with 40 m curvature radius (b).

The results of simulation presented above and the results of simulation carried out at other conditions (different length of crystals, different crystal axes and planes orientations concerning to the incident beam etc.) show that for positively charged particles mechanism of planar channeling is the most effective for beam deflection in a bent crystal. Due to this mechanism deflection on angles much greater than θ_c are possible. Due to multiple scattering by bent atomic strings splitting of positively charged particles beams is possible (see figure 1b). After all, when positively charged particles beam reflection on bent atomic planes conditions are fulfilled the deflection is possible only on angles on the order of θ_c.

For negatively charged particles the most effective mechanism of beam deflection is stochastic mechanism (see figure 1c). Due to this mechanism almost all beam particles are deflected on angles on the order of several critical angles of axial channeling ψ_c. Mechanisms of planar channeling and reflection on bent atomic planes leads to a weak beam deflection.

References

[1] Lindhard J 1965 *Mat. Fys. Medd. Dan. Vidensk. Selsk.* **34** 14
[2] Gemmell D S 1974 *Rev. of Mod. Phys.* **46** 129
[3] Akhiezer A I and Shul'ga N F 1996 *High Ehergy Electrodynamics in Matter* (Amsterdam: Gordon and Breach) p 143
[4] Tsyganov E N, 1976 *Preprint* Fermilab TM-682, TM-684
[5] Taratin A M and Vorobiev 1987 *Phys. Lett.* A **119** 425
[6] Greenenko A A and Shul'ga N F 1991 *JETP Lett.* **54** 524
[7] Bak J F, Jenesen P R *et al.* 1984 *Nucl. Phys.* B **242** 1
[8] Biryukov V M, Chesnokov Yu A and Kotov V I 1997 *Crystal Channeling and its Application at High-Energy Accelerators* (Berlin: Springer) p 60
[9] Taratin A M 1998 *Phys. Part. Nucl.* **29**, 437
[10] Akhiezer A I, Shul'ga N F, Truten' V I, Greenenko A A and Syshchenko V V 1995 *Physics-Uspekhi* **38** 1119
[11] Greenenko A A and Shul'ga N F 2002 *Nucl. Instr. Meth.* B **193** p 133
[12] Akhiezer A I, Truten' V I and Shul'ga N F 1991 *Phys. Rep.* **203**, 289
[13] Scandale W, Vomiero A *et al.* 2008 *Phys. Rev. Lett.* **101** 164801
[14] Scandale W, Vomiero A *et al.* 2009 *Phys. Lett.* B **680** 301

Electron acceleration in a helical waveguide

X. Artru, C. Ray

Institut de Physique Nucléaire de Lyon, Université de Lyon,
Université Lyon 1 and CNRS/IN2P3, F-69622 Villeurbanne, France

E-mail: x.artru@ipnl.in2p3.fr, c.ray@ipnl.in2p3.fr

Abstract. Travelling wave in a helical wave guide is considered for electron acceleration. A first determination of the travelling wave modes using a partial wave expansion (PWE) and a point matching method is presented. It gives a rapid solution for moderate deformation of the guide relative to a straight cylinder. Strong deformations will give higher accelerating gradient but the PWE diverges. Two methods overcoming this difficulties are suggested.

1. Introduction

An electromagnetic wave in a straight wave guide cannot accelerate particles because its phase velocity $v_{\rm ph}$ is larger than c. In conventional linear accelerators, $v_{\rm ph}$ is reduced by a chain of resonant cavities along the guide. A scheme not using a resonance phenomenon is the travelling wave tube (TWT). One type of TWT is a cylindrical tube with a conducting helical wire inside [1, 2]. Another type is a tube with a thick dielectric wall [3].

In this paper we study another kind of helical device : the helically bent tube. The main "photon path" is the helix passing trough the centres of the local cross sections of the tube. For a strong enough bending, the difference in length between this path and a straight path is such that the effective $v_{\rm ph}$ is smaller than c, although the local $v_{\rm ph}$ is larger than c.

One may expect several advantages in using helically bent tube in place of accelerating cavities : easy construction, easy cooling, easy maintainance of the wall surface using an helical arm, large frequency spacing of the modes, large group velocity which allow a fast filling of the guide by the wave. Alternatively, the wave can be generated by a driving beam. Like with the dielectric wall in [4], several modes can cooperate to transfer the energy from one beam to the other.

The helical wave guide is defined by its axis Oz, its period $\Lambda = 2\pi/q$ and the *base curve*, which is the section of the wall with the $z = 0$ plane, of parametric equation $\mathbf{r}_b(s) = (x_b(s), y_b(s))$. The point $(x, y) = (0, 0)$ is supposed to be inside the base. The wall surface is generated by helices H_s of parametric equation

$$(x, y, z) = (\mathcal{R}[qz]\mathbf{r}_b(s),\ z) \qquad (1)$$

where the 2×2 matrix $\mathcal{R}[qz]$ describes the rotation of angle qz about $0z$. One method to make a helical waveguide is to draw helical grooves on the internal face of an initially cylindral tube. The case of rectangular grooves has been studied in [5, 6]. The corresponding wall base curves have indentations. Here we will restrict to base curves with smooth curvatures.

This work is the continuation of a first study [7], where the electromagnetic wave modes were calculated using a partial wave expansion (PWE) associated to a point matching (PM) method.

Since that time, we realized that PWE is limited to weak enough helical bending and sufficiently "round" base curves ; if it is not the case, the PWE converges only inside a cylinder which does not contain the whole guide.

In this paper we will recall the PWE-PM method, some of our old results, present some new ones and discuss the limitations of this method. Two alternative methods will be evoked.

2. Propagating modes in the scalar field case

We first present the simpler case of a scalar field $\Phi(t,x,y,z)$, which obeys the Klein-Gordon equation (in units $c=1$)

$$(-\partial_t^2+\partial_x^2+\partial_y^2+\partial_z^2)\,\Phi(t,x,y,z)=0\,, \tag{2}$$

with the boundary condition $\Phi(t,x,y,z)=0$ on the wall. These two equations are invariant under a time translation and under a *helical displacement*

$$\{\text{translation by } \Delta z\}\times\{\text{rotation by } q\Delta z\}\,. \tag{3}$$

In the transverse plane we introduce a *"rotating frame"*, the coordinates (X,Y) of which are related to the fixed frame ones by $x+iy=e^{iqz}\,(X+iY)$. On can therefore look for a solution of the form

$$\Phi(t,x,y,z)=e^{iPz-i\omega t}\,\Psi(X,Y) \tag{4}$$

which has definite reduced frequencies ω and definite eigenvalue P of the operator

$$\mathcal{P}_{\text{helic}}=p_z+q\mathcal{J}_z=-i\partial/\partial_z-iq\partial/\partial_\varphi\,. \tag{5}$$

2.1. The partial wave expansion (scalar field case)

Let us temporarily assume that a mode can be expanded in free cylindrical partial waves which are eigenstates of $\mathcal{J}_z$ and p_z with eigenvalues l and p respectively. A single partial wave writes

$$\Phi_{\omega,p,l}(t,x,y,z)=e^{-i\omega t}\,e^{ipz}\,e^{il\varphi}\,\psi_{l,k}(r)\,, \tag{6}$$

where (r,φ,z) are the cylindrical coordinates and

$$\psi_{l,k}(r)=J_{|l|}(kr)/k^{|l|}=I_{|l|}(\kappa r)/\kappa^{|l|}\,. \tag{7}$$

J_l and I_l are the regular Bessel functions of first and second kind, $k=(\omega^2-p^2+i0)^{1/2}$ is the radial momentum and $\kappa\equiv -ik=(p^2-\omega^2-i0)^{1/2}$. The two expressions of (7) are equivalent. If one whishes to work with real quantities, one takes the first one for $|p|\le|\omega|$, the second one for $|p|>|\omega|$. The denominator $k^{|l|}$ is introduced to make $\psi_{l,k}(r)$ finite at $k\to 0$. The partial wave expansion reads

$$\Phi(t,x,y,z)=e^{-i\omega t}\sum_{l=-\infty}^{+\infty}a_l\,e^{ip(l)z+il\varphi}\,\psi_{l,k(l)}(r)\,, \tag{8}$$

with

$$p(l) \text{ or } p_l=P-lq\,,\quad k(l) \text{ or } k_l=\left(\omega^2-p_l^2+i0\right)^{1/2}\,, \tag{9}$$

so that each partial wave is an eigenstate of $\mathcal{P}_{\text{helic}}$ with eigenvalue P.

It is sufficient to write the boundary condition at $t=z=0$, thanks to the invariances under time translation and helical displacement. For every point $\mathbf{r}_b(s)$, of polar coordinates $(r_b(s),\varphi_b(s))$, of the base curve one has the constraint

$$\sum a_l\,\exp[il\varphi_b(s)]\,\psi_{l,k}[r_b(s)]=0\,. \tag{10}$$

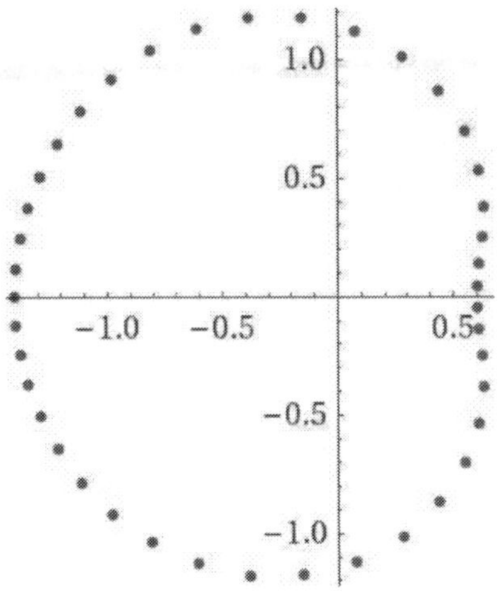

Figure 1. Matching points on the base curve.

2.2. The point matching method in the scalar field case

One keeps only a finite set $\mathcal{L}$ of partial waves and selects an equal number of matching points $\mathbf{r}_1, ...\mathbf{r}_N$ of the base curve (see Fig.2.2). For a base without discrete rotational symmetry, $\mathcal{L} = [l_{\rm inf}, l_{\rm sup}]$ with $l_{\rm sup} = l_{\rm inf} + N - 1$. The choice of $\mathbf{r}_n$'s and $\mathcal{L}$ should be guided by the following conditions :

- the $\mathbf{r}_n$'s should give a precise enough definition of the base curve,
- $\mathcal{L}$ should contain l's of both signs, however with more l's of the same sign as q.
- $l\,(\varphi_{n+1} - \varphi_n)$ for $l \in \mathcal{L}$ should not be too large compared to unity,
- $k_l\,(r_{n+1} - r_n)$ for $l \in \mathcal{L}$ should not be too large compared to unity.

The a_l are approximatively determined by the system of linear equations

$$\sum_{l\in\mathcal{L}} a_l\, e^{il\varphi_n}\, \psi_{l,k}(r_n) = 0\,, \quad 1 \le n \le N\,, \tag{11}$$

(r_n, φ_n) being the polar coodinates of $\mathbf{r}_n$. A nonzero solution exists if the determinant est zero:

$$\det\{C_{nl}\} = 0\,, \text{ where } C_{nl} = e^{il\varphi_n}\, \psi_{l,k}(r_n)\,. \tag{12}$$

This equation, which relates ω and P, is the dispersion relation in the PWE-PM approximation. One may impose ω and look for the roots of (12) as a function of P, or vice-versa. Alternatively, in view of accelerating highly relativistic electrons one may impose $\omega/P \equiv v_{\rm ph} = v_e \simeq 1$ and look for the roots of (12) as a function of ω. Using a graphical software, the roots are easily seen as dips in the curve of $\ln|\det\{C_{nl}\}|$ versus the free variable. If $(\det\{C_{nl}\})$ is $e^{i\alpha}$ times a real number, the roots are precisely located as jumps of the curve $\arg\left(ie^{-i\alpha}\det\{C_{nl}\}\right)$ from $\pm\pi/2$ to $\mp\pi/2$.

Provided that the partial wave expansion converges, the quality of the approximation is improved by increasing N. The convergence is obtained first for the lowest modes (which have smallest ω at fixed P or $v_{\rm ph}$, or largest P at fixed ω), then for higher and higher modes. A first criterium is the stability of the root when N is increased by one or two units, when $\mathcal{L}$ is translated by one or two units or when one moves the matching points. The ultimate criterium is the stability of the coefficients a_l, at least for l not close to $l_{\rm inf}$ or $l_{\rm sup}$.

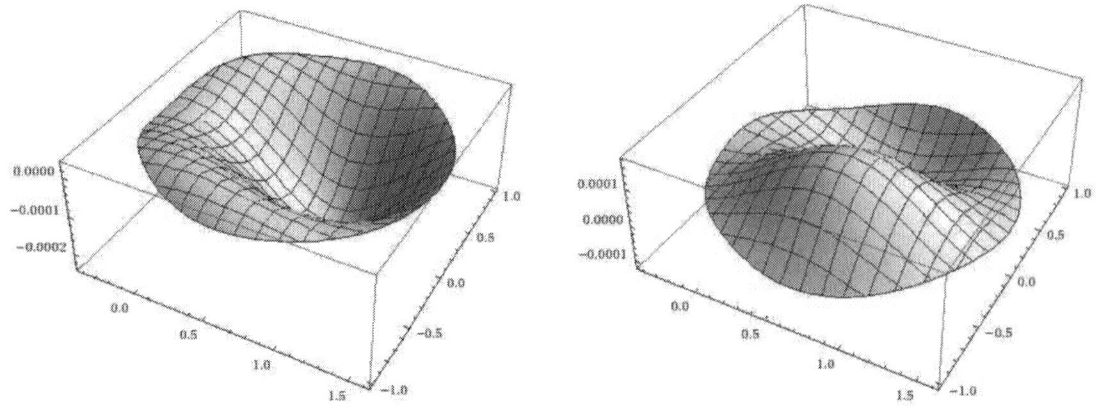

Figure 2. 3-dimensional representation of the scalar field $\Psi(x, y) = \Phi(0, x, y, 0)$ of the $\omega a = 5.0$ mode, for a circular base of radius a, off-centred by $\varepsilon = 0.4$ from the axis, and $q = 1$. Left: real part; right: imaginary part.

Case of a symmetric base. If the base is symmetrical about the x-axis, one can choose the $\{\mathbf{r}_n\}$ to be symmetrical by pairs, except for those which are on the x-axis. Then $i^\nu \det\{C_{nl}\}$ is real, ν being the number of pairs. The a_l can be taken all real, which gives

$$\Phi(0, x, -y, 0) = \Phi^*(0, x, y, 0)\,. \tag{13}$$

If the base is symmetrical about Oz, the modes are either even or odd under the parity symmetry $(x, y) \to (-x, -y)$, therefore $\mathcal{L}$ has to contain either even or odd l's. More generally, if the base has a rotational symmetry of order S, the l's should be equal modulo S. To avoid redundant equations in (12), the points $\mathbf{r}_n$ should be all different modulo the symmetry. It suffices to choose them all on one sub-period of the base curve.

Numerical exemple with the scalar field. The simplest base curve is a circle or radius a which is de-centered by εa with respect to the z-axis. We took $\varepsilon = 0.4$, $q = 1$ and imposed the phase velocity $v_{\text{ph}} = 0.99999$. The lowest modes are at $\omega a = 5.0$ and 7.0. They are stable at the 0.1 precision when $\mathcal{L}$ contains at least the set $[-1, +5]$. Fig.2 displays the cross section $\Psi(X, Y)$ of the field at $z = t = 0$, for $\omega a = 5.0$ mode. Note the symmetry (13). The maximum of $|\mathrm{Re}\,\Psi|$ is not at the centre of the base, but farther from the helix axis. This is a manifestation of the centrifugal effect.

3. Propagating modes of the Maxwell field

We gather the fields $\mathbf{E}$ and $\mathbf{B}$ in a 6-component vector $\mathbf{F} = \{\mathbf{E}, \mathbf{B}\}$. In analogy with (4) we look for solutions of the form

$$\mathbf{F}(t, x, y, z) = e^{iPz - i\omega t}\,\mathcal{R}[qz]\,\mathcal{F}(X, Y) \tag{14}$$

with $\mathcal{F}(X, Y) = \{\mathcal{E}(X, Y), \mathcal{B}(X, Y)\}$. $\mathcal{R}[qz]$ is the rotation matrix of (1) applied to the vectors $\mathcal{E}$ and $\mathcal{B}$. The physical field is the real part of $\mathbf{F}$.

Boundary conditions. We introduce the vectors

$$\vec{\sigma}(s) = \mathbf{r}'_b(s) = \begin{pmatrix} x'_b(s) \\ y'_b(s) \\ 0 \end{pmatrix}, \quad \vec{\tau}(s) = \hat{\mathbf{z}} + q\,\hat{\mathbf{z}} \times \mathbf{r}_b(s) = \begin{pmatrix} -q y_b(s) \\ q x_b(s) \\ 1 \end{pmatrix}, \tag{15}$$

which are tangent to the wall at $z = 0$, and the normal vector $\vec{\nu}(s) = \vec{\sigma}(s) \times \vec{\tau}(s)$. The boundary conditions are

$$\begin{cases} \vec{\mathcal{E}} \cdot \vec{\sigma} = 0 \\ \vec{\mathcal{E}} \cdot \vec{\tau} = 0 \\ \vec{\mathcal{B}} \cdot \vec{\nu} = 0 \end{cases} \quad \longrightarrow \quad \begin{cases} \mathcal{E}_r \, \sigma_r + \mathcal{E}_\varphi \, \sigma_\varphi = 0 \\ qr \, \mathcal{E}_\varphi + \mathcal{E}_z = 0 \\ \mathcal{B}_r \, \sigma_\varphi + (qr\mathcal{B}_z - \mathcal{B}_\varphi)\sigma_r = 0 \end{cases} \tag{16}$$

The right set is written in cylindrical coordinates. These 3 conditions are not independent. Only two suffice.

Partial wave expansion. **F** can be expanded in free partial waves like Φ in (8). However, since the photon has two polarisation states, there are two waves for each l, which now denotes the total (spin + orbital) angular momentum. We take the states of right and left *chirality*, or *photon helicity* $\chi = \pm 1$. The analogue of $\psi_{l,k}(r)$ in (6,8) is $\mathcal{F}_{\omega,p,l,\chi}(r) = \left\{ \vec{\mathcal{E}}_{\omega,p,l,\chi}(r), \vec{\mathcal{B}}_{\omega,p,l,\chi}(r) \right\}$. For $l = 0$, the cylindrical components are

$$\vec{\mathcal{E}}_{\omega,p,0,\chi}(r) = \begin{pmatrix} \mathcal{E}_r \\ \mathcal{E}_\varphi \\ \mathcal{E}_z \end{pmatrix}_{\omega,p,0,\chi} = \begin{pmatrix} -ip\,\chi\,\psi_1(r) \\ \omega\,\psi_1(r) \\ \chi\,\psi_0(r) \end{pmatrix}, \qquad \vec{\mathcal{B}}_{\omega,p,0,\chi}(r) = -i\chi\,\vec{\mathcal{E}}_{\omega,p,0,\chi}(r)\,. \tag{17}$$

For $l \neq 0$,

$$\vec{\mathcal{E}}_{\omega,p,l,\chi} = \begin{pmatrix} -i\left[\psi_{\lambda-1} + (p - \chi'\omega)^2\,\psi_{\lambda+1}\right] \\ s_l\,\left[\psi_{\lambda-1} - (p - \chi'\omega)^2\,\psi_{\lambda+1}\right] \\ 2(p - \chi'\omega)\,\psi_\lambda \end{pmatrix}, \qquad \vec{\mathcal{B}}_{\omega,p,l,\chi} = -i\chi\,\vec{\mathcal{E}}_{\omega,p,l,\chi}\,. \tag{18}$$

Here $s_l = \mathrm{sign(l)}$, $\chi' = s_l\chi$, $\lambda = |l|$ and $\psi_\lambda = J_\lambda(kr)/k^\lambda$. These states have finite and linearly independent limits when $p^2 \to \omega^2$. The analogue of (8) is

$$\mathbf{F}(t, x, y, z) = e^{-i\omega t} \sum_l e^{il\varphi}\, e^{ip(l)z} \left[a_l\, \mathcal{F}_{\omega,p(l),l,\chi=+s_l}(r) + b_l\, \mathcal{F}_{\omega,p(l),l,\chi=-s_l}(r) \right]\,. \tag{19}$$

We made the choice to associate a_l and b_l with positive and negative χ' instead of χ. For $l = 0$ we set $s_l = +1$ in (19), but we can alternatively take the (TM, TE) basis as in [7].

The point matching method in the Maxwell field case. Generalising the PWE-PM method of Section 2.2 we keep only N values of l and write the first two conditions of (16) for N points $\mathbf{r}_n$ of the base curve. It gives $2N$ linear equations linking N coefficients a_l and N coefficients b_l. The $2N \times 2N$ matrix of the system is given in [7]. The dispersion relation is the vanishing of its determinant.

Numerical results for the Maxwell field. Taking the same de-centered circular base curve as in the scalar field case $\varepsilon = 0.4$, but $q = 1$ and imposing $v_{\rm ph} = 0.999$, we obtain the lowest modes at $\omega a = 2.28$, 3.6, 4.8 and 5.0. They remain the same for various $\mathcal{L}$ sets like $[-6, +8]$, $[-5, +9]$ and $[-3, +11]$. The TM/TE character, measured by $|E_z(0)/B_z(0)|$, are 0.3, 0.25 and 0.4. and 5.4 respectively. The $\omega a = 5.0$ mode seems therefore more appropriate for electron acceleration. It has a *figure of merit*

$$f.o.m. = |E_z(0)|/[\max(|\mathbf{E}|) \text{ on the wall}] \simeq 1/8 \tag{20}$$

and a group velocity $v_{\rm g} = d\omega/dP = 0.68$. Plots of the fields of this mode can be seen in [7]. Let us mention that $|\mathcal{B}_r|$ and $|\mathcal{B}_\varphi|$ look very similar to $|\mathcal{E}_\varphi|$ and $|\mathcal{E}_r|$ respectively, which is reminiscent of a plane wave.

Increasing the guide periods $\Lambda = 2\pi/q$ while imposing fixed $v_{\rm ph} < 1$ makes the ω of each mode increase until the mode disappear at $\omega = \infty$. For $q = 0.7$, the two lowest modes are at $\omega a = 2.7$ and 4.2 with $|E_z(0)/B_z(0)| = 0.25$ and 0.26 respectively. The lowest mode has $f.o.m \simeq 1/9$.

With a stronger helical bending of the waveguide (larger ε and q), the mode frequencies decrease, but the convergence of the PWE-PM method becomes questionable. For $\varepsilon = 0.6$, $q = 1.5$ and $v_{\rm ph} = 0.999$, three first modes are seen at $\omega a = 2.2$, 3.3 and 4.0 with $|E_z(0)/B_z(0)|$ $= 0.35$, 0.3 and 0.8 respectively. About twenty matching points were sufficient to obtain stable roots. However the field amplitudes were not stable.

We also considered a non-circular base curve of the Pascal's limaçon type (figure 1), parametrized by

$$r_b = a\,(1 - \varepsilon \cos s)\,, \quad \varphi_b = s - \varepsilon' \sin s\,, \quad s \in [0, 2\pi]\,. \tag{21}$$

The inward bump is expected to give a stronger $E_z(0)$. Indeed when a large part of the base curve is close to the origin, the boundary condition inhibits $E_z(0)$. This is avoided with the bump. For $\varepsilon = \varepsilon' = 0.3$, $q = 0.7$ and $v_{\rm ph} = 0.999$, two modes are clearly seen at $\omega a = 2.4$ and 6.7. A state at $\omega a = 3.5$ was also seen for odd $l_{\rm sup}$, but it appeared only as a smooth dip of $\ln|\det\{C_{nl}\}|$ for even $l_{\rm sup}$.

Finally we considered a base curve with two inside bumps and a rotational symmetry of order 2, paramerized by

$$r_b = a\,(1 - \varepsilon \cos s)\,, \quad \varphi_b = (s - \varepsilon' \sin s)/2\,, \quad s \in [0, 4\pi]\,. \tag{22}$$

The modes are either even or odd under this symmetry. Odd modes are not interesting for particle acceleration since $E_z(0) = 0$. Even modes have the property $E_x(0) = E_y(0) = 0$, which guaranties that electrons on the z-axis are not deflected and emit no synchrotron radiation. For $\varepsilon = \varepsilon' = 0.1$, $q = 0.7$ and $v_{\rm ph} = 0.999$, even modes are seen at $\omega a = 4.0$ and 6.3, with the different l truncations $[l_{\rm inf}, l_{\rm sup}] = [-4, 8]$, $[-8, 8]$ and $[-10, 10]$. With more pronounced bumps ($\varepsilon = \varepsilon' = 0.2$ or 0.3), no stable root was found.

4. Limits of the PWE-PM method

We have encountered difficulties when the deformation of the wave guide relative to a straight cylindrical tube becomes too strong. This may be attributed not to the point matching method but to a lack of convergence of the partial wave expansion at r larger than a critical radius r_c.

Let us indeed consider the expansion (8), in the scalar field case, and suppose that it converges at some point of polar coordinates (r_0, φ_0). Then it converges at all points (r, φ) with $r < r_0$. It implies that the domain of convergence is bounded by a cylinder.

Demonstration : The convergence of (8) at (r_0, φ_0) implies that $|a_l\, \psi_{l,k(l)}(r_0)|$ is bounded by some number M. Using (7), we have

$$|a_l\, \psi_{l,k(l)}(r)| \le M \left| \frac{I_{|l|}(\kappa_l r)}{I_{|l|}(\kappa_l r_0)} \right| \tag{23}$$

for any r. At large $|l|$, we have $\kappa_l \simeq |l|q - s_l P$ with $s_l = \text{sign}(l)$ and the asymptotic form of the modified Bessel functions gives $I_{|l|}(\kappa_l r)/I_{|l|}(\kappa_l r_0) \simeq \exp[(|l|q - s_l P)\,(r - r_0)]$. Then

$$|a_l\, \psi_{l,k(l)}(r)| \le M\, \exp[s_l P (r_0 - r)]\, e^{|l|q(r-r_0)} \quad \text{at large } l\,. \tag{24}$$

If $r < r_0$, the expansion (8) is bound in modulus by a convergent geometric series in $|l|$, therefore converges at $l \to \pm\infty$.

If the base curve has a singularity, *e.g.*, an anglar point of polar coordinate (r_1, φ_1), the field is non-analytic at this point, therefore $r_c \leq r_1$. On the other hand, for a strong enough deformation relative to the straight cylinder, part of the wave guide may be outside the cylinder of convergence, even if the base curve is analytic. A similar situation occurs when one uses the Rayleigh expansion for the wave diffracted by a grating : this expansion does not converge down to the bottom of the grooves if these ones are too deep [8, 9, 10].

5. Conclusion

We have seen that an increase in the strength of the helical bending (increase of ε and q in the case of a de-centered circular basis) leads to a decrease of the dimensionless frequencies ωa of the low modes at imposed phase velocity $v_{\rm ph} < c$ (a characterizes the transverse size of the guide). One can also expect that it leads to a larger accelerating field $E_z(0)$, at imposed power of the wave or imposed maximum field on the wall. We have indeed, in order of magnitude

$$E_z/E_T \sim p_T/p_z\,, \quad p_T \sim 1/a\,, \quad p_z \sim \omega \quad \rightarrow \quad E_z/E_T \sim 1/(\omega a)\,. \tag{25}$$

Inward bumps of the base curve also favor large $E_z(0)$. On the other hand, the PWE-PM method is limited to weak helical bending and sufficiently "round" base curves. Here we propose two numerical methods to find the propagating modes for larger bending and base curves with more pronounced structures.

The finite element method. This is the standard way to solve numerically the Maxwell equations. In the rotating frame these equations take the 2-dimensional form at fixed ω and P :

$$\begin{aligned}
i\omega\mathcal{B}_X &= +\partial_Y\mathcal{E}_z - iP\mathcal{E}_Y - q\mathcal{E}_X + q(X\partial_Y - Y\partial_X)\mathcal{E}_Y\,,\\
i\omega\mathcal{B}_Y &= -\partial_X\mathcal{E}_z + iP\mathcal{E}_X - q\mathcal{E}_Y - q(X\partial_Y - Y\partial_X)\mathcal{E}_X\,,\\
i\omega\mathcal{B}_Z &= +\partial_X\mathcal{E}_Y - \partial_Y\mathcal{E}_X\,,\\
-iP\mathcal{B}_Z &= \partial_X\mathcal{B}_X + \partial_Y\mathcal{B}_Y + q(Y\partial_X - X\partial_Y)\mathcal{B}_z\,,
\end{aligned} \tag{26}$$

and analogous equations obtained with the duality transformation $\mathcal{B} \rightarrow \mathcal{E}$, $\mathcal{E} \rightarrow -\mathcal{B}$.

Method of virtual currents. Instead of expanding the field in cylindrical partial waves, one does it with eigenfunctions of $\mathcal{P}_{\rm helic}$ (equation 5)which are regular inside the waveguide but singular at some outside places. A candidate is the retarded or advanced field produced by the charge-current distribution

$$j^\mu(t, \mathbf{r}, z) = \begin{pmatrix} \rho \\ \mathbf{j}_T \\ j_z \end{pmatrix} = e^{iPz - i\omega t}\,\delta^{(2)}(\mathbf{r} - \mathcal{R}[qz]\mathbf{d}) \begin{pmatrix} 1 \\ qv_{\rm ph}\,\hat{\mathbf{z}} \wedge \mathbf{r} \\ v_{\rm ph} \end{pmatrix}, \tag{27}$$

with $v_{\rm ph} = \omega/P$. $\mathcal{R}[qz]$ is the rotation matrix of (1). The x and y components of any vector are gathered in a bi-vector. The point $(x, y, z) = (\mathcal{R}[qz]\mathbf{d}, z)$ draws an helix of path $\Lambda = 2\pi/q$ intersecting the $z = 0$ plane at point $(\mathbf{d}, 0)$. $\rho(t, x, y, z)$ is a periodic distribution of charges running along this helix with the longitudinal velocity $v_z = v_{\rm ph}$ and $\mathbf{j}$ is the associated current. j^μ is periodic in time with frequency ω and is an eigenfunction of $\mathcal{P}_{\rm helic}$ with eigenvalue P, therefore the retarded or advanced field has the same property. Let us denote, for instance, by $\mathbf{F}^{(\mathbf{d})} = \{\mathbf{E}^{(\mathbf{d})}, \mathbf{B}^{(\mathbf{d})}\}$ the retarded field. Different $\mathbf{d}$'s give different independent solutions. To take into account the fact that there are two polarisations, we also consider the dual field $\tilde{\mathbf{F}}^{(\mathbf{d})} = \{\mathbf{B}^{(\mathbf{d})}, -\mathbf{E}^{(\mathbf{d})}\}$, which would be produced by a distribution of magnetic charge and current. In the $z = 0$ plane these fields have a singularity in $1/|\mathbf{r} - \mathbf{d}|$ at $\mathbf{r} = \mathbf{d}$; they are Green's functions of (26). Choosing N singular points $\{\mathbf{d}_1, \mathbf{d}_2, \cdots \mathbf{d}_N,\}$ external to the base curve, we may approximate the field of a mode by

$$\mathbf{F} = \sum_{n=1}^{N} a_n \, \mathbf{F}^{(\mathbf{d}_n)} + b_n \, \tilde{\mathbf{F}}^{(\mathbf{d}_n)} \, , \tag{28}$$

The coefficients a_n and b_n should be choosen to satisfy the boundary conditions on N matching points $\mathbf{r}_1, ...\mathbf{r}_N$ of the base curve.

Near a sharp inward bump of the base curve the field is strong and, by analytic continuation, one probably meets a singularity not far behind the curve, analogous to a point-like electrical image in electrostatics. This singularity may be simulated by placing one or a few $\mathbf{d}_n$'s on it or near to it. It amounts to replace the continuous surface charge and currents on the bump by charges and currents on virtual helical wires.

Whichever method will be used, it will be usefull to compare it with the PWE-MP method in the case of weak helical deformation, as a check.

[1] Klein, H. ; Junior, P. ; Schempp, A. ; Finke, P. ; Herminghaus, H. ; Klabunde, J. ; Lehr, L. *Proceedings of the 1972 proton linear accelerator conference,* October 10-13, 1972.

[2] J. Aron *et al.*, Stanford 1974, *Proceedings, Ninth International Conference On High Energy Accelerators,* Springfield 1975. Published in *IEEE Trans.Nucl.Sci.* **20**, (1975) 76.

[3] Zhang T B and Marshall T C, 1996 *Nucl. Instr. in Phys. Res.* A **375** 614-617.

[4] Park SY and Hirshfield J L, 2000, *Phys. Rev.* E **62**, 1266-83.

[5] Wenxiang Wang, Yanyu Wei, Guofen Yu, Yubin Gong, Minzhi Huang and Guoqing Zhao, *Journal of infrared, millimeter and Terahertz waves* **24**, 9, pp 1469-84.

[6] T.M. Wallet, K.R. Vaden, J. Freeman, A. Haq Qureshi, Technical characteristics of a novel helical-groove traveling-wave tube structure. *IEEE Transactions on Magnetics* **34**, issue 4, pp. 1408-10.

[7] Artru X and Ray C 2009, *Proceedings of Channeling 2008*, in press

[8] Petit R and Cadilhac M. 1966 *Compt. Rend. Acad. Sci.* (Paris) **262B** 468

[9] Hill N R and Celli V 1978 *Phys. Rev.* B **17** 2478-81

[10] Agassi G and George T F 1986 *Phys. Rev.* B **33** 2393-2400

Cherenkov Radiation and Dielectric Based Accelerating Structures: Wakefield Generation, Power Extraction and Energy Transfer Efficiency.

Alexei Kanareykin

S.Petersburg Electrotechnical University "LETI", Prof.Popov 5, St.Petersburg, Russia; Euclid Techlabs LLC, 1375 Piccard Dr., # 150, Maryland, USA

alexkan@euclidtechlabs.com

Abstract. We present here our recent results of the Euclid Techlabs LLC/Argonne National Laboratory/St.Petersburg Electrotechnical University "LETI" collaboration on wakefield high energy acceleration of electron bunches in dielectric based accelerating structures. This program concentrates primarily on Cherenkov radiation studies providing efficient high energy generation aimed at a future 1 TeV collider. We report here on recent experiments in high power Cherenkov radiation and corresponding dielectric material developments and characterizations. Progress in diamond, quartz and microwave low-loss ceramic structure development in GHz and THz frequency ranges is presented. Beam Breakup effects and transverse bunch stability are discussed as well. We e report on recent progress on tunable dielectric based structure development. A special subject of our paper is transformer ratio enhancement schemes providing energy transfer efficiency for the dielectric based wakefield acceleration.

1. Introduction

Since the invention and implementation of a photomultiplier general properties of Cherenkov radiation such as energy and momentum angular dependence [1,2] have become a well known and widely used analytical methods for particle diagnostics in high energy physics [1,3]. Threshold Cherenkov counters and Differential and RICH (Ring Imaging Cherenkov) or CRID (Cherenkov Ring Imaging) detectors have been developed and successfully applied for identifying the type of particles in nuclear physics and elementary particle science [3]. It should be note that IR and light emissions of charged particle are primarily used in Cherenkov detectors of high energy physics application.

At the same time, a new application of microwave and THz Cherenkov radiation has been proposed and studied in last decade to be used for linear accelerators and colliders in high energy physics as well [4-8]. It is a Dielectric Loaded Accelerator, or DLA concepts [4]. With this paper, we describe an experimental DLA program and present our latest results obtained by the Argonne National Laboratory/Euclid Techlabs LLC/St.Petersburg Electrotechnical University "LETI" collaboration [4-8,10,12-15].

2. Dielectric Based Accelerator

Dielectric loaded accelerator (DLA) structures using low-loss microwave ceramics or quartz and excited by a high current electron beam or an external high frequency high power RF source have been under extensive study recently [4-8]. The basic wakefield RF structure is very simple - a cylindrical, dielectric loaded waveguide with an axial vacuum channel is inserted into a conductive sleeve, Fig.1. A high charge, (typically 20 – 40 nC), short, (1 – 4 mm) electron drive beam generates TM_{01} mode electromagnetic Cherenkov radiation (wakefields) while propagating down the vacuum channel [4,10]. Following at a delay adjusted to catch the accelerating phase of the wakefield is a second electron (witness) beam. The witness beam is accelerated to high energy by the wakefield produced by the drive beam. A series of proof of principle experiments have been successfully performed in microwave frequency range at Argonne's Advanced Accelerator [4-8]. THz range wakefields have been successfully generated by the UCLA-SLAC collaboration [11].

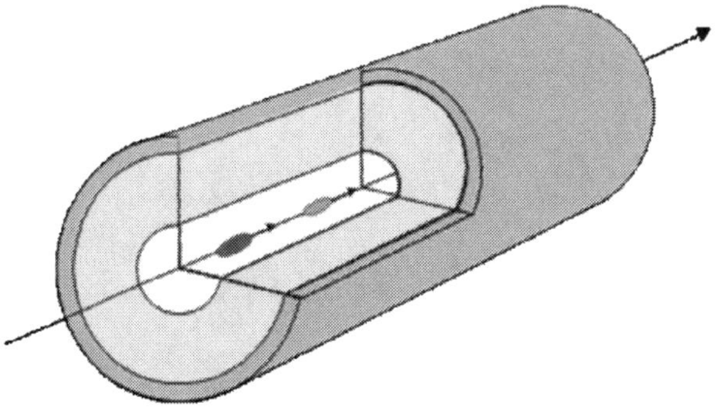

Figure 1. Partially Dielectric Loaded Accelerating structure excited by the drive beam. Inner region is vacuum; outer layer is dielectric; and the outermost layer is metal.

The advantages and potential problems of using dielectric for loading an accelerating structure are discussed in the above references [4-8] and are only summarized here. The advantages are: (1) Simplicity of fabrication: The device is simply a tube of dielectric surrounded by a conducting cylinder. This is a great advantage for high frequency (> 30 GHz) structures compared to conventional structures where extremely tight fabrication tolerances are required. The relatively small diameter of these devices also facilitates placement of quadrupole lenses around the structures. (2) Dielectrics can potentially exhibit high breakdown thresholds relative to copper, and high shunt impedance. (3) Reduced sensitivity to the single bunch beam break-up (BBU) instability: The frequency of the lowest lying HEM_{11} deflecting mode is lower than that of the TM01 accelerating mode. (4) Easy parasitic mode damping.

Potential challenges of using dielectric materials in a high power RF environment are breakdown and thermal heating, same as for all-metal conventional accelerating structures. Microwave DLA required high brightness high charge drive beam providing >200 MV/m accelerating gradient that has to be sustained by the dielectric material of the structure loading [10]. Currently we consider fused silica and CVD diamond as our primarily candidates for the high gradient DLA structure loadings [10,12].

$$\sigma_r = \left(\frac{\varepsilon_N}{\gamma}\beta\right)^{1/2}; \qquad W_Z(z) \approx \frac{Q}{a^2}\exp\left[-2\left(\frac{\pi\,\sigma_z}{\lambda_n}\right)^2\right]\cos(kz) \qquad (1)$$

Formula (1) presenting longitudinal wake function below clearly shows that the high gradient can be reached by using the small aperture structure (~ a^2) that, in turn, requires low emittance $_N$, high charge Q drive beam (or bunch train) to be generated and transferred through the DLA structure [9,10]. The power extraction has to be provided for the two-beam accelerator design. 26 GHz wakefield power extractor has been successfully demonstrated in 2009 [13], and 30 MW power level has been reached for the ~ 10 ns pulse [13]. Beam handling is always a special issue for the decelerating section of a high charge wakefield accelerator [14]. We also proposed an original method of DLA structure tuning based on additional ferroelectric layer [15] introduce into the DLA loading.

It should be also noted that expected problems with dielectric charging are easily mitigated by using a dielectric with a small dc conductivity.

3. Dielectric Based Accelerator Development.

3.1. Diamond DLA Structure.

Quartz and cordierite structures have been recently beam tested, and accelerating gradient exceeding 100 MV/m has been demonstrated [10]. Low-loss (tan $\sim 1\times10^{-4}$) fine microwave ceramics have been high power tested recently at 1.9 MW power with the gradient corresponding to surface field more than 100 MV/m at the coupler joints [10]. Amorphous dielectrics, such as glass, are known to exhibit high breakdown limits (~100's of MV/m), but most have a loss tangent that is too high for high-gradient accelerator applications.

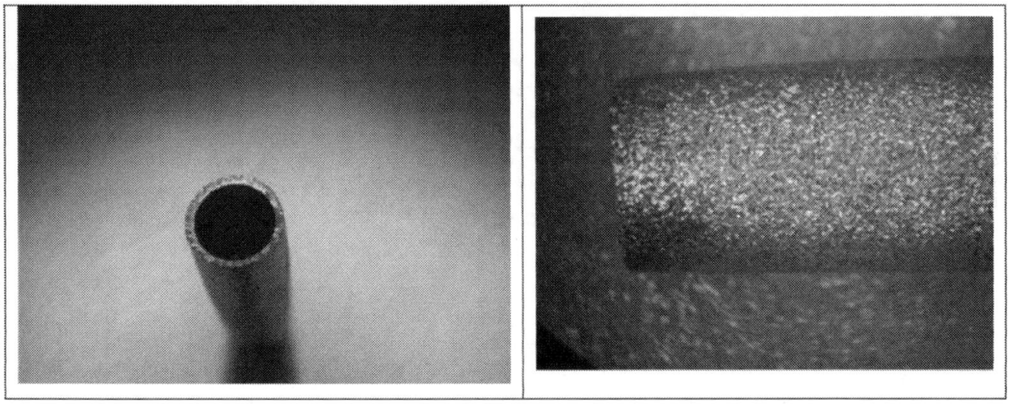

Figure 2. Photographs of CVD diamond tube developed: (a) top view; (b) axial view. Tube parameters are: 5 mm inner diameter, 2.5 cm long and ~ 500 μm thick. Light reflects off the naturally smooth individual facets of diamond crystals comprising the polycrystalline aggregate.

Low-loss microwave ceramics and quartz are not the only materials that are being intensively studied as potential DLA loading. An alternative is to use polycrystalline artificial diamond produced by CVD [12], which shows promise for use in a high-gradient, dielectric-loaded accelerator (DLA) structure. It has a very high breakdown field up to 2 GV/m at DC field (no data available for RF frequency range yet), low loss tangent ($<10^{-4}$), and the highest known thermal conductivity (2×10^{3} $W\times m^{-1}\times K^{-1}$) [12].

In Ref. [16], a diamond-based rectangular DLA was discussed and all-metal and dielectric-based accelerating structures were compared. For cm- and mm-wavelength linear accelerators with normal-conducting structures as studied by the NLC/JLC collaboration and the CLIC study group [16], the main limitation to the achievement of high accelerating gradient is rf breakdown. It is known that the field limit is proportional to an inverse fractional power of the pulse length, and will vary with material and surface processing. Recent experiments on 30GHz accelerating structures at the CLIC Test Facility with an rf pulse width of 16 ns showed the usable average gradient is lower than these values; e.g. <130 MeV/m for molybdenum [17].. Therefore, a clear objective is to increase acceleration gradient without incurring an unacceptable rate of breakdown events in the accelerating structure.

A DLA structure, unlike a conventional metallic one, admits the unique possibility of sustaining extremely high gradients by using a diamond-based material as the dielectric loading of the guiding structure [12,16]. These structures are based on cylindrical diamond dielectric tubes that are manufactured via a relatively simple and inexpensive chemical vapor deposition (CVD) process, plasma assisted CVD. Use of the CVD process is a much simpler method to achieve high quality rf microcavities compared to other microfabrication techniques. Our initial work was based on 100 μm scale tubes with fundamental frequencies in the 0.1–1.0 THz range; promising results were obtained using the plasma assisted and hot-filament CVD process. For the larger structures required for Ka band (34 GHz) and longer wavelength applications, the use of microwave plasma-enhanced CVD (PECVD) was determined to have a greater likelihood of success based on its larger rate of diamond deposition and the ability to control surface temperatures on the substrate during deposition. The 5 mm inner diameter, 2.5 cm long and 500 μm thick diamond tube has been fabricated and characterized with SEM, micro-Raman and micro-photoluminescence spectrum analysis. Fig.2 shows the finished free standing diamond tube.

Inner Diameter	no vacuum gap	
	1.5 mm	3 mm
R_{outer}, cm	0.19	0.24
P_{attn}, dB/m	-4.7	-4.5
r_s, MΩ/m	152	85
$E_{z.diel}/E_{z.accel}$	1	1
$E_{r.diel}/E_{z.accel}$	0.37	0.68
$E_{r.metal}/E_{z.accel}$	0.17	0.25

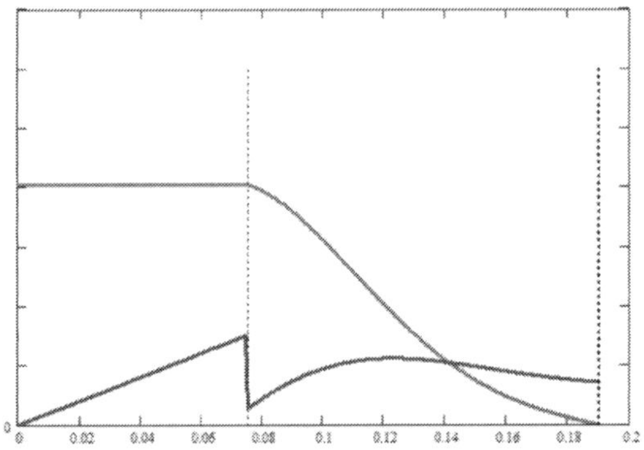

Figure 3. (a) diamond-based cylindrical DLA structure parameters in case of, no vacuum gap. Surface field ratio $E_{metal}/E_{acc} > 0.17$ for a 1.5 mm beam channel aperture; (b) longitudinal (red) and transverse (blue) electric field profiles normalized to the acceleration gradient E_{acc} for the structure parameters: (a) - inner diameter 2a = 1.5 mm, outer diameter 2b = 3.79 mm, diamond thickness is 1.15 mm.

3.2. Numerical Simulations.

Numerical simulations of completed and planned experiments with these structures have been reported in [5,7,9,12-15]. Field analysis for diamond-based DLA structures has been carried out analytically [9]. The dispersion equation solutions and field magnitude expressions for TM_{0n} modes of the cylindrical dielectric loaded waveguides can be found in [9]. Parameters of the diamond-based DLA structures with the 1.5 mm and 3 mm apertures presented in Fig. 3(a) [12]. The structure operates at the TM_{01} mode in the Ka band frequency range where its axial wave number corresponds to a phase velocity of c; the diamond has dielectric constant of 5.7 and loss tangent tan $<1\times10^{-4}$.

The radial field profiles normalized to the accelerating gradient for the 34 GHz diamond-based structure are presented in Fig. 3(b). The accelerating gradient is equal to the maximum E_z field magnitude on the inner dielectric (diamond) surface. The transverse E_r/E_{acc} ratio on this surface does not exceed 0.37. If the dielectric breakdown limit is indeed in the range of 0.5-1.0 GV/m [32, 33], the maximum acceleration gradient also would exceed 500 MV/m, if dielectric breakdown is the ultimate limitation. It should be noticed that for relatively short RF pulse length, the accelerating gradients to be available with the diamond-based DLA structure are evidently well in excess of those for iris-loaded all-metal accelerating structures [12,16].

From Fig. 3(a), the shunt impedance is equal R= 152 MV/m for an aperture of 1.5 mm. Fig. 5b shows the calculated shunt impedance R vs. the vacuum beam channel radius as it was calculated for the cylindrical diamond-loaded traveling-wave (TW) structure, Fig.3a. Simulations showed that the maximum shunt impedance R_{max} increases with $1/R_c$ along with the accelerating gradient. Note that for the rectangular DLA structure the larger aperture decreases the shunt impedance [16], another advantage of the cylindrical design presented in this paper.

3.3. Beam Break-Up.

The dynamics of the beam in structure-based wakefield accelerators leads to beam stability issues not ordinarily found in other machines [5,14]. In particular, the high current drive beam in an efficient wakefield accelerator loses a large fraction of its energy in the decelerator structure, resulting in physical emittance growth, increased energy spread, and the possibility of head-tail instability for an off axis beam, all of which can lead to severe reduction of beam intensity. Beam breakup effects resulting from parasitic wakefields provide a potentially serious limitation to the performance of dielectric structure based wakefield accelerators as well.

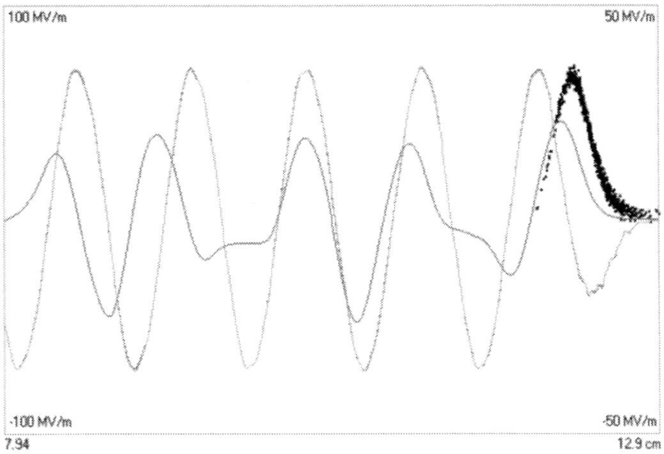

Figure 4. Wakefields (longitudinal: blue, transverse: green) and beam profile for the Ka band dielectric wakefield experiment (diamond based DLA). The beam at the time of this snapshot has been propagating for ~430 ps (13 cm). The bunch tail has been deflected by the transverse wakefield almost to the inner radius of the vacuum channel [14].

We used a dielectric structure BBU code derived from the software described in Ref. [5]. The longitudinal and transverse wakefields of a bunch of macroparticles (1000 in this case) are propagated down a dielectric structure using the analytic Green's functions [14]. For this calculation the first 5 TM and HEM modes were used to compute the wakefields of the bunch. The magnitude of the longitudinal single bunch wakefield is ~70 MV/m, while the transverse force magnitude for the initial offset as given is ~20 MV/m. The bunch propagates about 13 cm through the structure before particle losses from deflection of the tail begin to occur, adequate for measurement of the signal without using external focusing to control the beam, Fig.4 [14]. The wakefield signal can be detected using radial field probes inserted into the copper sleeve of the diamond structure.

3.4. Frequency Tuning.

The frequency of a metallic accelerating structure is defined by its geometry. DLA structures, on the other hand, have another important parameter that determines the frequency spectrum -- the dielectric constant of the loading material. There are two classes of materials that can be tuned, in other words, materials with electromagnetic properties that can be controlled by external fields: ferrites, controlled by magnetic fields, and ferroelectrics, controlled by electric fields. Ferrites do not appear to be a practical solution for use in high frequency high gradient accelerators because of the high loss factor and also because a magnetic field could interfere with the electron beam optics. We proposed a new technique that allows control of the dielectric constant (and consequently the frequency spectrum) for dielectric waveguides by incorporating ferroelectric layers [15].

The most noteworthy feature of the tunable DLA is the replacement of a single ceramic by a composite of 2 layers as shown in Fig. 5a. The inner layer is ceramic, with permittivity $_1$ typically in the range of 4 - 36 [15]. The outer layer is a film made of BST ferroelectric, of permittivity $_2$, placed between the ceramic layer and the copper sleeve. The DLA structure tuning is achieved by varying the permittivity, $_2$, of the ferroelectric film by applying an external DC electric field across the ferroelectric. This allows us to control the effective dielectric constant of the composite system and therefore, to control the structure frequency during operation [15].

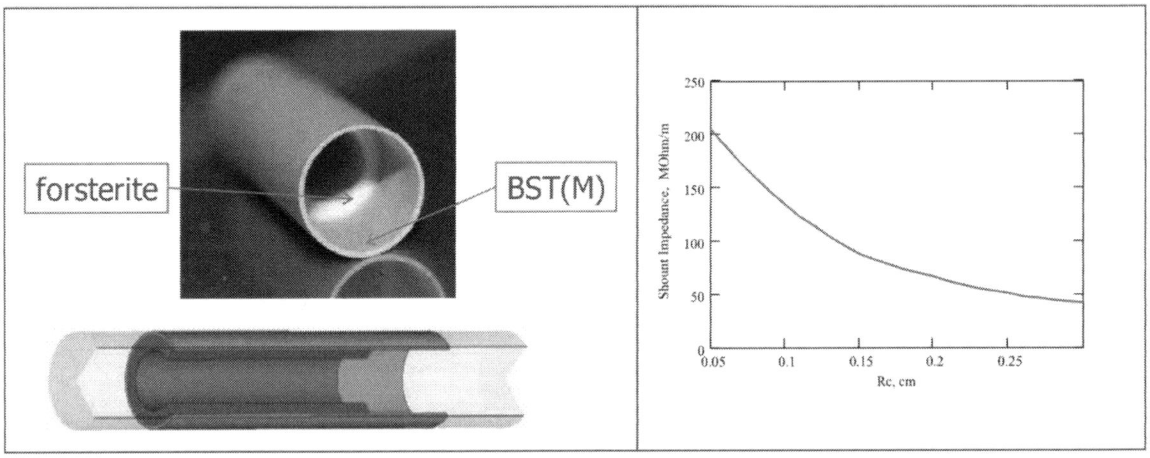

Figure 5. (a) Tunable double-layer DLA structure loading. The nonlinear layer is made of BST(M) ferroelectric, thickness is of 400 μm, dielectric constant is of 450. One can see the inner forsterite layer (white) inserted into the ferroelectric one; (b) shunt impedance R vs beam channel radius for the 34. 272 GHz cylindrical diamond-based DLA structure.

Using a combination of ferroelectric and ceramic layers permits tuning of a composite ceramic–ferroelectric waveguide while keeping the overall material loss factor in the (4 - 5)×10-4 range. Two sets of dual layer waveguides were fabricated. The ferroelectric components had 400 and 800 μm thicknesses and dielectric constants of ~550 and ~450. The tunable DLA resonator with the double layer loading was assembled with the following parameters: inner layer made of forsterite ceramic, dielectric constant 6.8, inner diameter 6mm, thickness 1.35 mm, length 25 mm; outer ferroelectric layer made of BST(M) ferroelectric, dielectric constant 450/550, thickness 400 μm, and length 34.7 mm. This double layer was inserted into a cylindrical copper sleeve. The geometry of the resonator is shown in Fig. 6 [15].

Unlike a simple copper jacketed resonator with dielectric loading, the ferroelectric based device showed a positive frequency response to temperature variation. In general, the ordinary ceramic material has very small thermal expansion and stable dielectric constant over a wide temperature range but the copper housing will expand and increase the volume so that the resonant frequency of a regular dielectric loaded resonator will shift down with increasing temperature. Here, since the dielectric constant of the ferroelectric tube will decrease significantly with a temperature increase that will lead to an increase of the resonance frequency, the overall effect is a positive slope over a rising temperature. The results from Fig. 4b show a very good linearity and sensitivity of the thermal tuning of this ferroelectric material: ~14MHz/ ^{0}K has been measured for both materials, Fig. 7(a). We also applied a high dc voltage to the DLA resonator that we used for the thermal testing. In these measurements, a tuning range of 6 MHz at 25 kV/cm biasing dc field has been demonstrated [15].

3.5. Energy Transfer Efficiency.

One approach to future high energy particle accelerators is based on the wakefield principle: a leading high-charge drive bunch is used to excite fields in an accelerating structure or plasma that in turn accelerates a trailing low-charge witness bunch. Energy transfer efficiency for the driver-witness bunches is a critical issue for this type of accelerator technique. The transformer ratio R is defined as the ratio of the maximum energy gain of the witness bunch to the maximum energy loss of the drive bunch. A number of techniques have been proposed to overcome the transformer ratio limitation. We reported recently the first experimental study of the ramped bunch train (RBT) technique in which a dielectric loaded waveguide was used as the accelerating structure. A single drive bunch was replaced by two bunches with charge ratio of 1:2.5 and a separation of 10.5 wavelengths of the fundamental mode. An average measured transformer ratio enhancement by a factor of 1.31 over the single drive bunch case was obtained in this experiment [8].

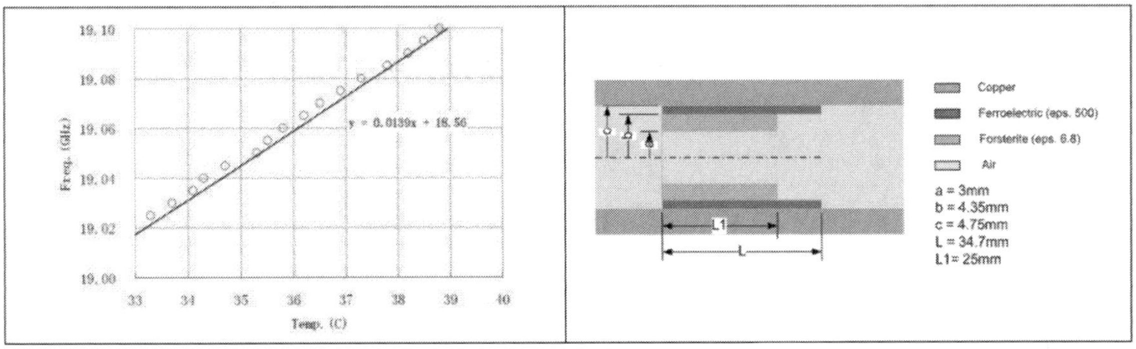

Figure 6. (a) Temperature dependence of the $TM0_{12}$ mode of the dielectric loaded resonator with ferroelectric tunable layer; (b) the tunable DLA resonator with double layer loading. Inner layer (forsterite ceramic), dielectric constant is 6.8, inner diameter 6mm, thickness 1.35 mm, length is of 25 mm; outer ferroelectric layer made of BST(M) ferroelectric, dielectric constant 450/550, thickness 400 μm, length 34.7 mm.

Acknowledgments. This work is supported by the SBIR Program of the US Department of Energy and RFBR award.

References

[1] V.L.Ginzburg. Phys. Usp. 39, 973 (1996)
[2] B.M. Bolotovskiĭ. Phys Usp. 62, 201 (1957); 75, 295 (1961)
[3] Akimov Y.K. Physics of Atomic Nuclei, 67, 1385 (2004).
[4] W. Gai, P. Schoessow, B. Cole, R. Konecny, J. Norem, J. Rosenzweig and J. Simpson, Phys. Rev. Lett. 61, 2756 (1988).
[5] W. Gai, A.D. Kanareykin, A.L. Kustov, J. Simpson, Phys. Rev. E55(3), 3481 (1997).
[6] P. Zou, W. Gai, R. Konecny, X. Sun et al. Review of Scientific Instruments, 71, 6, 2301 (2000).
[7] J.G. Power, M.E. Conde, W. Gai, A. Kanareykin, R. Konecny and P. Schoessow. Physical Review ST-AB, 3, 101302, (2000).
[8] C. Jing, A. Kanareykin, J. Power, M. Conde et al. Phys. Rev. Lett., 98, 144801 (2007).
[9] M. Rosing and W. Gai, Phys. Rev. D42, 1829, (1990).
[10] W. Gai, AIP Conference Proceedings 1086, 3, (2009).
[11] M. C. Thompson et al. AIP Conference Proceedings 877, 903 (2006).
[12] P. Schoessow, A. Kanareykin, and R. Gat, AIP Conference Proceedings, 1086, 398 (2008)
[13] C.-J. Jing, A. Kanareykin, A. L. Kustov, P. Schoessow et al. Proceedings Particle Accelerator Conference PAC-2009 , Vancouver, CA 2009 to be published.
[14] A. Kanareykin, C.-J. Jing, A. L. Kustov, P. Schoessow et al. Proceedings Particle Accelerator Conference PAC-2009 , Vancouver, CA 2009 to be published.
[15] A. Kanareykin, C. Jing, E. Nenasheva, P. Schoessow et al. AIP Conference Proceedings, v. 1086, 386, (2009).
[16] C.Wang, V.P.Yakovlev and J.L.Hirshfield. Proceedings Particle Accelerator Conference PAC-2005, 1282 (2005).
[17] I. Syratchev, CLIC Note 505, CERN (2001).

Radiation from Relativistic Electrons in a Thin Crystal Target as Complementary X-Ray and Gamma Source at Synchrotron Light Facilities

O V Bogdanov, A A Evdokimov, K B Korotchenko , Yu L Pivovarov and TA Tukhfatullin

Tomsk Polytechnic University, Lenin Ave 30, 634050 Tomsk, Russia

Email: TTA@tpu.ru

Abstract. Concept of using crystal targets as complementary X-ray and gamma ray sources at synchrotron light facilities is considered. Calculations of channeling radiation, parametric X-radiation and bremsstrahlung from relativistic electrons in the crystal targets are performed for parameters of synchrotrons operating in Moscow (Kurchatov Center for Synchrotron Radiation, KCSR) and Karlsruhe (ANKA).

1. Introduction

The relativistic electrons penetrating through a thin crystal target emit several kinds of radiation (in X-ray and gamma region): coherent and ordinary bremsstrahlung (CBS and BS), transition radiation (TR), parametric X-radiation (PXR), channeling radiation (CR).

The BS and TR produced in internal amorphous thin target have been considered recently as principal EUV and hard X-ray generation mechanisms in the table-top synchrotron storage rings [1–2], operating at beam energy 6–20 MeV.

Similar concept probably can be extended to more high energy storage rings, e.g. operating at 2.5 GeV:

- Kurchatov Center for Synchrotron Radiation in Moscow (KCSR) [3],
- ANKA at FZ Karlsruhe [4].

The crystal target can be used both as internal target in the main storage ring and photon production target in the booster synchrotrons or injectors (80-100 MeV linac at KCSR and 53 MeV microtron at ANKA).

So, the goal of the present work is to perform calculations of CR, PXR and BS for beam energies indicated above to optimize the production of photons in the thin Si, C, Ge crystal targets.

2. Theory

Calculation of channeling radiation (CR) from relativistic electrons under (111) planar channeling in a Si and LiF crystal is performed using both classical and quantum methods. The quantum approach is used when the number of allowed transverse energy levels of channeled electron is not large. The classical treatment can be applied for the greater energies of electrons. We developed the new computer code for solving these problems. It is called "Basic Channeling with Mathematica (BCM)" and will include a set of packages to make symbolic calculation of the planar channeling phenomena [5].

2.1. Classical description of CR

In the classical description, for every individual electron trajectory $x(t)$ ($x(t)$ is the electron transverse motion coordinate perpendicular to the channeling planes), the CR intensity spectrum is calculated using [6]:

$$\frac{dW}{d\omega dz} = \frac{e^2\omega}{c^4T^2}\sum_{l=1}^{\infty}\Theta[1-\eta_l]\cdot\left[\eta_l^2-\eta_l+\frac{1}{2}\right]\cdot|\dot{x}_{\tilde{\omega}}|^2;$$
$$\eta_l = \frac{T\omega}{4\pi\gamma^2 l};\tilde{\omega} = \frac{2\pi l}{T};\dot{x}_{\tilde{\omega}} = \int_{-T/2}^{T/2}\dot{x}e^{i\tilde{\omega}t}dt. \tag{1}$$

Here, Θ is the Heaviside theta function, γ is the relativistic factor, T is oscillation period of electron in the planar channel, which depends on the initial transverse energy, $\dot{x}_{\tilde{\omega}}$ is the Fourier component of electron velocity, l is the harmonic number. To obtain electron trajectory $x(t)$ we numerically solved classical equation of motion of relativistic electron. The periodic planar potential used in the equation of motion was obtained as in our previous works [7-8]. Equation (1) is obtained assuming: $\upsilon_\perp \equiv \dot{x} << \upsilon_\parallel$; $\upsilon_\perp / c << 1/\gamma$, $\upsilon_\perp$ is the velocity of transverse motion, $\upsilon_\parallel$ is the average velocity of electron along channeling planes. The Fourier components of velocity $\dot{x}_{\tilde{\omega}}$ have been defined numerically using the calculated trajectories.

2.2. Quantum description CR

In the quantum description, the spectral distribution of CR intensity is described by the formula [9]

$$\frac{dW}{d\omega dz} = \hbar\omega\frac{dW_{if}}{d\omega} = \sum_i\sum_{f<i}\frac{dI_{if}}{d\omega} = \alpha\hbar\omega\sum_i\sum_{f<i}|M_{if}|^2\left[1-2\frac{\omega}{\omega_m}+2\left(\frac{\omega}{\omega_m}\right)^2\right]P_i(k_j,k_0), \tag{2}$$

α is the fine-structure constant, M_{if} are the transition matrix elements, and $\omega_m = \hbar\Omega_{if}(1 - \beta_\parallel)$, $\beta_\parallel = \upsilon_\parallel / c$, $\hbar\Omega_{if} = E_i(k_j) - E_f(k_j)$, $E_i(k_j)$ is the transverse energy of electron, k_j is transverse wave vector of electron in allowed band, and $P_i(k_j,k_0)$ is the initial population of ith band:

$$P_i(k_j,k_0) = \frac{1}{d_p}\left|\int \exp(-ip\theta x/\hbar)\varphi_i(k_j,x)dx\right|^2. \tag{3}$$

Here, p is the incident electron momentum, d_p is the interplanar spacing. The wavefunctions $\varphi_i(k_j,x)$, which describe the transverse motion, have been calculated for planar periodic potential from the Schrodinger equation with relativistic mass $m^*=\gamma m_e$ and using the Bloch boundary conditions. The matrix elements M_{if} were calculated without use of dipole approximation.

2.3. Parametric X-ray radiation

The parametric X-ray radiation is emitted in the direction satisfying the Bragg conditions, i.e. at large angle with respect to the direction of particle momentum. The angular distribution of emitted photons per unit crystal length is described by the formula obtained in Ref. [10], using 2-wave theory of X-Ray diffraction:

$$\left(\frac{d^3N}{d\theta_x d\theta_y dz}\right)_{PXR} = \frac{\alpha\omega_B}{4\pi c\sin^2\theta_B}\left(\frac{\theta_x^2}{4(1+W_{\upsilon\parallel}^2)}+\frac{\theta_y^2}{4(1+W_{\upsilon\perp}^2)}\right), \tag{4}$$

$$W_{\omega\sigma} = \frac{1}{2|\chi_g|P_\sigma}\left[\theta_x^2 + \theta_y^2 + \theta_{kin}^2 - \frac{|\chi_g|P_\sigma}{\theta_x^2 + \theta_y^2 + \theta_{kin}^2}\right], \quad (\sigma = \|, \perp), \tag{5}$$
$$P_\perp = 1, P_\| = \cos(2\theta_B).$$

Here, θ_x and θ_y are the angles of emission measured from the direction satisfying the Bragg condition, θ_B is the Bragg angle, ω_B=|**g**|/(2sin θ_B) is the frequency of virtual phonons satisfying the Bragg diffraction condition, **g** – is the reciprocal lattice vector, $\theta^2_{kin}=\gamma^{-2}+|\chi_0|$, χ_g is the Fourier component of the dielectric susceptibility χ and $\chi_0=\varepsilon_0-1$.

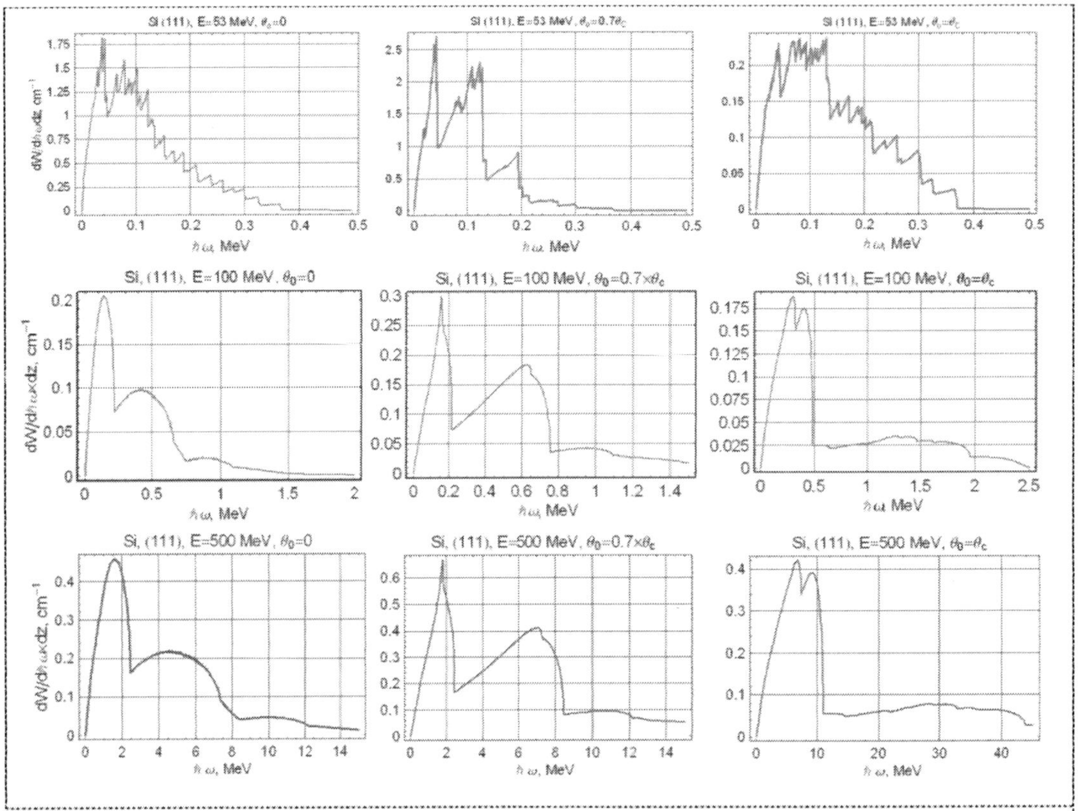

Figure 1. The CR spectra calculated for Si crystals using quantum approach (electron beam energy E=53 MeV), and classical approach (electron beam energy E=100 MeV and E=500 MeV), and for different incident angles θ_0 with respect to the (111) planes (θ_C is the Lindhard angle).

2.4. Bremsstrahlung

The bremsstrahlung spectra were calculated using well known free software package GEANT4 [11]. This package provides the energy loss of electrons and positrons due to the radiation of photons in the field of a nucleus. In the amorphous target the production of a photon of energy $\hbar\omega$ is occurred due to bremsstrahlung. The probability of this process is estimated according to the appropriate cross section. For calculation of electrons transportation through target Method Monte-Carlo is used.

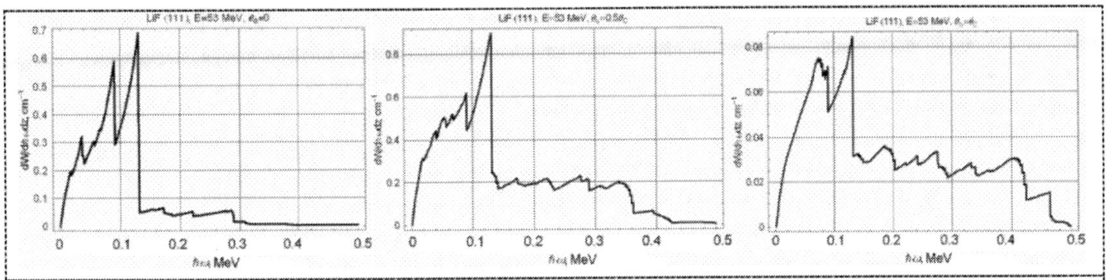

Figure 2. The CR spectra calculated for LiF crystals using quantum approach for electron beam energy E=53 MeV, for different incident angles θ_0 with respect to the (111) planes (θ_C is the Lindhard angle).

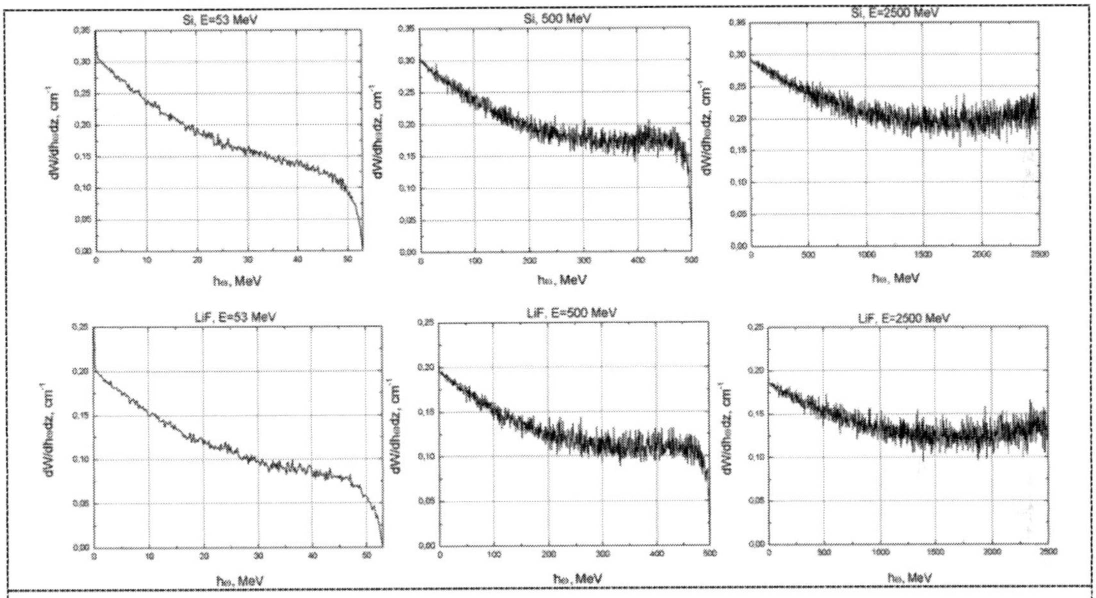

Figure 3. The BS spectra calculated for Si and LiF crystals (electron beam energy: E=53, 500, 2500 MeV). Total number of incident particles 10^9. Crystal thickness 10 μm.

3. Results and discussion

We calculated of CR and BS spectra and PXR angular distribution using the computer code mentioned above.

- It is seen that the CR spectrum strongly depends on electron beam energy and the beam alignment with respect to the crystallographic planes [10].
- The CR spectra consist of large number of peaks which correspond to transition between large numbers of transverse energy levels if we use quantum approach (Fig. 1-3). For classical approach CR spectra contain two maxima which correspond to electron motion in two potential wells (Fig. 1). CR spectrum strongly depends on the beam alignment with respect to the crystallographic axes or planes (Fig. 1, 2).

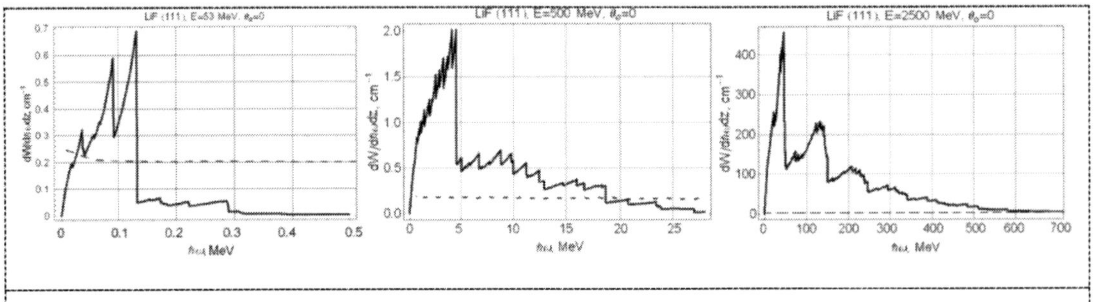

Figure 4 The CR spectra (solid lines) and the BS spectra (dashed lines) calculated for LiF crystals using quantum approach and electron beam energies E=53, 500, 2500 MeV, for incident angles θ_0=0 with respect to the (111) planes.

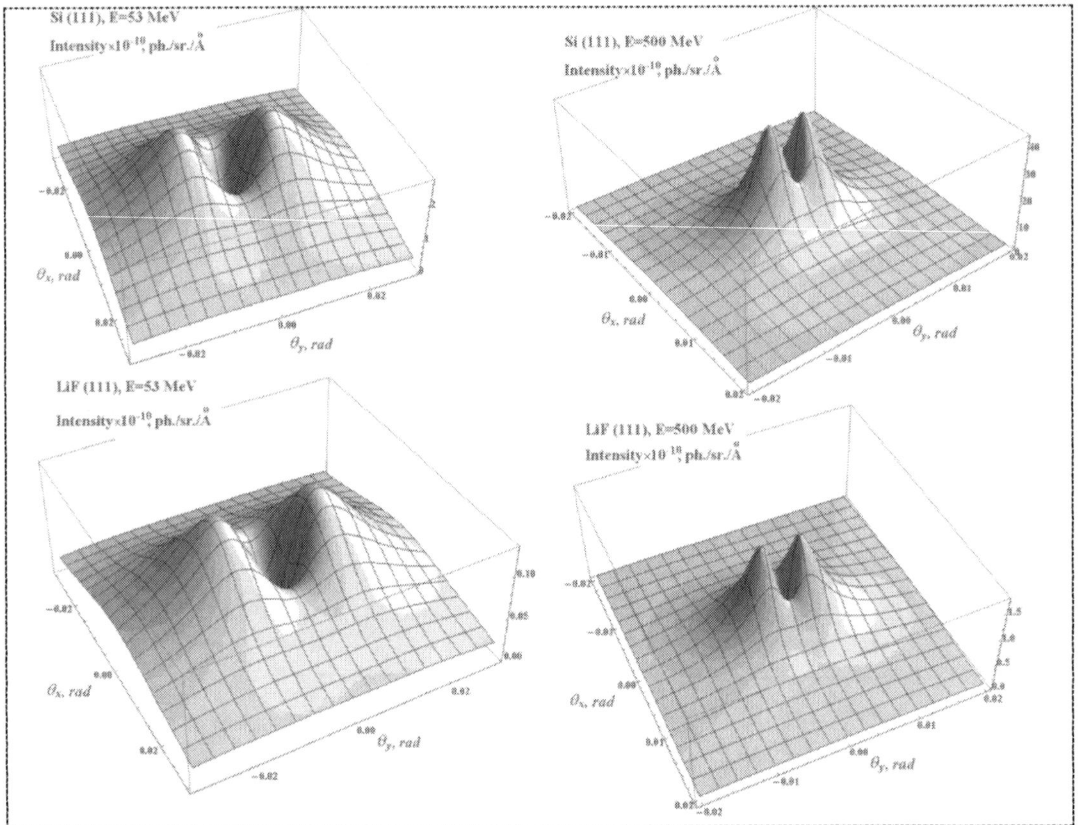

Figure 5. The angular distribution of PXR for Si and LiF crystals (electron beam energy 53 MeV and 500 MeV) for the angle of incidence with respect to the crystallographic planes $\theta_0 = 0$, in vicinity of the Bragg angle (emission angle) $\theta_B = \pi/8$. The photon energy is for Si 5.17 keV, for LiF 6.97 keV.

- The intensity of BS does not depend on the energy of incident electron beam and shape of the spectra changes with increasing energy of primary electron beam, mainly due to appearance of hard part of radiation spectrum (Fig 3.).
- Intensity of CR in contrast to BS increases with increasing of the electron energy and BS becomes negligibly small for energy electron beam greater than 500 MeV.

- As one can see from Fig. 3,4 the intensity of BS much less then intensity of CR for low energy photons so these two kind of radiation may be separate from each others.
- PXR in contrast to CR is emitted at large (Bragg) angle with respect to reflecting crystallographic planes and consists of several spectral peaks with photon energies ranging from several keV up to several decades keV. PXR peak positions do not depend on the electron beam energy while the shape and width of angular distributions of PXR is very specific in vicinity of the Bragg angle (see Fig.5).

4. Conclusions

In this paper we have presented the results of calculation of PXR, CR and BS from relativistic electrons passing through the crystal. for electron energies typical for the KCSR and ANKA injector and booster synchrotrons. For calculation of CR and PXR a new computer code BCM 1.0 was used [5]. The results of calculations show that CR, PXR and BS from relativistic electrons penetrating through the crystal indicate on possibility of creation of X-ray and gamma-ray source complementary to synchrotron radiation at existing of synchrotron light facilities. The PXR and CR can be used for generation of the photons from keV to MeV regions while BS can be used for generation of more hard photons. The crystal target can be used in the booster synchrotrons or injectors during beam lifetime in the main storage ring (about 20 hours) or as internal target in the main storage ring at the end of beam life-time.

5. Acknowledgments

The authors are grateful to Dr. I.S. Tropin for valuable comments on calculations with GEANT4.
The work is partially supported by the Russian Science and Innovations Federal Agency under contract No 02.740.11.0238 and by the Russian Federal Agency of Education under contract No P2596.

References

[1] Minkov D, Yamada H, Toyosugi N, Yamaguchi T, Kadono T, Morita M 2006 *Journal of Synchrotron Radiation* **13** 336
[2] Toyosugi N, Yamada H, Minkov D, Morita M, Yamaguchi T, Imai S 2007 *Journal of Synchrotron Radiation* **14** 212
[3] http://ankaweb.fzk.de
[4] http://www.kcsr.kiae.ru
[5] Korotchenko K B, Fiks E I, Pivovarov Yu L, Tukhfatullin T A 2009 *Radiation from Relativistic Electrons in Periodic Structures (RREPS-09): Book of Abstracts of VIII International Symposium (Zvenigorod, Russia, 7-11 September 2009)* pp 97-98
[6] Baier V N, Katkov V M, Strakhovenko V M Electromagnetic 1998 *Processes at High Energies in Oriented Single Crystals* (Singapore: World Scientific Publishing Co)
[7] Bogdanov O V, Korotchenko K B, Pivovarov Yu L 2008 *J Phys* B **41** 055004
[8] Bogdanov O V, Korotchenko K B, Pivovarov Yu L 2007 *JETP Letters* **85** 555
[9] Beloshitskiy V V, Kumakhov M A *JETP* 1978 **74** 1244
[10] Yabuki R, Nitta H, Ikeda T, and Ohtsuki Y H 2001 *Phys Rev* B **63**, 174112
[11] http://geant4.web.cern.ch/geant4/support/index.shtml

Tunable X-ray source based on mosaic crystals using for practical applications

D A Baklanov, Duong Thi Giang, R A Shatokhin, I E Vnukov and Yu V Zhandarmov

Belgorod State University, 14 Studencheskaya str., Belgorod 308007, Russian Federation

E-mail: vnukov@bsu.edu.ru

Abstract. The prospect of creation of the X-ray source with tunable wavelength on the base of a middle energy accelerator and oriented crystals are analyzed. It is proved, that due to the contribution of diffracted bremsstrahlung, the mosaic crystals provide the essentially greater yield of hard radiation, than perfect crystals. The method of choosing crystals with necessary characteristics in contribution ratio of parametric X-ray and diffracted bremsstrahlung in radiation yield is proposed.

1. Introduction

Generation mechanisms of X-ray, appeared when fast charged particles interact with periodic structures, have been extensively studied in the past two decades [1]. The interest to this study is caused by the search for new tunable X-ray sources, capable to offer an alternative to storage rings. One of such mechanisms is parametric X-ray (PXR) of fast electrons in crystals [2, 3]. Advantages of PXR over other sources, based on the use of radiation of fast particles in substance, is the opportunity to vary the energy of photons by the change of orientation of a crystal or a position of the radiated object and smaller doze loadings as PXR is let out at the big angles to the direction of an electron beam. It is supposed that the most perspective is the use of this type of radiation in medical purposes, where small sources of hard quasimonochromatic X-ray ($\omega \geq$20 keV, $\Delta\omega/\omega \leq$5%) [4, 5, 6] are demanded.

Measurements of the PXR characteristics, performed for a range of conventional crystals with perfect structure: diamond, silicon, germanium, tungsten, quartz, lithium fluoride (see [1] and references therein) showed that the radiation yield weakly depends on the crystal and is not enough for the practical realization of the source, based on this mechanism of radiation. The ways of radiation yield increase, offered recently (see, for example, [6, 7] and references therein) also hasn't found any practical application.

In works [8, 9] it is noted that the use of ideal mosaic crystals permits to increase significantly the radiation yield due to the contribution of diffracted photons of the bremsstrahlung (DB) or transition radiation (DTR). In quoted works the DB (DTR) contribution to the yield of the registered radiation in some times exceeded the PXR contribution and was well described by the theory of X-ray diffraction in mosaic crystals [10]. Accordance of the measuring results for the mosaic crystals of pyrolytic graphite [8] and diamond [9] with the calculation allows to compare perfect and mosaic crystals from the point of view of their use for generation of hard X-ray emission and appreciate possible advantages and disadvantages.

2. General expressions

As it is noted above mosaic crystals are able to provide higher intensity of quasimonochromatic radiation by means of additional contribution of DB and DTR mechanisms, than perfect crystals. However, as for these crystals PXR yield remains high, it is necessary to consider the contribution of this mechanism to the radiation observed. The kinematic PXR theory describes the measuring results with the fault not worse than 10-15% (see, for example, [11] and references therein), that is why to count the PXR yield the formula of spectral and angular distribution, obtained by the kinematical approach in the work [12] is used:

$$\frac{d^2N}{dZd\Omega} = \frac{\sum_\nu \alpha\omega^3|\chi_{\vec{g}}|^2}{2\pi\varepsilon_0^{3/2}\beta(1-\sqrt{\varepsilon_0}\vec{\beta}\vec{n})}\left[\frac{(\omega\vec{\beta}-\vec{g})\vec{e}_{\vec{k}\nu}}{(\vec{k}_\perp+\vec{g}_\perp)^2+\frac{\omega^2}{\beta^2}\{\gamma^{-2}+\beta^2(1-\epsilon_0)\}}\right]^2 \tag{1}$$

Here and further the system of units used is $\hbar = m_e = c = 1$. Here $\varepsilon_o = 1 - \omega_p^2/\omega^2$, where ω_p -plasma frequency of medium, $\vec{\beta} = \beta\vec{n}_0$ - vector of electron speed, $\vec{n}_0$, $\vec{n}$ - unit vectors in the line of the flying electron and emitted photon (with energy ω and momentum $\vec{k}$), $\vec{g}$ is the reciprocal lattice vector, $\vec{e}_{\vec{k}\nu}$ are the polarization vectors, $\vec{e}_{\vec{k}1} = \frac{[\vec{n},\vec{n}_0]}{|[\vec{n},\vec{n}_0]|}$, $\vec{e}_{\vec{k}2} = [\vec{e}_{\vec{k}1}, \vec{n}]$, $\perp$ - index, indicated the projections of vectors on the plane, which is parallel to $\vec{n}_0$. The other notations are conventional. The term $|\chi_{\vec{g}}|^2$ denotes the following value:

$$|\chi_{\vec{g}}|^2 = |S(\vec{g})|^2 \exp(-2W)\left[-\frac{\omega_p^2}{\omega^2}\frac{f(\vec{g})}{z}\right]^2 \tag{2}$$

Here $|S(\vec{g})|^2$ is the structure factor, $\exp(-2W)$ is the Debye-Waller factor, $f(\vec{g})$ is the Fourier component of the spatial distribution of electrons in a crystal atom ($f(0) = z$, where z is the number of electrons in an atom).

As it is shown in [13] the total PXR yield doesn't depend on crystal mosaicity. During the radiation collimation mosaicity causes the broadening of the orientational dependencies (OD) and PXR angular distributions, and in case of comparability of Θ_{ph} and mosaicity factor σ_m it decreases radiation yield at the expense of PXR photons re-reflection [8]. The crystal mosaicity was taken into account by the calculation of the spectrum for different mosaic blocks, taking into account their distribution in the target.

It is known (see [10] and quoted literature) that by the perfection degree crystals can be classified by two factors: sizes of regular blocks or areas in a crystal and the extent of their mutual disorientation. According to the first factor all the crystals can be divided into to classes a and b. In the a class crystals separate areas are large enough for the primary extinction influence was visibly revealed, i.e. their linear size is comparable with the primary extinction length l_{ex}. In the class b crystals the regular block size is small that is why the effect of primary extinction is not observed. According to the second factor crystals also can be divided into two classes - α and β. In the α class crystals the blocks are almost parallel to each other, their mutual disorientation is small, that is why the contribution of the secondary extinction is big. In the class β crystals blocks are distributed irregularly, that is why the contribution of the secondary extinction is small.

Combining these factors we can classify crystals by their perfection degree into 4 groups from $a\alpha$ to $b\beta$. Class limit when possible disorientation of mosaic blocks is less than the area of total X-ray reflection (Darwin's table) is an ideal perfect crystal, and the class $b\beta$ limit is an ideal mosaic crystal. It should be noted that belonging to these classes isn't prescribed once and for all, as it depends on the extinction length, which in its turn depends on the order of reflection and photon energy [10].

It is known that the X-ray reflectivity of crystals is directly related to the perfection of their structure. Class $a\alpha$ crystals provide the least narrow rocking curve (FWHM ~ 20-30″), which is well described by dynamical theory of X-rays diffraction in perfect crystals. For this reason, their integral reflectivity is small. Crystals with a damaged structure have bigger reflectivity. Class $b\beta$ mosaic crystals have maximum reflectivity and the broadest rocking curve.

Class $a\alpha$ crystals can be represented sometimes as a set of crystallites, i.e. perfect crystals of small sizes, each of which reflects X-rays in accordance with the dynamic theory of diffraction. Reflectivity of such a crystal as a whole is close to the reflectivity of a perfect crystal with the same sizes, if the size of the crystallite is greater than the radiation absorption length. Otherwise, reflectivity of such crystal is bigger than of perfect one, since the radiation can be reflected from the crystallites located deeper in the course of the beam [10].

Characteristic parameters of the above classification of crystals - the width of the Darwin "table" $\Delta\Theta$ and the length of the primary extinction l_{ex} depend on the reflection order and photon energy [10]. For unpolarized radiation and the lack of absorption it can be written:

$$\Delta\Theta = 2 \cdot \gamma \Delta\theta_0, \tag{3}$$

where $\Delta\theta_0 = 2 \cdot \delta / \sin 2\Theta$ - allowance to the Bragg angle Θ because of the wave refraction in a crystal, $\delta = (\omega_p/\omega)^2/2$ - the difference of refraction index from 1, $\gamma = \frac{1}{2}\frac{F(\vec{g})}{F(0)}(1+\cos(2\Theta))$, and $F(\vec{g}) = S(\vec{g})f(\vec{g})$ - a structure factor.

As an estimation of the primary extinction length one can use the expression:

$$l_{ex} = d/(2\bar{\xi}\sin\Theta), \tag{4}$$

where d - interplanar distance, and $\exp(-2\bar{\xi})$ - weakening of primary wave intensity when passing through one plane:

$$2\bar{\xi} = \frac{\pi d^2 N F(\vec{g})}{n}\frac{e^2}{mc^2}. \tag{5}$$

Here N is scattering centers concentration, n - the order of reflection.

If you use crystals to generate intense beams of X-ray a divergent photon beam with continuous spectrum falls on it or occurs in it, some of photons of which are then reflected and recorded by the detector with the fixed angular location. In such experimental set up the imperfect structure causes broadening of the detected radiation spectrum and its intensity increase. That is, for the a class $a\alpha$ crystal the detected radiation is of the least intense, and the spectrum width and orientational dependence is mainly determined by the radiation collimation angle ϑ_c. For the class $b\beta$ crystal radiation intensity is maximum and the width of the orientation dependence and the radiation spectrum are determined basically by the crystal mosaic structure.

The main advantage of mosaic crystals is the additional contribution to the registered spectrum of diffracted real photons. For the electrons of medium energy ($E_e \leq 100$ MeV) the main photon source is the bremsstrahlung (BS). For the relativistic electron and soft photons ($\omega \ll E_e$) the spectral and angular distribution of BS intensity produced per unit of path length in an amorphous substance can be represented in the form [14]:

$$\frac{d^2 I}{d\omega d\Omega} = \frac{\gamma^2}{\pi L}\frac{1+\gamma^4\theta^4}{(1+\gamma^2\theta^2)^4}, \tag{6}$$

where L is the radiation length, θ - angle of photon outlet.

In the analyzed case in the crystal the divergent photon beam with continuous spectrum occurs. To reflect a monodirect and monochromatic photon beam with the energy ω and direction $\vec{n}$ from the element of mosaic crystal with volume ΔV can be wrote as [10]:

$$\int P(\theta)\, d\theta = Q\Delta V, \tag{7}$$

where $P(\theta)$ - reflecting power of the crystal element at the angle θ, which is proportionate to the distribution of mosaic blocks in the crystal [15]. $Q\Delta V$- integral reflection from the element ΔV, Q - the X-ray reflectivity depends on the crystal parameters and the radiation energy as:

$$Q = \left(\frac{e^2}{mc^2}\right)^2 \frac{N^2\lambda^3}{\sin(2\Theta)} |F_p| \left|F(\vec{g})\right|^2 \cdot \exp(-2W), \tag{8}$$

where λ - wavelength, $|F_p|$ - polarization multiplier. For an unpolarized photon beam $|F_p| = (1 + \cos^2 2\Theta)/2$.

Let the radiation with the spectral and angular intensity distribution $I(\omega, \vec{n})$ be spread in a mosaic crystal possessing the distribution of reciprocal lattice vectors $P(\vec{g})$. Here $\vec{g} = |\vec{g}|\vec{\alpha}$, where $\vec{\alpha}$ is the unit vector specifying the deflection of crystal mosaic blocks from the average direction $\vec{g}_0 =< \vec{g} >$. Starting from Bragg law for the photon with energy ω and direction $\vec{n}$ we can obtain the direction of vector $\vec{g}$ of the block, on which this photon can diffract, and the driving direction of the reflected photon $\vec{n}'$. Then probability density of the reflection of photon with fixed ω and $\vec{n}$ in the block of mosaic crystal with the thickness Δt along the photon driving direction can be written as:

$$f(\omega, \vec{n}) = q(\omega, \vec{n})Q(\omega)\Delta t, \tag{9}$$

where $q(\omega, \vec{n})$ - a factor of the crystal mosaicity:

$$q(\omega, \vec{n}) = \int P_m(\alpha_x(\omega, \vec{n}, \alpha_y), \alpha_y)d\alpha_y. \tag{10}$$

Here $P_m(\alpha_x,\alpha_y)$- distribution of crystal mosaicity, expressed through $\omega, \vec{n}, \alpha_y$ [8].

To calculate the yield of the diffracted photons in the aperture of the collimator for each order of the reflection i, spectral and angular distribution of the radiation intensity, taking into account multiple electron scattering in the target $\frac{d^2I(\omega,\vec{n})^*}{d\omega d\Omega}$, was reduced with the probability density of the diffraction by all variables, including energy, angles of photon outlet θ_x and θ_y and the crystal thickness T:

$$Y_i = \int\limits_{\Delta\omega_i} d\omega \int\limits_{T} dt \iint\limits_{\Delta\Omega} \frac{d^2I(\omega, \vec{n})^*}{d\omega d\Omega} S(\omega, \vec{n}, \vec{n}', t)Q(\omega)q\left(\omega, \vec{n}\right) d\theta_x d\theta_y, \tag{11}$$

here $S(\omega, \vec{n}, \vec{n}', t)$ - factor considering the radiation absorption in the target material and the experiment geometry, $\Delta\Omega$ angular acceptance, and $\Delta\omega_i$ energetic capture of the detector for i reflection order.

According to the kinematic PXR theory for the definitely Bragg direction the intensity of radiation is equal to zero [2, 3, 12]. That is why in perfect crystals diffraction of PXR photons usually is not viewed. In mosaic crystals for the photon, produced in one of the mosaic blocks, the diffraction conditions in other block can be fulfilled. To consider this effect for each PXR photon with the energy ω and momentum $\vec{k}$ from Bragg law in accordance with (9)-(10) the blocks of mosaic, on which this photon can diffract, and the probability of reflection were defined. From here, taking into account the distribution of the mosaic blocks, PXR photon beam attenuation on its way from the birth point to the outlet out of crystal was defined. More detailed the method of calculation of the PXR and DB yields in the mosaic crystals was described in [8].

3. Comparison perfect and mosaic crystals

The PXR yield linearly depends on the thickness of the crystal, that is why the natural way to increase the intensity is to enlarge the thickness of a target. In Figure 1 there is a calculated dependence of PXR yield for the reflection (220) and $\Theta_B = \Theta_D/2$ from the thickness of the silicon crystal for the following conditions: E_e=45 MeV; electron beam divergence ϑ_e=1.5 mrad; observation angle Θ_D=16.7°($\omega \sim$ 33 keV); ϑ_c =0.85 mrad.

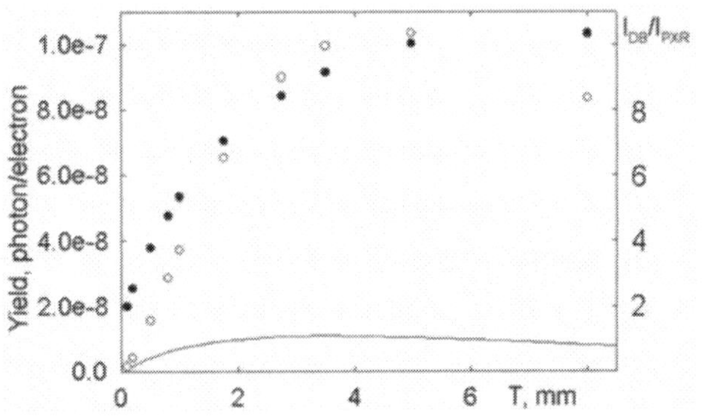

Figure 1. The dependence of PXR and DB intensity on the silicon crystal thickness. Curve - PXR, ∘ - DB, • - Y_{DB}/Y_{PXR}

The use of class *b* mosaic crystals permit to get essentially bigger radiation yield due to the DB contribution [8, 9]. In this figure there is also a dependence of DB yield for the mosaic sample with σ_m=1 mrad (circles) and the ratio of DB and PXR yields (points). It is clear from the figure that even for T=0.1 mm the DB yield from the mosaic crystal is bigger than the PXR yield from perfect one. As the PXR yield little depends on the sample mosaicity [13], and the DB yield from the perfect crystal is less than PXR yield [16], a mosaic crystal almost always provides the bigger radiation yield than perfect one. It should be noted that fulfilling the condition $\sigma_m < \vartheta_c$ the DB spectrum is narrower than PXR spectrum [9], e.g. for σ_m=0.6 mrad and T=3.5 mm the full width of the spectrum (on level 10% from max) $\Delta\omega_{DB}$=1.2 keV is less than for the PXR spectrum $\Delta\omega_{PXR}$=1.95 keV, and the relation of intensities ~8.

As it is seen from the figure, the maximum of the PXR and DB yield is reached for the crystal thickness is about the absorption length of photons. That is why with the decrease of the photon energy the advantage of the mosaic crystals over perfect ones becomes lesser. E. g. for E_e=45 MeV, of a silicon crystal and photon energy $\omega \approx$18 keV the relation of the DB and PXR yields for the target with optimal thickness decreased to ~4 [8].

The main problem when we use the crystals with optimal thickness to generate radiation is not the spectrum width but the big BS background in the location of the radiated object. As was shown in Ref.[17] the number of BS and DB photons, fallen on the radiated object when using the crystals of optimal thickness, as a rule is comparable and the full energy is in some extent bigger than the full energy of useful radiation. Apparently this is the main reason why the authors' idea [5] about using PXR beam from a thick pyrolytic graphite crystal for mammography was not realized.

One of the solutions for this problem is a double-crystal scheme. Recently in the work [18] the system of two perfect crystals for the PXR generation in a thin crystal and its subsequent diffraction in another, thicker one was realized. The main advantage of such scheme is that the narrow width of spectrum leads to low radiation intensity. If the width of the spectrum is not a critical parameter, e.g. for the radiography, where $\Delta\omega/\omega \sim$ 2–3% may be enough, than the use of mosaic crystals instead of the perfect ones can essentially enlarge the radiation yield [9].

To realize the merits of mosaic crystals for the obtaining of intensive X-rays the samples with low Z and σ_m <0.4-0.6 mrad are necessary. The low limit of σ_m and the required sizes of blocks are determined by the limits of the application of the class b mosaic crystal model $\sigma_m \gg \Delta\Theta$ and $l \ll l_{ex}$. These crystals can be used on the beams of synchrotron radiation instead of perfect crystals with asymmetric cut or curved ones [19], when high intensity is demanded and there are no strict requirements for its monochromaticity.

4. Choice of crystals

For most applications, which require high monochromaticity of radiation, the use of pyrolytic graphite crystals is limited because of the large value of mosaicity factor. As shown above, and in the work [9], the other crystals, which meet the requirement that $\sigma_m \gg \Delta\Theta$ and $l \ll l_{ex}$ should have the same merits. The choice of crystals with small mosaicity is easily done with the help of conventional methods of X-ray analysis. Estimation of mosaic blocks sizes and the angles of their relative misalignment is a more complex experimental task, which is successfully solved by conventional methods only for the surface layers, for example, see [20].

Characteristic parameters of the classification $\Delta\Theta$ and l_{ex} depend on the reflection order and photon energy. In this regard, we analyze the measuring results for significantly different photon energies. Figure 2 shows the measuring results of the vertical angular distribution of radiation yield for the reflection (220) from the crystal of a natural diamond with the sizes $6 \times 10 \times 0.35$ mm^3 in the experiment [21] (circles) for the following experimental conditions: E_e=900 MeV, Θ_D=90°, ϑ_c=1.88 mrad, $\sigma_m \approx$0.2 mrad. The crystal is installed in Bragg geometry (Θ_B=45°) so, that the thickness of the target along the electron beam was $\approx$ 0.5 mm.

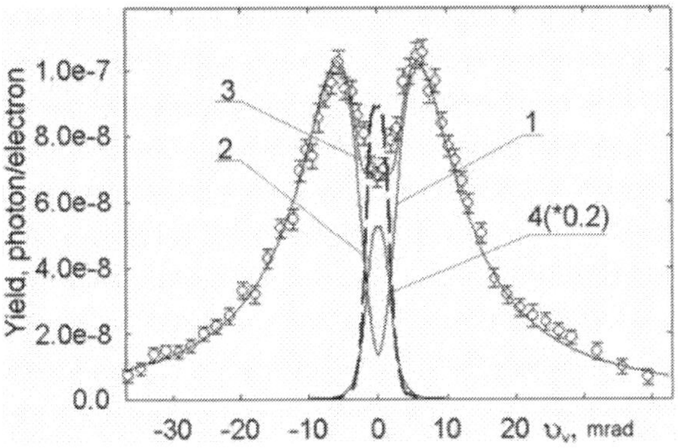

Figure 2. The vertical angular distribution for the experiment [21]. o - experiment. Lines - calculation: 1- PXR for the perfect crystal; 2 - DTR for the perfect crystal; 3- PXR+DTR for the perfect crystal; 4 - DTR for the class $b\beta$ mosaic crystal.

Dependence 1 in this figure is calculation results according to the kinematic PXR theory [12] in the assumption that the crystal is perfect. As it is shown in the works [8, 9, ?], under the condition $\sigma_m \ll \Theta_{ph}$, as in this case, mosaicity has almost no effect on the spectrum and angular PXR distribution. As in the quoted experimental work, because of the normalization fault (~20-30%) measurements are normalized to the calculation results of photons emission angle $\vartheta_v \geq 5$ mrad. The figure shows that the calculated dependence soundly describes the experimental results only for the angles $\vartheta_v \geq$ 5-7 mrad. For smaller angles the experimental points are significantly higher than the calculated curve, indicating the contribution of radiation with narrower angular distribution than PXR, i.e. the contribution of real photons diffraction In this case (ω ~6.97 keV $\ll \gamma\omega_p$ ~67 keV) it is DTR.

Curve 2 shows the results of the calculation of vertical dependence of DTR yield carried out according to the method of work [22]. Curve 3 is the resulting angular distribution of PXR + DTR, which satisfactorily describes the experimental results. Nevertheless, in the center of the angular distribution experimental and calculated dependencies are somewhat different, which is apparently caused by a crystal mosaicity (see below).

Mosaicity factor σ_m ~0.2 mrad is comparable with the width of total reflection area for this photon energy $\Delta\Theta$ ~0.02 mrad. Therefore, by the degree of blocks disorientation this crystal should be attributed to class α. To test the applicability of a mosaic class b crystal approach to this model, the angular DTR distribution in this approximation by the method of the work [22](curve 4) was calculated. It is seen from this figure that the angular DTR distribution for the perfect and mosaic crystals are similar, but differ in intensity almost by an order of magnitude.

Curve 3 is closer to the experimental points than the sum of curves 2 and 4, therefore the diamond crystal, used in the experiment [21], should be classified as $a\alpha$ class, i.e. the sizes of its constituent crystallites are greater than l_{ex} ~7 μm. The difference between curve 3 and the experimental results is caused by the fact that for photons with ω= 6.97 keV, the absorption length $l_a \sim 0.43$ mm $\gg l_{ex}$. Therefore blocks, located in the depths of the crystal and disoriented relative to the blocks on the surface $\Delta\Theta$, can contribute to the detected radiation yield. The value of this contribution is very small in comparison with a ratio of values σ_m and $\Delta\Theta$ (see above and Fig. 2), therefore, the crystallite size is significantly bigger l_{ex}.

In Figure 3 the points show the dependence of the first order reflection photons yield from the angle of (110) plane misalignments relative to the direction of Bragg reflection obtained in the experiment [9].

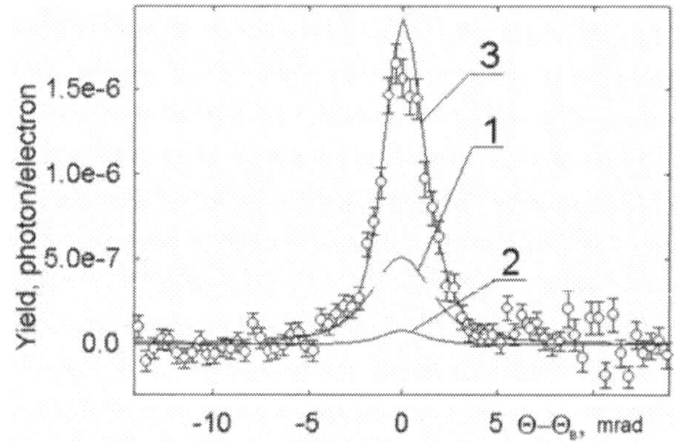

Figure 3. The orientation dependence for the experiment [9]. ○ - experiment. Lines - calculation: 1- PXR for the perfect crystal; 2 - DB for the perfect crystal; 3- PXR+DTR for for the class $b\beta$ mosaic crystal.

Experiment is made in Laue geometry on the natural diamond crystal with the surface mosaicity σ_m ~0.2 mrad, the orientation of <110>, and sizes $6 \times 8 \times 2$ mm ^{3}, cut from the same sample as the target, used in the experiment [21]. Experimental conditions: E_e= 500 MeV, Θ_D = 4° ($\omega \sim$ 145 keV), ϑ_c=1.9 mrad.

Curve 1 shows the calculated dependence obtained within kinematic PXR theory. Measured and calculated OD are close enough in form, but differ significantly in amplitude. The difference in the widths (FWHM) of calculated and measured dependencies $\Delta\Theta_{calc}$=3.97 mrad and $\Delta\Theta_{exp} = (2.9 \pm 0.2)$ mrad exceeds the experimental fault (measurement pitch OD $\approx$0.4 mrad) and witnesses the radiation contribution with a narrower angular distribution than PXR. For the conditions of the quoted work ($\omega \approx$ 145 keV $\gg \gamma\omega_p \sim$ 35 keV) it is DB.

Evaluation of DB contribution by the method of the Ref. [16] in the assumption, that the crystal is perfect, (curve 2) showed that in this case DB yield does not exceed 25% of the PXR one and cannot explain the experimental results. Accounting of the crystal mosaicity by the method [8] in the assumption of a homogeneous distribution of mosaic blocks with sizes smaller than the primary extinction length l_{ex} = 148 μm in the crystal thickness, i.e. its belonging to the mosaic crystals of class $b\alpha$, showed that the difference is indeed caused by mosaicity. The resulting dependence of PXR+DB, calculated with due account for the mosaicity for both components (curve 3) is close to the experimental one. The calculated values of the yield $Y_{calc} = 1.94 \cdot 10^{-6}$ phot./electr. and the width of the OD of $\Delta\Theta_{calc}$ = 2.67 mrad are satisfactory agreed to the measuring results $Y_{exp} = (1.63 \pm 0.008) \cdot 10^{-6}$ phot./electr. $\Delta\Theta_{exp} = (2.9 \pm 0.2)$ mrad [9].

Both targets are cut from the same sample of natural diamond. Consequently, their microstructure has to be approximately the same. However, the measurements for them are described in different models. For Θ_D=90° the crystal appears as a perfect or $a\alpha$ class mosaic crystal. For Θ_D=4° the results are described in the approximation of $b\alpha$ class mosaic crystal.

Different manifestations of the same crystal structure is caused by the energy difference of the detected radiation. With photon energy increasing l_{ex} increases and $\Delta\Theta$ decreases. Therefore, with photon energy increasing any imperfect crystal becomes a class b mosaic crystal. That is, changing the observation angle one can trace the transition from one description to another, and, therefore, estimate the sizes of mosaic blocks. PXR theory in the kinematic approximation describes the results of measurements carried out in the perfect crystals with an error not worse than 10-15% [11]. With same accuracy the theory of X-ray diffraction in mosaic crystals describes the yield of DB and DTR from class b mosaic crystals [8]. Therefore having measured the radiation yield at Bragg angles for several observation angles and compared it with the predictions of the kinematic PXR theory one can register a contribution of real photon diffraction and make a conclusion about the average sizes of blocks which form the sample.

These values behave this way with of the reflection order increasing. Consequently, having measured the angular distribution of radiation for one observation angle and different orders of reflection, and compared it with the calculation results of PXR and DB+DTR yields for the perfect crystal and the mosaic class b crystal, we can determine for which reflection order the transition from one description to another will start, i.e. to estimate the average size of blocks.

5. Conclusion

The results of the conducted researches can be formulated in the following way:

1) Use of the class $b\beta$ mosaic crystals instead of the perfect ones in a great extent increases the quasimonochromatic X-radiation yield. Mosaicity $\sigma_m < \vartheta_c$ provides the increasing of the radiation yield without noticeable worsening of radiation monochromaticity.

2) Registration of the radiation yield at Bragg angles (DTR+PXR+DB) for different observation angles (different reflection orders) permit to obtain the information about characteristic sizes of blocks, which form the test sample, and its applicability for getting intensive beams of quasimonochromatic X-ray.

This work is supported in part by the internal grant of Belgorod State University.

References

[1] Rullhusen R, Artru X and Dhez P 1999 *Novel Radiation Sources Using Relativistic Electrons* (Singapore: World Scientific)

[2] Garibian G Shi Yang 1971 *Sov. Phys. JETP* **61** 930

[3] Baryshevskii V, Feranchuk I 1971 *Sov. Phys. JETP* **61** 944

[4] Baldelli P, Taibi A, Tuffanelli A, Gambaccini M 2004 *Physics in Medicine and Byology* **49** 4125

[5] Peistrup M A *et al* 2001 *Review of Scientific Instruments* **72** 2159

[6] Takashima Y *et al* 1998 *NIM* B **145** 25

[7] Kaplin V V *et al* 2002 *Applied Physics Letters* **80** 3427

[8] Bogomazova E A *et al* 2003 *NIM* B **201** 276

[9] Baldin A N, Vnukov I E, Shatokhin R A 2007 *Technical Physics Letters* **33** 625

[10] James R *The optical principles of the diffraction of X- rays* 1958 (London: G. Bell and Sons)

[11] Brenzinger K-H *et al* 1997 *Phys. Rev. Lett.* **79** 2462

[12] Nitta H 1991 *Phys. Lett.* A **158** 270

[13] Afanas'ev A M and Aginian M A 1978 *Sov. Phys. JETP* **74** 300

[14] Bazylev V A, Zhevago N K *Radiation of Relativistic Particles in External Fields and in Matter* 1987 (Moscow: Nauka Pub.) (in Russian).

[15] Chabot M *et al* 1991 *NIM* B **61** 377

[16] Baklanov D A, Baldin A N, Vnukov I E, Nechaenko D A, Shatokhin R A 2007 *The Journals of Kharkiv National University* **763** No.1(33) 41 (in Russian)

[17] Shatokhin R A, Vnukov I E, Zhandarmov Yu V 2009 *Problems of Atomic Science and Tekhnology, Series "Nuclear Physics Investigatios"* **51** 113

[18] Hayakawa Y *et al* 2006 *NIM* B **252** 102

[19] Arfelli F 2000*NIM* A **454** 11

[20] Ohler M, Baruchel J, Moore A W, Galez Ph, Freund A 1997 *NIM* B **129** 257

[21] Baldin A N, Vnukov I E, Kalinin B N, Karataeva E A 2006 *Poverkhnost'* No. 4, 72 (in Russian)

[22] Adischev Y N *et al* 2003 *NIM* B **201** 114

On electron beam motion near the SPARC photoinjector cathode

A S Podlesnaya[1], A V Dik[2], S B Dabagov[2,3] and M Ferrario[3]

[1]MEPhI, Kashirskoe shosse 31, 115409 Moscow, Russian Federation
[2]LPI PAS, Leninsky prospect 53, 119991 Moscow, Russian Federation
[3]INFN-LNF, Via E. Fermi, 40-00044 Frascati, Rome, Italy

E-mail: alinapodlesnaya@mail.ru

Abstract. Description of the electron beam behavior at the moment of its generation inside a radio–frequency gun of the SPARC[1] photo–injector is presented by this research work aiming to show the influence of laser radiation presence on the beam accelerated by the electric field. The electron motion inside the photo–injector RF–gun has been analytically solved within the reasonable approximations. The partial qualitative analysis of both electron beam distribution in a space as well as pulse photoelectron distribution at the moment of beam generation are given. The electron redistribution inside the beam was analyzed and visualized by means of the developed MATLAB code. The obtained estimations have been performed using the parameters of electron beam defined during the SPARC commissioning.

1. Introduction

The Free-electron lasers (FELs) have developed rapidly over the past 30 years as a powerful source of coherent radiation [1]. One of the European FEL projects, the SPARC, has been realized as a facility within the collaboration of the Italian research Institutes such as INFN and ENEA (CNR Frascati) [2]. One of the main goals of the SPARC project is to develop a photoinjector facility for high brightness electron beams generation to drive a self-amplified spontaneous emission FEL (SASE FEL) experiment [3] and other applications [4-9]. SASE FEL regime, which is characterized by the beam microbunching, requires a high-quality electron beam and long undulators. There are several ways to form a beam microbunching that, in turn, increases the emitted radiation intensity in SASE FEL regime. In this work we aim to consider the influence of laser radiation on electron beam motion near the SPARC photoinjector cathode as a possibility for microbunching of electron beam before the undulator entrance.

2. Motion of electron beam inside the SPARC photoinjector

At generation, the electron beam is being affected by both incident $\vec{E}_l^{in}$ (at 70 degs) and reflected $\vec{E}_l^{ref}$ laser beams (190 MV/m amplitude, ~10 ps pulse duration, 266 nm wavelength), as well as by

[1] SPARC is a FEL project at Laboratori Nazionali di Frascati, Istituto Nazionale di Fisica Nucleare

the accelerating electric field $\vec{E}_{ac}$ (120 MV/m). The cathode quantum efficiency is about 10^{-4} [10]. The problem with similar conditions has been previously reported [11].

Let us consider simplified case of normal laser incidence on the cathode. In our case the penetration intensity change with respect to nonlinear absorption process is determined by the following

$$I(z) = I_0 \frac{\exp(-\mu_1|z|)}{1 + I_0 \frac{\mu_2}{\mu_1}(1 - \exp(-\mu_1|z|))},$$

where I_0 is the absorbed radiation intensity [12]. The estimates done according to the SPARC parameters show that the term $I_0\mu_2/\mu_1$ is negligible and $\mu_1 \geq 10/\lambda$ [13]. Therefore, the linear optics calculations can be applied [14]. The laser beam distribution is determined as a distribution with the Gaussian-beam profile, $I_i(x,y) = C\exp(-x^2/2\sigma_x^2 - y^2/2\sigma_y^2)$, the mean intensity of a step-shaped temporal distribution is $\overline{E}_0 \sim 10^7$ V/m [11]. Taking into account the dependence $I(z) = I_0\exp(-\mu_1|z|)$, we admit that all unreflected radiation is absorbed on the cathode surface. The mean energy absorbed per one period at the depth z can be written as $d^2\overline{E}/dt = \mu_1 I(x,y,z)dV$ that defines the number of electrons per one vibration period of a laser pulse

$$dN_e = \frac{d^2\overline{E}}{n\hbar\omega_l} = \frac{\mu_1 I(x,y,z)}{n\hbar\omega_l}dV \quad (1)$$

According to our calculations the total number of electrons is $N_e \sim 10^{10}$ that corresponds to the SPARC experimental data [15]. The photoelectron concentration per one period can be defined as

$$n_e(x,y,z) = \frac{\mu_1}{n\hbar\omega_l}(1-R)C\exp(-\frac{x^2}{2\sigma_x^2} - \frac{y^2}{2\sigma_y^2} - \mu_1|z|) \quad (2)$$

A simple approximate result can be obtained on the basis of the partial qualitative analysis of the momentum distribution function $f(\vec{p}) = B\exp(-p^2/2a)p^2\sin\theta$ where unknown parameters can be determined from the electron escape maximum angle and the mean energy of the electron beam [16]. The photoelectron momentum distribution can be determined as follows

$$\frac{dN}{dpd\Omega} = \frac{N_0}{\overline{\theta}^2\sqrt{2}}\sqrt{\left(\frac{3}{2\pi\overline{E}m}\right)^3}\exp\left(-\frac{3p^2}{4m\overline{E}}\right)p^2, \quad (3)$$

where $\overline{\theta}$ is the beam divergence angle with respect to the axis z.

Figure 1 shows the electron bunch at the time $t = 0$, supposing that the electron beam has normal distribution on the plane of ellipsoid symmetry axes.

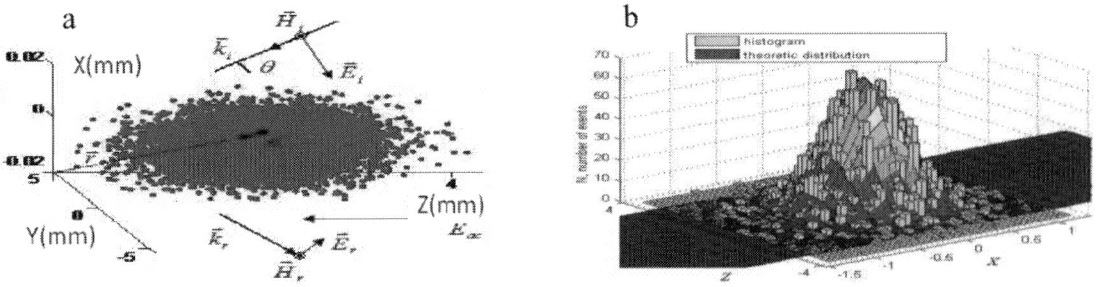

Figure 1. (a) Motion of electron experiences a field shaped by incident and reflected waves, and by accelerating field (The electron bunch with the shape of an ellipsoid (x– direction: a~1 mm, z–direction: b~2 mm, t=0). (b) The electron beam distribution: three-dimensional image (theory), histogram (simulation).

The distribution of electrons performed by the simulation in MATLAB is shown in Figure 1(b). We can use such distribution as a distribution at the initial position of the electron beam, see Figure 4(a).

3. Motion of electron beam under the action of electron beam

Photoemitted electrons experience a field formed by both incident and reflected waves and by the accelerating field. The reflection coefficient of electromagnetic wave R is defined from the relation: $\left|\vec{E}_{ref}\right| = R\left|\vec{E}_{inc}\right|$. As shown below, the magnetic component of laser field is negligible, see Figure 3. The total electromagnetic field is determined by the following: $\vec{E}_r = RE_0(\cos\theta, 0, \sin\theta)\sin(\omega_l t - k(z\cos\theta - x\sin\theta))$, $\vec{E}_{ac} = (0,0,-E_{ac})$, and $\vec{E}_i = E_0(-\cos\theta,\ 0,\ \sin\theta)\sin(\omega_l t + k(z\cos\theta + x\sin\theta))$. The electron trajectory can be represented as a smooth curve $\{X(t), Z(t)\}$, moving along of which the electron executes very fast small-amplitude oscillations $\{\xi(t), \varsigma(t)\}$ in a rapidly alternating laser field [17]. Then the general solution of the electron motion is described by the formulas

$$\begin{Bmatrix} x(t) \\ z(t) \end{Bmatrix} = \begin{Bmatrix} X(t) \\ Z(t) \end{Bmatrix} - f_{\left\{ 1 \atop 3 \right\}} \sin(\omega_l t) - f_{\left\{ 2 \atop 4 \right\}} \cos(\omega_l t), \quad f_i = \frac{f_i(X,Z)}{m\omega_l^2}, \tag{4}$$

where $X(t) = C_0 t + C_1$, $Z(t) = \dfrac{eE_{ac}}{2m} t^2 + C_2 t + C_3$ and $f_i(X,Z)$ function is given by

$$f_{\left\{ 1 \atop 2 \right\}} = eE_0 \cos\theta \begin{pmatrix} \cos(k(Z\cos\theta + X\sin\theta)) - R\cos(k(Z\cos\theta - X\sin\theta)) \\ \sin(k(Z\cos\theta + X\sin\theta)) + R\sin(k(Z\cos\theta - X\sin\theta)) \end{pmatrix}$$

$$f_{\left\{ 3 \atop 4 \right\}} = -eE_0 \sin\theta \begin{pmatrix} \cos(k(Z\cos\theta + X\sin\theta)) + R\cos(k(Z\cos\theta - X\sin\theta)) \\ \sin(k(Z\cos\theta + X\sin\theta)) - R\sin(k(Z\cos\theta - X\sin\theta)) \end{pmatrix}.$$

Figure 2(a, b, c) shows the motion of electron beam as a function of time and coordinate. There is an insignificant difference (~10 Å) between electron trajectories under the action of the magnetic component (curve 1, see Figure 3) and without its action (curve 2, see Figure 3).

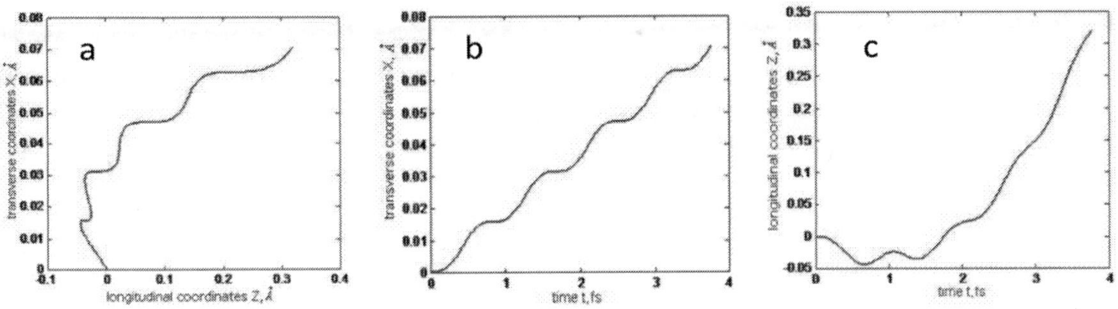

Figure 2. (a) Electron trajectory in the plane of polarization of the laser wave. (Longitudinal displacement Z (Å) versus transverse displacement X (Å)). (b) Transverse displacement X (Å) versus time t (fs). (c) Longitudinal displacement Z (Å) versus time t (fs).

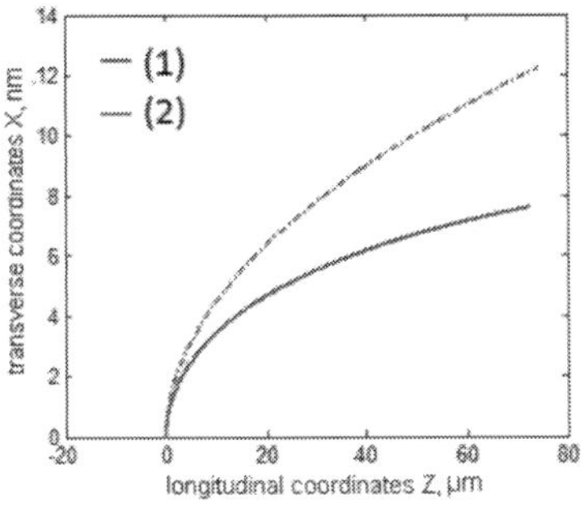

Figure 3. Curve 1 – the numerical solution of electron motion equation taking into account the magnetic component. Curve 2 – the analytical solution of the equation of electron motion without magnetic component. Longitudinal displacement Z (μm) versus transverse displacement X (nm).

The simulation of electron beam motion near the SPARC photoinjector cathode have been performed by means of the MATLAB soft, see Figure 4.

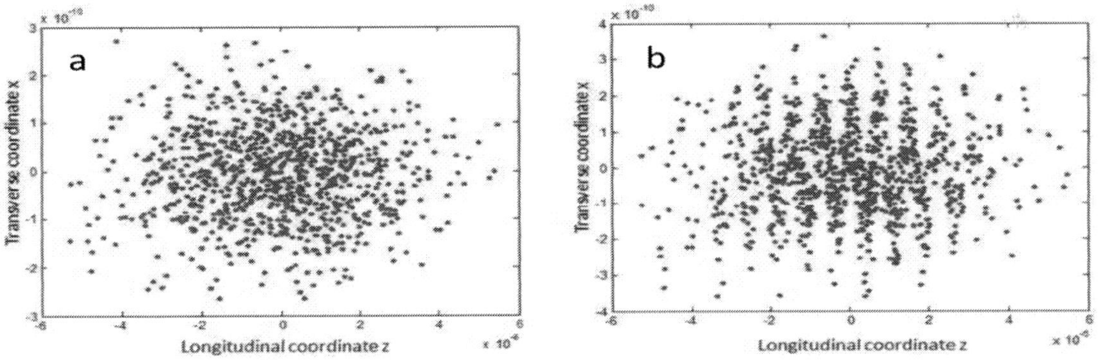

Figure 4. (a) The motion of electron beam in the SPARC photoinjector (the number of electrons $N \sim 1000$) at the initial time $t = 0$. (b) The motion of electron beam in the SPARC photoinjector (the number of electrons $N \sim 1000$) at the time $t \sim 1 fs$.

Examination of the beam shaping process in the SPARC photoinjector has showed that, while the electron beam moving in the laser field ($\sim 10^{-12}$ s), the beam size remains practically constant, see Figure 4(a,b). The obtained periodic structure of the electron beam as a result of high-frequency laser field action can be applied for the development of microbunching of electron beam before the undulator entrance.

4. Conclusion

The presented estimations have been performed using the parameters of electron beam measured during the SPARC commissioning. According to our calculations the total number of electrons is $N_e \sim 10^{10}$ that corresponds to the SPARC experimental data [15].

Numerical solutions of the equation for electron motion inside the RF–gun shows that the dependence of electron trajectories on action of the magnetic component of a laser is negligible (the absolute value is ~10 Å, while the beam size is ~1 mm). The electron redistribution inside the beam was analyzed and visualized by means of the MATLAB software. The beam size, as resulted from the examination of the beam shaping process, remains practically constant in the laser field ($\sim 10^{-12}$ s). At the same time one appears a periodic structure of the electron beam (microbunching) as a result of high-frequency laser field action (the beam modulation) that can be applied as the possibility for microbunching of electron beam before the undulator entrance. The technical solutions for the beam modulation can be different.

Acknowledgments

The present work has been resulted in Diploma theses [16, 18]. The authors (A.P. and A.D) wish to acknowledge the National Research Nuclear University "MEPhI" (Moscow, Russia) and the Frascati National Laboratories LNF INFN (Frascati, Italy) for a possibility to perform the studies at LNF.

References

[1] G. Dattoli, A. Renieri, and A. Torre, Lectures on Free Electron Laser Theory and Related Topics (World Scientific, Singapore, 1993).

[2] Valeria F, Beam dynamics and collective effects in the SPARC project Tesi di Dottorato in Elettromagnetismo, Universita degli Studi di Roma "La Sapienza".

[3] Alesini D, Bertolucci S, Biagini M E, et. al., 2003 The SPARC project: a high-brightness electron beam source at LNF to drive a SASE-FEL experiment *Nucl. Instr. and Meth. Phys. Res. Sect. A* **507** 345

[4] Cianchi A, Alesini D, Bacci A et. al., 2008 High brightness electron beam emittance evolution measurements in an rf photoinjector *Phys. Rev. Special Topics: Accelerators and Beams* **11** 032801

[5] Bacci A, Broggi F, DeMartinis C et. al., 2009 Status of Thomson source at SPARC/PLASMONX *Nucl. Instr. and Meth. Phys. Res. Sect. A* **608**, 90

[6] Giannessi L, Alesini D, Biagini M , et. al., 2008 Seeding experiments at SPARC *Nucl. Instr. and Meth. Phys. Res. Sect. A* **593** 132

[7] Dabagov S B ed., 2007 International conference on charged and neutral particles channeling phenomena, "Channeling 2006" *SPIE* 6634

[8] Dabagov S B, Zhevago N K 2008 On radiation by relativistic electrons and positrons channeled in crystals *La Rivista del Nuovo Cimento* 3

[9] Alberto R 2003 SPARC project and SPARX proposal *Nucl. Instr. and Meth. Phys. Res. Sect. A* **507** 507

[10] Biagini M, Boscolo M, Ferrario M, et. al., 2003 Beam dynamics studies for the SPARC project *Proceedings of the 2003 Particle Accelerator Conference IEEE 2077*

[11] Ferrario M, Gatti G, Serafini L, Rosenzweig J B 2006 Electron beam-laser interaction near the cathode in a high brightness photoinjector *Published in Edinburgh 2006*, EPAC 2805-2807

[12] Delone N B 1989 Interaction of laser radiation with matter *a Lecture Course,* Nauka Moscow (in Russian)

[13] Blombergen L 1966 *Nonlinear Optics* Mir, Moscow (Russian translation)

[14] Schubert M and Wilhelmi B 1973 Introduction to nonlinear optics *part 1* Mir, Moscow (Russian translation)

[15] Ferrario M, Alesini D, Bacci A 2007 Direct measurement of the double emittance minimum in the beam dynamics of the SPARC high-brightness photoinjector *Phys. Rev. Lett.* **99** 234801

[16] Dik A V 2009 Generation and forming of electron beam inside the SPARC photoinjector *Diploma Thesis* National Research Nuclear University "MEPhI" (in Russian)

[17] Landau L D and Lifshits E M 1988 Mechanics *Volume 1 Course of theoretical physics* Nauka, Moscow

[18] Podlesnaya A S 2009 Simulation of electron beam close to photocathode surface of the SPARC photoinjector *Diploma Thesis* National Research Nuclear University "MEPhI" (in Russian)

AUTHOR INDEX

Institute of Physics
Dirac House, Temple Back
Bristol BS1 6BE UK

CURRAN ASSOCIATES INC
proceedings.com